AF692468

Reimagining the Globe and Cultural Exchange

Studies in the History of Christianity in East Asia

Editors-in-Chief

M. Antoni J. Ucerler, s.j.
Wu Xiaoxin 吳小新
Ricci Institute for Chinese-Western Cultural History–Boston College

Editorial Board

Donald L. Baker (*University of British Columbia, Canada*)
Anthony Clark (*Whitworth University, USA*)
Kiri Paramore (*Leiden University, The Netherlands*)
Stephen J. Roddy (*University of San Francisco, USA*)

VOLUME 9

The titles published in this series are listed at *brill.com/hcea*

Reimagining the Globe and Cultural Exchange

The East Asian Legacies of Matteo Ricci's World Map

Edited by

Laura Hostetler

BRILL

LEIDEN | BOSTON

This is an open access title distributed under the terms of the CC BY-NC-ND 4.0 license, which permits any non-commercial use, distribution, and reproduction in any medium, provided no alterations are made and the original author(s) and source are credited. Further information and the complete license text can be found at https://creativecommons.org/licenses/by-nc-nd/4.0/

The terms of the CC license apply only to the original material. The use of material from other sources (indicated by a reference) such as diagrams, illustrations, photos and text samples may require further permission from the respective copyright holder.

Cover illustration: Cropped detail from a 1604 Ricci map reproduced in Japan in color. The image is taken from the Kano Collection at the Tohoku University Library in Japan and is available open source. https://commons.wikimedia.org/wiki/File:Kunyu_Wanguo_Quantu_(坤輿萬國全圖).jpg.

Library of Congress Cataloging-in-Publication Data

Names: Hostetler, Laura, editor.
Title: Reimagining the globe and cultural exchange : the east asian legacies of Matteo Ricci's world map / edited by Laura Hostetler.
Description: Leiden ; Boston : Brill, 2024. | Series: Studies in the history of Christianity in east Asia, 25423681 ; volume 9 | Includes index.
Identifiers: LCCN 2023055072 (print) | LCCN 2023055073 (ebook) | ISBN 9789004682665 (hardback) | ISBN 9789004684782 (ebook)
Subjects: LCSH: Asia—Maps. | Asia—Historical geography—Maps. | Cartography–History. | Ricci, Matteo, 1552–1610. | LCGFT: World maps. | Early maps.
Classification: LCC G2201.S1 R4 2024 (print) | LCC G2201.S1 (ebook) | DDC 912.5—dc23/eng/20231226
LC record available at https://lccn.loc.gov/2023055072
LC ebook record available at https://lccn.loc.gov/2023055073

Typeface for the Latin, Greek, and Cyrillic scripts: "Brill". See and download: brill.com/brill-typeface.

ISSN 2542-3681
ISBN 978-90-04-68266-5 (hardback)
ISBN 978-90-04-68478-2 (e-book)
DOI 10.1163/9789004684782

Copyright 2024 by Laura Hostetler. Published by Koninklijke Brill NV, Leiden, The Netherlands.
Koninklijke Brill NV incorporates the imprints Brill, Brill Nijhoff, Brill Hotei, Brill Schöningh, Brill Fink, Brill mentis, Vandenhoeck & Ruprecht, Böhlau and V&R unipress.
Koninklijke Brill NV reserves the right to protect this publication against unauthorized use.

This book is printed on acid-free paper and produced in a sustainable manner.

Contents

PART 3
Reverberations of Ricci's Maps in East Asia

POSTLUDE
Reflections on the Curation of Cartographic Knowledge

Foreword: Maps, Missionaries, and the Global Exchange of Knowledge in the Early Modern World

The Ricci Institute for Chinese-Western Cultural History at Boston College is pleased to be able to present this volume to readers interested in the history of cartography as a vehicle of intercultural dialogue and the global exchange of knowledge over the past centuries, with a specific focus on the contribution of missionaries during the early modern and modern periods, from the sixteenth through the eighteenth centuries. This book is part of our Institute's monograph series, *Studies in the History of Christianity in East Asia*, which we are privileged to publish with Brill Academic Publishers in Leiden and Boston. We asked Prof. Laura Hostetler, from the University of Illinois at Chicago, to edit the volume. She had previously served as our EDS-Stewart Chair at the Ricci Institute in 2016, when the Institute was still part of the University of San Francisco; and she was involved in helping the Institute with the various events that we organized around this theme, including an international conference with experts on cartography entitled *Reimagining the Globe and Cultural Exchange: From the World Maps of Ricci and Verbiest to Google Earth*. We invited a group of scholars from across the world to share their insights about these dynamics of intercultural exchange with a particular focus on cartography.

We regret that it has taken much longer than we had anticipated for this book to appear in print on account of inevitable delays tied to the outbreak of the coronavirus pandemic as well as the recent move of the Ricci Institute from San Francisco to Boston. Fortunately, these obstacles have now been overcome. In this foreword, I believe it will be useful to provide some context and to explain some of the ideas that helped to frame and define the contours of this multifaceted project and its related events, while mentioning some aspects of the history of cartography that served as its scholarly inspiration. The resulting volume may not conform completely to what some readers might initially expect from a more technical volume on the history of cartography; but it is our hope that many will find something of interest in the pages that follow and will discover some of the unexpected links between the different topics covered therein.

I should note from the outset that it was not our aim to provide an exhaustive treatment of the topic, but rather to engage in an exploration—from different perspectives—of cartography as a medium for the transmission of knowledge across cultures, continents, and languages, beginning with the early modern period, a time of exploration, navigation, colonization, and Christian missionary activity beyond the shores of Europe. It was also an exciting time marked by the invention of the Gutenberg handpress, a technology that revolutionized the ability to transmit knowledge at a cost that was no longer prohibitive. More specifically, our aim was to explore the specific role played by missionaries in the development of cartography as intermediaries; and how this, in turn, stimulated further cultural interaction and exchange of ideas with the peoples of East Asia. It was most definitely a period marked by the mutual *discovery* of the "Other"; and perhaps nothing was more concrete than the questions Europeans relate they were asked upon their arrival. We can well imagine that, after the initial query: "Who are you and why did you come here?" what soon followed were questions informed by curiosity: "Where did you come from?," "How did you get here?," and "What do your lands look like?" This final question called for detailed descriptions of mountains, rivers, lakes, sea passages, as well as the fauna and flora of different regions.

What soon becomes clear from both a perusal of the written reports as well as the material culture, as expressed in maps, is that knowledge travelled in both directions; and while Europe gained knowledge of Asia, so too East Asian scholars in China, Japan, and Korea gained knowledge of world geography. The fifteenth to eighteenth centuries were a fluid time in terms of cartographic knowledge; and not all cartographers or schools of cartography agreed upon the physical details they were describing, sometimes even debating the contours of entire countries or even continents. The island of Hokkaido in northern Japan, formerly referred to as Ezo (or Yezo), for example, did not appear at all on a European map until the early seventeenth-century manuscript map drawn by the Jesuit missionary, Geronimo De Angelis (1567–1623), who would later die as a martyr in 1623. He was also the first non-Japanese to describe the Tsugaru straits between Honshu and Hokkaido. From his correspondence, we know that he visited Matsumae in northern Japan between 1618 and 1622 and studied the entire region.

Thus, sometimes unwittingly, missionaries ended up exploring places they had never imagined they would see. The result was that they found themselves at new cultural and scientific crossroads, as they traveled to new parts of the world to carry out their primary goal, namely to preach the Christian faith. This is not to say that others were not involved in these explorations. But we have chosen to explore their varied activities and how these led to new knowledge or the exchange thereof. Even a perfunctory study of the European presence in East Asia in the sixteenth through eighteenth centuries reveals that they often ended up dominating the scene as unintentional brokers of information. Soon after their arrival in East Asia, Christian missionaries began to engage in the compilation of geographical data, including the actual drafting of localized maps. At first, these were produced primarily with a European audience in mind. But once the Jesuit missionaries established themselves at the Chinese imperial court as scientists, they were also called upon to share their knowledge of the world with their East Asian interlocutors. Thus, the information was not exclusively traveling in one direction; nor was the story of missionary endeavors in East Asia limited simply to describing their work of evangelization or their involvement with the merchants on whose ships they traveled.

The Italian Jesuit, Matteo Ricci (1552–1610), aided by the Chinese literatus, Li Zhizao (1565–1630), undertook the momentous project of creating for the Chinese a map of the entire whole world, with place names of all known countries in their own language. After several earlier attempts, beginning in 1584, the 1602 *Kunyu wanguo quantu* 坤輿萬國全圖 or *A Complete Map of the Myriad Kingdoms of the Whole World*, printed with woodblocks on six large panels, became one of Ricci's most celebrated achievements. It did not fail to astonish Ricci's Chinese interlocutors. An early adoption of a Mercator-type projection, it provided detailed visual information about most of the continents. It also included other scientific information regarding various natural phenomena and astronomical data. Many place names transliterated into Chinese characters for the first time by Ricci on his 1602 map are still used to this very day in China, more than four hundred years later. Keen to make this cartographic treasure born of the collaboration between Chinese and European scholars working together available to a wider audience, we set out to borrow one of the rare original copies of this map, which is part of the extensive map collections of the James Ford Bell Library's collections at the University of Minnesota. We wished to include it as part of an exhibit we proposed to the Asian Art Museum of San Francisco. This came to fruition after three years of preparation in the spring of 2016 as part of the Museum's 50th anniversary celebrations. I co-curated this exhibit with Natasha Reichle of the AAM. But Ricci's map was not the only map in this exhibition. We thought that it would be more interesting also to display the aforementioned 1674 map created by Ferdinand Verbiest (1623–88), namely his *Complete Map of the World* or *Kunyu quantu* 坤輿全圖.

Even larger than Ricci's six-panel *mappamondo*, Verbiest's map of the world in two hemispheres, represented stereographically on six large panels, with the Eastern hemisphere on the left, thereby placed China visually close to the center rather than at the edge of the world as the "Far East." In this respect, Verbiest followed what Ricci had done in his 1602 map. The various cartouches on the additional two outer panels (for a total of eight altogether), provide short treatises in classical Chinese on astronomic phenomena, the tides of the oceans, earthquakes, rivers, and mountains, among other themes.

It also includes detailed drawings of animals from various parts of the world unknown to inhabitants of the Middle Kingdom, including, for example, the North American beaver and turkey, as well as the African rhinoceros, referred to by missionaries on some maps as a "sea horse" 海馬 or "cavallo marino" (in Italian manuscripts). There are even mythological creatures, such as a unicorn, a mermaid, and a merman.

During the years when I was teaching history at Georgetown University, I discovered to my surprise that the Library of Congress possessed an original copy of Verbiest's map on eight scrolls. When we were debating how best to put together a proposal for an exhibit, I approached Ralph Ehrenberg, the former Head of the Geography and Map Division, to see whether we could ask the Library of Congress to authorize the loan of this map. Thanks to his generous efforts, we were able to display the two large Jesuit maps together at the Asian Art Museum. As far as we know, this may well have been the first time in several centuries that the two maps came together for a public exhibit. Other materials were also included in this exhibit entitled: *China at the Center: Ricci and Verbiest World Maps*. Among these additional artifacts were two large hand-painted portraits of Matteo Ricci and his first convert and colleague, Xu Guangqi 徐光啓 (1562–1633), painted by orphans at the Tushanwan (Tou-sè-we) 土山灣 orphanage painting workshop, within the Zikawei (Xujiahui 徐家匯) Jesuit complex in Shanghai, ca. 1910–1914. These paintings were originally exhibited at the Republic of China pavilion during the Panama Pacific International Exposition in San Francisco in 1915; and they were once again on exhibit in San Francisco a century thereafter. These artifacts were requested by the newly established government in China following the collapse of the Qing dynasty in 1911.

In conjunction with the exhibit at the Asian Art Museum, the Ricci Institute also sponsored its own exhibit on the campus of the University of San Francisco at the Manresa Gallery of St. Ignatius Church, with early modern European maps that we borrowed from the extensive collections of the Kirishitan Bunko (Christian Archives) of Sophia University in Tokyo. Entitled, *Mapping "The East": Envisioning Asia in the Age of Exploration*, we worked with the Art History and Arts Management Program of USF. The exhibit was curated by Prof. Catherine Lusheck and Madeline E. Warner '15. A second exhibit on the University's campus was held at the Thacher Gallery. This included the response of two contemporary artists, Amanda Hughen and Jennifer Starkweather, through works on wood and paper, to a Japanese version of Ricci's map produced in 1785 by the Edo scholar, Nagakubo Sekisui 長久保赤水 (1717–1801). This copy was also part of the Thacher Gallery exhibit. Finally, in collaboration with the ¡Sacabuche! Ensemble, two original concerts with a repertoire of early European and

Chinese music, were held at St. Ignatius Church in San Francisco and at the Mission Church on the campus of Santa Clara University. These highly original audiovisual performances, centered on Ricci's 1602 map, retell the story of Ricci's time in China through images taken from his map, with a narrative script in English and Chinese, edited by Ann Waltner, Professor of Chinese history at the University of Minnesota.

Why then should we make such a fuss over these maps? These unique artifacts represent much more than mere curiosities from a bygone age, inferior to ours in scientific terms. They became so popular in East Asia itself that they were reproduced numerous times not only in China but also in Edo Japan and in Joseon Korea. In fact, they continued to be reprinted, and many were then hand-colored, until the mid-nineteenth century. This is historically significant, as both Japan and Korea in the seventeenth and eighteenth centuries were not otherwise receptive to Western Learning mediated by Christian missionaries. Official Korean records reveal the interactions between the envoys that the royal Joseon court regularly sent to Beijing and Jesuit missionaries at the Qing court. Among others, Crown Prince Sohyeon (1612–45), returning from his exile as a hostage in Shenyang, met Johann Adam Schall von Bell, Verbiest's senior colleague, in Beijing in 1644. Korean envoys, such as Jeong Duwon 정두원 (鄭斗源) (1581–?), also speak of their encounters in 1631 with João Rodrigues Tçuzzu (1561/2–1633/34), known as the "Interpreter," following the latter's exile from Japan to China in 1614. This encounter is recorded in the *Veritable Records of the Joseon Dynasty* under the entry for the seventh month, twelfth day (in the lunar calendar) of the ninth year of the reign of King Injo 인조 (仁祖) (1595–1649).

This and other Korean sources explicitly mention, for example, that the envoys acquired and brought back with them copies of Ricci's 1602 map. It was most probably during this period that the exceedingly rare 1603 reprint of Ricci's map, with the title, *Liangyi xuanlantu* 兩儀玄覽圖, or *A Mysterious Map of the Two Forms*, made its way to Korea. The only known extant copy of this rare treasure is preserved in the Korean Christian Museum at Soongsil University in Seoul. Even the highly conservative Joseon court readily came to recognize the importance of the Ricci map; so much so, that in 1708 the king ordered it to be reproduced anew. Further copies of Verbiest's 1674 map were also reprinted in Korea until the late nineteenth century, and notably in 1860. But they possessed authority *not* on account of their "Western" origin, but rather because these Jesuits produced their work as Chinese court scientists. Thus, Joseon scholars respected this knowledge because it had originated in the Middle Kingdom and its text was composed in classical Chinese. What is curious in this regard is initial reluctance on the part of Korea to acknowledge the Qing "barbarians" as legitimate heirs to the "enlightened" Ming throne, whom they had overthrown. These negative perceptions were only confirmed by the second Qing invasion of the Korean peninsula in 1636 that saw the court humiliated and subjugated by the Manchus. What has thus far proven to be unique to Korea are the round world maps, i.e., the *Chonhado* 천하도 (天下圖) or *Map of All under Heaven*, which were produced in the seventeenth and eighteenth centuries. They are of great interest, as they represent a curious hybrid between earlier Chinese Buddhist cosmological maps, which include mythical islands or kingdoms (e.g., the country of women, the country of giants, etc.), and the new knowledge that had been introduced from the West through maps like those of Ricci, Aleni, and Verbiest.

In another historical instance that illustrates the weight given to the cultural authority of China, the Jesuits were commissioned to take part in a larger project, literally to map "all under heaven" or the entire Chinese empire governed by the Qing rulers. This involved the Jesuits' participation in the Chinese imperial survey, underwritten by the Kangxi Emperor (1654–1722) between 1708 and 1718 with the help of the Jesuits who had been sent to China several decades earlier by Louis XIV. The atlas that they produced was part of the broader portfolio of scientific work that they were engaged in at the imperial court centered primarily around astronomy and calendar reform.

As in the case of Korea, which eventually accepted even the cultural authority of the Qing court, the results of such efforts were keenly followed in other countries beyond the borders of China. In Edo Japan Ricci's 1602 map was reprinted with woodblocks and hand colored as early as 1604. There were numerous reiterations of this map that continued to be printed for several centuries. What is particularly notable is that this took place despite the ban on Christianity, which had been decreed by the Edo shogunate in 1614. As soon as the Japanese discovered that Christian books or books written by Jesuit missionaries in Classical Chinese were being brought into Japan by Chinese merchants who were coming to trade at Nagasaki, they put an immediate halt to the importation even of scientific works produced by the Jesuit missionaries in China. Having set up an Inspector or Inquisitor for Books (*shomotsu aratameyaku* 書物改役) as early as 1630, it was sufficient for Ricci's Chinese name, Li Madou 利瑪竇 (or that of other missionaries) to appear as the author of a work, or even simply to be mentioned in an unrelated work (e.g., a seventeenth-century guidebook of Beijing that happened to mention the Jesuit Church and Ricci tomb as notable sites to visit) for it to be strictly banned. As for the unfortunate Chinese merchants who were wittingly or unwittingly responsible for the importation of such books, they were faced with severe punishment, either death or, if leniency was shown, a permanent ban from ever trading again in Japan.

And yet many copies of Ricci's world map continued to circulate in various forms throughout the Edo period. In fact, archival research seems to indicate that there are more extant copies of Ricci's map (or numerous versions thereof, including those by Aleni and Verbiest) produced in Japan than in China itself. We find some of these hand-drawn and colored versions (rather than re-printed with woodblocks) still being copied in the middle of the nineteenth century by curious scholars. Other more elaborate copies were painted directly onto traditional folding screens (*byōbu*), some of which were even decorated with gold leaf and became collectible works of art. Thus, their aesthetic qualities complemented or even superseded their primary function of transmitting scientific knowledge. Perhaps for some it was their appeal as decorative works of art that lent an air of sophistication to their collectors; and in some cases, this quality may even have been more prized than the actual knowledge conveyed by the cartographic data itself. With regard to the latter, however, Edo scientists were urging the shogunate to ease the rigid ban on the importation of scientific books produced by the missionaries in China, lest they fail in their own task of reforming the Japanese calendar and gaining sufficient up-to-date knowledge about the world. As a result of these entreaties, in 1720, the eighth shogun, Tokugawa Yoshimune (1684–1751), made provisions to allow the purchase in China and reproduction in Japan of such works, with the proviso: "insofar as they did not promote the Christian religion." This opened the doors to another century of copying, distribution, and study of these treaties and maps.

Of particular interest to Edo scholars was the compilation on world geography or *The Record of Foreign Lands* or *Zhifang Waiji* 職方外紀 by the Italian Jesuit, Giulio Aleni (1582–1649), first published in 1623, which included detailed descriptions of different parts of the world interspersed with foldout maps. We find copies of this work preserved in various important archives of the time, including the shogun's library, the Momijiyama bunko (紅葉山文庫), and the Confucian temple in Edo. First established as the Shinobugaoka Sacred Hall (忍岡聖堂) in 1632 with the support of the head of the Owari branch of the shogun's family, Tokugawa Yoshinao (1601–50), it was led by the renowned Confucian scholar, Hayashi Razan (1583–1657). His family would be responsible for the government sponsored school of Confucianism in Edo, the Shōheizaka Gakumonjo (昌平坂学問所) or Shōheikō (昌平黌), which was on the grounds of the Confucian temple, which was transferred to Yushima Sacred Hall (湯島聖堂) in 1691. Evidence of the existence of these cartographic treatises in the archives of these institutions reveals a great deal about the importance the shogunate placed on the acquisition of global knowledge. The fact that they had been written by missionaries in China was either forgotten or perhaps no longer considered relevant, for they did not include explicit references to Christianity, and they came from China.

Another influential map in Japan was produced by eighteenth-century Japan's most distinguished cartographer, Nagakubo Sekisui. As mentioned earlier, he reproduced in 1785 a somewhat simplified version of Ricci's 1602 map, with names written in both Chinese characters and Japanese kana. He may have also had access to an earlier version of this world map compiled by Ricci in 1584 and 1600. The latter bears the title, *Shanhai yudi quantu* 山海輿地全圖 or *Complete Terrestrial Map*. In the preface printed on Sekisui's map, the Japanese geographer refers to the original cartographer as "Master Li" 利氏 and uses the title of Ricci's earlier 1600 map while also referring to "six panels that together come to form the complete map," presumably a reference to Ricci's 1602 *mappamondo*. It is almost certain that, by this time, almost no one would have known that "Master Li" had been a foreigner, and a missionary of the banned "Kirishitan" religion. As mentioned earlier, as the map had originally come from Ming China, it had the authority of the Chinese court, which was still considered an important source of accurate scientific information. This version of Sekisui's map would also be reproduced numerous times, well into the nineteenth century, both in print and as hand-drawn manuscripts.

Missionaries were also a privileged conduit of information about East Asia in general and about its geographical features in particular. Detailed descriptions and drawings were sent to Europe on a regular basis, beginning in the mid-sixteenth century. One of the best-known maps of Japan from this period is the 1595 version produced by Jodocus Hondius (1563–1612), based on the information taken from the work of Luís Teixeira, whose *Iaponiae insulae descriptio* or *Description of the Island of Japan*, appeared in Abraham Ortelius's (1527–98) monumental atlas, *Theatrum orbis terrarum* or *Theatre of the Orb of the World* (literally "of the lands"), which was first printed in Antwerp in 1570. These maps went hand in hand with various reports and other treatises that missionaries sent back to Europe recounting their activities and important events that had taken place. The 1595 map of Japan, mentioned earlier, was published during Hideyoshi's invasion of Joseon Korea, which commenced in 1592. The Spanish Jesuit, Gregorio de Céspedes (1551–1611), accompanied some of the Christian lords fighting for Hideyoshi to Korea as a chaplain of sorts and thereby became an important source of information. In an annual letter from the Japanese mission for 1591 and 1592, published in Rome by Luigi Zannetti, we find one of the first European references to "Il Regno di Corai" or the

"Kingdom of Korea." For an earlier reference, we must look to a handwritten report (composed ca. 1580) on China written in Latin by Alessandro Valignano (1539–1606), the Visitor of the Jesuit Superior General for the missions of India and East Asia. He includes a short paragraph on "the kingdom of Korea" or *De regno Coreae*. It was thus during the late sixteenth century that reports about Joseon Korea became available for the first time to European readers.

The thirst for accurate scientific knowledge had the surprising effect of breaking down ideological and confessional divides that otherwise seemed insuperable. When the Italian Jesuit missionary in China, Martino Martini (1614–61), made his way back to Europe to solicit funds and men for the Chinese mission, he brought back with him an elaborate set of maps, which would become the source for the first complete atlas of China, the *Novus atlas sinensis* or *New Chinese Atlas*, printed in Europe. It is noteworthy that his atlas was printed not in Rome, Madrid, or Lisbon, but in Amsterdam in 1655 by the famous Dutch printer, Joan Blaeu (1596–1673). This was true of other maps as well, including those of Japan by Jodocus Hondius (1563–1612) and Willem Janszoon Blaeu (1571–1638). Yet another example is a 17th-century Danish map of Tartary, i.e., *Kort over det ostlige Tartari til den almindelige Reise beskrivelse af Jesuiternes Kort* or *A Map of Eastern Tartary with a General Description of Travel of the Jesuit Map*, which makes explicit reference to its source as the Jesuit maps or *Jesuiternes Kort* written between 1708 and 1715, the very years the Jesuits were engaged in the aforementioned imperial survey of all of China. This is highly significant, as it reveals the extent to which the *res publica litterarum* was able to overcome such ideological and/or specifically confessional conflicts in the common pursuit of reliable knowledge about the world—a world that seemed to expand and change its contours with each new geographical discovery. It appears that the same dynamic can account for the reaction of scholars in Edo Japan, who took risks to put knowledge of the world and science ahead of political orthodoxy.

Another curious feature of these early maps that merits further exploration is how secondary information was transmitted via these maps. A good case in point is Jodocus Hondius's 1606 Map of China. While this map is primarily, as indicated by its title, a map of China, it includes a colorful cartouche showing the execution of the first twenty-six Christians by crucifixion on Nishizaka Hill in Nagasaki on February 5, 1597. This involved being stabbed to death with two long swords or *naginata*, while placed on a wooden cross with arms, neck, and legs fastened firmly to the wood with metal clamps. This information was taken directly from Jesuit and Franciscan reports sent from Japan and published in Europe shortly after the fateful event. In fact, even a summary perusal of reports from this period on the Japanese missions published in Europe in various languages reveals a wealth of editions in various languages that recount this fateful event. Yet another source of information were various engravings that depicted this martyrdom. One notable example is a broad sheet with an engraving with text, including the names of all twenty-six martyrs, published in Augsburg in 1628 by Wolfgang Kilian (1581–1663), a prominent engraver, publisher, and cartographer.

This event was considered of such importance to *all* of Christendom that Hondius felt compelled to include this new information from those regions on his map of China. In fact, this map was printed just nine years after this martyrdom in Japan. Hence, one might even say that maps sometimes functioned almost like modern websites, with hyperlinks, which provided the curious observer with the "breaking news" of the day. And just as one newspaper or tv channel will carry information first reported by another news outlet, so too this map, and the information it contained, was deemed too important to relegate to a single edition, published by a single cartographer in a particular country and only in one language. Hence, the English cartographer, John Speed (1552?–1629), would produce his own map of China in 1626 and include his own cartouche of this martyrdom with the English caption, "the [i.e., Japanese] manner of execution." At times these captions would be faithful translations of previous editions, whereas in other cases the text was edited, abbreviated, or lengthened according to what the publisher thought most appropriate for the target audience. For good measure, on both sides of the map Speed also includes a pictographic representation in color of "a solider of Japan," armed with an arquebus (first introduced to Japan in 1543) and a sword.

What these examples suggest is that there was not always a clear line of demarcation between professions, and that the producers of these maps could comfortably wear several hats at the same time. An engraver might also work on maps, while cartographers would be consuming and possibly producing other textual and pictographic media, all the while incorporating that data into the final version of their maps. The same held true for the works of Jesuit missionaries, including those by Giulio Aleni and Ferdinand Verbiest, where the distinctions between ethnographic treatises, complementary illustrations, and cartography proper were sometimes not as explicit as they would become in later centuries. The *Flora sinensis* or *Chinese Flora*, an illustrated book compiled by the Polish Jesuit, Michał Boym (1612–59), and printed in Vienna in 1656, is a good case in point. It visually represented the fauna and flora of different parts of the world and was an important extension of his own cartographic endeavors. He had produced three notable maps of China; but then, almost as if he were zooming into that region on Google Earth, as we can do today, he felt the need to provide his readers with a more localized and—literally—specific "picture" of the country's defining features. Hence, we may even speak of an "intertextuality" between maps and other printed and manuscript textual sources produced by the missionaries that complemented, elucidated, or otherwise provided further information about peoples and places. It is also important to note that the hand coloring of these maps or ethnographic treatises added to their visual impact and enhanced their value as desirable objects for purchase.

Their function was not only to convey knowledge but also to produce a desired visual rhetorical effect that would elicit its own emotional response. For Catholic Europe, these maps were sometimes viewed as a rebuttal to Protestant claims that the Roman Catholic Church was no longer the true Church. Rome would claim that missionary expansion was a clear sign of a vibrant Church in line with the ideal of the return to the primitive Church of the apostles—often cited as an ideal by Reformers. And this expansion was demonstrated by maps clearly indicating all the places where the Roman Catholic Church had newly preached the Gospel. A striking example is the *Iaponiae nova et accurata descriptio* by António Francisco Cardim (1596–1659), based on an early 1641 version by Bernardino Ginnaro. In the case of Cardim, he included it in his book on Christian martyrs in Japan, *Fasciculus e Iapponicis floribus, suo adhuc madentibus sanguine* or *A Bundle of Japanese Flowers, still moist with the Blood (of the Martyrs)*, printed in Rome in 1646.

In these and many other cases, we can identify clear visual references, or what we might term "inter-visual" or "inter-objective" cross references between different maps. We find, for example, Willem Janszoon Blaeu including numerous cartouches representing different men and women of countries in East and Southeast Asia, together with smaller representations of major cities in Asia, in his *Asia noviter delineata*, which was reprinted numerous times in the seventeenth century, following its first issue in 1630. Far from considering such copying or borrowing of texts and images as plagiarism, reproducing details from earlier maps could lend them even greater authority. Imitation conferred approval and acceptance of previously transmitted data. And for those who were still new to their trade, copying could also be a safe practice, as their work appeared to rely on accepted knowledge. Nevertheless, this imitation did not always have the intended effect. A good case in point is the depiction of both the Korean and Californian peninsulas as islands. Curiously, Matteo Ricci's famous world map of 1602 correctly shows California as a peninsula, whilst his fellow Jesuit, Ferdinand Verbiest, when creating his own monumental world map more than a generation later in 1674, reverted to drawing California as an island. In the latter case, i.e., of California, the issue would not be definitely resolved until the overland exploration by Eusebio Kino (1645–1711) between 1698 and 1701—an expedition that would finally provide irrefutable evidence that California was in fact a peninsula. Kino's map notwithstanding, it would take almost another half a century for the "island theory" to be definitively abandoned.

Similar uncertainty plagued the geographical status of Korea until the publication of maps in the eighteenth century, including Jacques Nicolas Bellin's (1703–72) *China nebst Corea und den benachbarten Laendern der Tartary aus denen Karten genommen, welche die Jesuiten von dem Jahre 1708 bis 1717 davon entworfen haben* or *China together with Korea, and the neighboring countries of Tartary, taken from the maps which the Jesuits have drawn of it from the year 1708 to 1717*, printed in Leipzig in 1749. This map explicitly refers to the Kangxi imperial survey with which the Jesuits were involved. Regardless of whether they were correct or committed blunders when drawing their maps, before the advent of modern techniques of surveying, which improved exponentially in the eighteenth century, not to mention our contemporary recourse to satellite imagery, we are often confronted with different "schools of cartography." Each presented data as a matter of received opinion or what they judged as the most reliable and verifiable knowledge for their own time, even if others argued the contrary (e.g., land being an island or a peninsula). In such cases where there was disagreement, the obvious risk for the early modern cartographer was copying what would turn out to be the wrong representation.

To sum up, the early modern age was a time when important developments in the history of cartography took place that would forever transform—both in Europe and in East Asia—the perception and understanding of our physical world and of its inhabitants across continents. Europeans had no idea where the mythical "Zipangu" or "Cyampagu," which Marco Polo had referred to in his famous travelogue, was located until the Portuguese washed up on the shores of Japan in 1543. Soon they began to send back information confirming both its existence, location, and geographical contours. On the other hand, the Chinese, Koreans, and Japanese, were astonished to lay eyes on Ricci's map and reimagine a world with entire continents well beyond the Middle Kingdom and the countries in its immediate vicinity. That said, as the scholar of Chinese history, Timothy Brook, has demonstrated, the Selden map preserved in the Bodleian Library in Oxford, reveals that Chinese merchants were independently mapping detailed commercial routes to numerous parts of both Southeast and South Asia. Thus, while the missionaries were certainly not the exclusive conduit of information regarding cartography, it goes without saying that they played a crucial historical role both in the actual transmission of knowledge and the material production of it, whether it involved the preparation of woodblocks in China or arranging the publication of these maps with cartographers and printing houses in Europe. These are limited examples of some of the trajectories that we chose to explore during our international symposium at the University of San Francisco in 2016. It is our sincere hope that this volume on cartography will allow our readers to embark on their own journeys of discovery of new places, all the while attentive to the visually embedded stories that these maps reveal to us about the intercultural encounters of centuries past.

M. Antoni J. Ucerler, S.J.
Director, Ricci Institute for Chinese-Western Cultural History
Boston College, April 2023

Preface and Acknowledgements

In 2008 the Ricci Institute for Chinese-Western Cultural History in San Francisco received a mysterious phone call. Wishing to remain anonymous, the speaker divulged only that she was calling on behalf of a trust that was considering the purchase of a rare exemplar of Matteo Ricci's 1602 *Kunyu wanguo quantu* 坤輿萬國全圖 [A Complete Map of the Myriad Countries of the World]. She was seeking to consult with experts in the field as part of the process, the Ricci Institute having been identified as among them. The sale moved forward, garnering much attention. On its arrival in the U.S., the newly acquired map was first displayed at the Library of Congress for several months, and then at the Minneapolis Institute of Art. The James Ford Bell Trust, which identified itself after the sale was complete, has since made a gift of the map to the James Ford Bell Library of the University of Minnesota where it will be permanently housed.

To celebrate the Trust's acquisition of the Ricci map and its commitment to making it accessible to the public, the Ricci Institute, under the direction of Fr. M. Antoni J. Ucerler, S.J., undertook to work together with the Asian Art Museum on bringing the 1602 Ricci map, as well as the later 1674 *Kunyu quan tu* 坤宇全圖 [Complete Map of the World] by Ferdinand Verbiest, held by the Library of Congress, to San Francisco. It was agreed that the museum would host *China at the Center: Rare Ricci and Verbiest World Maps*, displaying these rare maps as well as other supporting materials.[1] The Ricci Institute would plan a scholarly conference, *Reimagining the Globe and Cultural Exchange: From the World Maps of Ricci and Verbiest to Google Earth*, as well as a full series of additional events in conjunction. These included an exhibition of early modern European maps of East Asia: *Mapping "The East": Envisioning Asia in the Age of Exploration*; a cartographically-inspired art exhibit: *Adjacent Shores: Hughen/Starkweather*; and a concert and reading by the *¡Sacabuche! Ensemble: Matteo Ricci: His Map and Music*; I was brought in for a semester as the EDS-Stewart Chair in order to assist with planning the conference.

From the beginning our goal was to bring together an interdisciplinary and international group of specialists including curators, collectors, librarians, bibliographers, historians, art historians, historians of science, historians of religion, and early modernists. We also felt it essential to include scholars from different national origins in order to include their voices and observations on

1 Reichle, Natasha, ed. *China at the Center: Ricci and Verbiest World Maps* (San Francisco: Asian Art Museum, 2016).

the history and legacy of the Ricci world map and its successors in Asia. Each author represented in this volume brings their own interests, expertise, and observations—as well as a wealth of scholarly knowledge—to explore a different facet of the Ricci map, its history, and/or legacy. The volume has been immeasurably enriched by the breadth and depth of the individual contributions, as by the variety of approaches and various areas of expertise represented throughout. I would like to thank each of the contributors for their fine work and patience through what has been a lengthy editorial process during which each of us has also been juggling other commitments and finding our way through a global pandemic.

I would also like to express my appreciation to the Ricci Institute for hosting me as EDS-Stewart Chair in the Spring of 2016, and for trusting me with editorial oversight of this volume. In particular, I would like also to thank each of the staff members who made me welcome and assisted in a variety of ways with the production of this volume. Fr. M. Antoni J. Ucerler, S.J., Director of the Institute (now located at Boston College), welcomed me the first morning of my stay in San Francisco with a lovely card, handwritten in fountain pen, which he had placed on my desk. His vision for the conference as well as direction of the multiple associated activities, mastery of many languages, and attention to the importance of fostering human relationships, all attest to a recognition of the multiple facets of what constitutes a life well lived. Dr. Xiaoxin Wu, now Director of Research at the Institute, provided a model of consummate administration. His genuinely thoughtful and considered approach to the many details involved in running an institute—and a conference—demonstrated to me that *how* our work is approached is an essential measure both of what can be accomplished, and the meaning we find in it. Mark Mir, Librarian and Archivist, first earned a special place in my heart by apologetically telling me that the library would close at 4:00 PM every day, thereby freeing me up to get some much-needed regular exercise. He continues to amaze me with the breadth and depth of his knowledge and bibliographic recall, which are somehow matched by a remarkable generosity of spirit, wry wit, and ability to build almost anything from the ground up. I am grateful also to Craig Asberry for assiduous work as my graduate research assistant, and to May Lee for helping me feel welcome and attending to a variety of aspects of the details of my daily life at the Institute. Thanks go also to Steven Ford, who joined the Ricci Institute after my time there in 2016, for his early editorial assistance with the project, and to Virginia M. Greeley for stepping in with final aspects of the manuscript preparation.

Thanks are also due to the Department of History and College of Liberal Arts and Sciences at the University of Illinois at Chicago for providing leave during my stay at the Ricci Institute, and more recently for supporting my work with the Engaged Humanities Initiative (EHI). Involvement in the EHI has broadened my view of what constitutes scholarship—including the importance of inviting multiple perspectives into our studies whether of material culture, human experience, or of the past—and fostered in me a deep appreciation that these approaches are always, and productively, multiply inflected.

Finally, I would like to thank my daughter, Naomi Liechty, for encouraging me to accept the assignment that took me away to San Francisco during the last semester of her senior year of high school, and my husband, Mark Liechty, for accepting the challenge of serving as single parent of a teen while I was away.

Laura Hostetler

Illustrations

Figures

Tables

Abbreviations

AHSI	*Archivum Historicum Societatis Iesu*
ARSI	*Archivum Romanum Societatis Iesu*
DHCJ	*Diccionario Histórico de la Compañía de Jesús*
MHSI	*Monumenta Historica Societatis Iesu*

Notes on Contributors

Aoyama Hiro'o
is Professor Emeritus of the National Museum of Japanese History at the National Institutes for the Humanities in Japan. He received his Doctor of Letters at Kyoto University. His research interests include the historical geography of landscape in Medieval Japan and the history of cartography and geographical knowledge. His recent works include one on the topic of Geographical Names and Landscape and another titled The Map of Japan and Geographical Data of Japan Obtained by Siebold (Journal of the Japan Cartographers Association 2018).

Paul Begheyn, s.j.
is Founder and Director of the Netherlands Institute of Jesuit Studies, Archivist of the Dutch Jesuits, and Bibliographer of the Society of Jesus worldwide. He entered the Society of Jesus in 1963 and received a degree in Philosophy from Nijmegen, Berchmanianum. His research focuses on the Jesuits worldwide. His recent books include *Jesuit Books in the Dutch Republic and its Generality Lands 1567–1773: A Bibliography* (2014) and *Frans van der Lugt s.j. (1938–2014): Bruggenbouwer en martelaar in Syrië* (2015).

Jose Casanova
is Emeritus Professor of Sociology, Theology and Religious Studies at Georgetown University and Senior Fellow at the Berkley Center for Religion, Peace, and World Affairs. He received his PhD in Sociology from the New School for Social Research. His book, Public Religions in the Modern World, has become a modern classic and has been translated into many languages including Japanese, Arabic, and Turkish. Recently he co-edited The Jesuits and Globalization (2016).

Robert Danieluk, s.j.
is an archivist at the Roman Archives of the Society of Jesus (ARSI) and Collaborator of the Jesuit Historical Institute in Rome, where he has served since 2004. Fr. Danieluk is the author of many works on the history and bibliography of the Jesuit Order in a variety of languages, including his native Polish, French, Italian, and English. He has studied theology and ecclesiastical history at the Pontifical Gregorian University and holds a PhD in History from the École Pratique des Hautes Études.

Paola Demattè

is a Professor in the Department of Theory and History of Art and Design at the Rhode Island School of Design. She received a Laurea in Chinese from the Università degli Studi di Venezia Ca' Foscari (Venice, Italy) and a PhD in Archaeology from the University of California Los Angeles. Her research focus is on the Neolithic archaeology of China, prehistoric rock art, and East–West contacts and exchanges. Among her publications are *The Origins of Chinese Writing* (Oxford 2022) and *China on Paper* (Getty 2007) co-edited with Marcia Reed with whom she also co-curated the exhibition at the Getty Research Institute.

Ralph E. Ehrenberg

is the Chief of the Geography and Map Division of the Library of Congress and earlier directed the Cartographic Records Division of the US National Archives. He received formal training from the US Navy in aerial photography and received his BA and MA from the University of Minnesota. His goal in the Navy was "to film the world," and his research focus has been on mapping it in vivid detail. His most recent books include *Mapping the World* (National Geographic Society 2006), and *Mapping the West with Lewis and Clark* (Library of Congress and Levenger Press 2015).

Laura Hostetler

is Professor in the Departments of History and Global Asian Studies at the University of Illinois at Chicago. She received her PhD in Asian and Middle Eastern Studies at the University of Pennsylvania. Her research interests include the history of cartography, empire, and encounters between Europe and Asia. Her seminal work is *Qing Colonial Enterprise: Ethnography and Cartography in Early Modern China* (2001). She has also published *The Art of Ethnography: A Miao Album of Guizhou Province* with David Deal (2006), and *Qing Illustrations of Tributary Peoples (Huang Qing zhigong tu): A Cultural Cartography of Empire* with Wu Xuemei (2022).

LIM Jongtae

is Professor in the Department of Science Studies at Seoul National University, South Korea. He received his PhD in the History of Science from Seoul National University. His research interests cover topics in the history of early modern Korean science, such as Western learning, Sino-Korean scientific exchanges, and science and technology in the state bureaucracy. His representative works include *Impact of Western Geography on East Asian Geographical Tradition in*

Seventeenth-and Eighteenth-Century China and Korea (2012); "The Tributary Relations between the Chosŏn and Ch'ing Courts to 1800" (2016); and *Travel, Statecraft Reform, and the Science and Technology in Eighteenth Century Korea* (2021).

Mark Stephen Mir
is the Archivist/Librarian of the Ricci Institute for Chinese-Western Cultural History at Boston College. He received his BA from San Francisco State University and MA in Asia Pacific Studies from the University of San Francisco. His responsibilities include the Ricci Institute's library, collections, and reference support for visiting scholars and Fellows.

Marguerite Ragnow
is the curator of the James Ford Bell Library at the University of Minnesota, and a member of the graduate faculties of History, Medieval Studies, and Early Modern Studies. She received her PhD in Medieval and Early Modern History at the University of Minnesota. Her research interests include how maps were used in the pre-modern world, and what maps can and cannot tell us about the time they were made. Recent publications include *Religious Conflict and Accommodation in the Early Modern World* and *Tulips, Chocolate & Silk: Celebrating 65 Years of the James Ford Bell Library*.

Kären Wigen
is Professor in the Department of History at Stanford University. She received her PhD in Geography from the University of California at Berkeley. Her research interests include world history, the history of cartography, and Japanese history. Her current project is an edited volume on the theme of Re-Mapping Sovereignty.

Qiong ZHANG
is Associate Professor in the Department of History at Wake Forest University. She received her PhD in the History of Science at Harvard University. Her research interests include early modern Chinese intellectual and cultural history and the history of science and medicine. Her current projects focus on the cultural context and social networks in which Bowu learning and meteorological knowledge were produced and circulated. Her most recent book is *Making the New World Their Own: Chinese Encounters with Jesuit Science in the Age of the Discovery*.

Introduction: Reflections on Form and Content

Laura Hostetler

Much has been written on The *Kunyu wanguo quantu* 坤輿萬國全圖 [A Complete Map of the Myriad Countries of the World], more commonly known as Matteo Ricci's 1602 World Map in Chinese, and rightly so.[1] Innovative works of historical significance are always worth revisiting on a regular basis. Given changing times, new interests and perspectives, and developments in cartography itself, scholars continually bring fresh sets of questions to the map that first introduced coordinate mapping to Asia, and with it a new way of viewing and representing the world.[2] Not only are scholars continually returning to the map, each time finding more to learn about it, the general public has also become fascinated with it—as attested by publicity and exhibits that followed the James Ford Bell Trust's acquisition of a rare exemplar in 2010.

When Matteo Ricci's 1602 World Map in Chinese (Figure 0.1) was exhibited in several venues in the United States during the following decade, one wonders what the various twenty-first-century visitors may have made of it. Those attuned to thinking about form might have noted the parallel vertical scrolls on which the contents are printed by woodblock. They might have commented

1 We have chosen not to standardize the translation of the title of Ricci's 1602 World Map in Chinese in this volume, preferring to allow authors to choose their own rendering.

2 The literature on Ricci's World Map in Chinese is copious. For an entry into the more recent literature see: Timothy Brook, *Completing the Map of the World: Cartographic Interaction between China and Europe* 全圖：中國與歐洲之間的地圖學互動 (Taibei: Institute of Modern History, Academia Sinica, Lecture Series no. 5, 2020); Mario Cams and Elke Papelitzky, *Remapping the World in East Asia: Toward a Global History of the "Ricci Maps"* (forthcoming from University Hawai'i Press); Cheng Fangyi, "Pleasing the Emperor: Revisiting the Figured Chinese Manuscript of Matteo Ricci's Maps," *Journal of Jesuit Studies* 6, no. 1 (March 11, 2019): 31–43; Michela Fontana, *Matteo Ricci a Jesuit in the Ming Court* (Lanham, MD: Rowman & Littlefield, 2011); Florin-Stefan Morar, "China's Great West: Matteo Ricci's World Map and the Quandaries of European Identity in the Late Ming Dynasty," *Journal of Jesuit Studies* 6 (2019): 14–30; Song Gang, "Re-Locating the 'Middle Kingdom': A Seventeenth-Century Chinese Adaptation of Matteo Ricci's World Map" in *Mapping Asia: Cartographic Encounters between East and West*, eds. Martijn Storms, Mario Cams, Imre Josef Demhardt, and Ferjan Ormeling (Cham: Springer International Publishing, 2019), 185–206; and Qiong Zhang, *Making the New World Their Own: Chinese Encounters with Jesuit Science in the Age of Discovery* (Leiden: Brill, 2015). For Ricci's journals see Matteo Ricci, *China in the Sixteenth Century: The Journals of Matthew Ricci, 1583–1610*, ed. Louis J. Gallagher, s.j. (New York: Random House, 1953). For an overview of the historiography of Jesuit cartography including that of Ricci see Robert Batchelor, "Introduction: Jesuit Cartography," *Journal of Jesuit Studies* 6, no. 1.

© LAURA HOSTETLER, 2024 | DOI:10.1163/9789004684782_002
This is an open access chapter distributed under the terms of the CC BY-NC-ND 4.0 license.

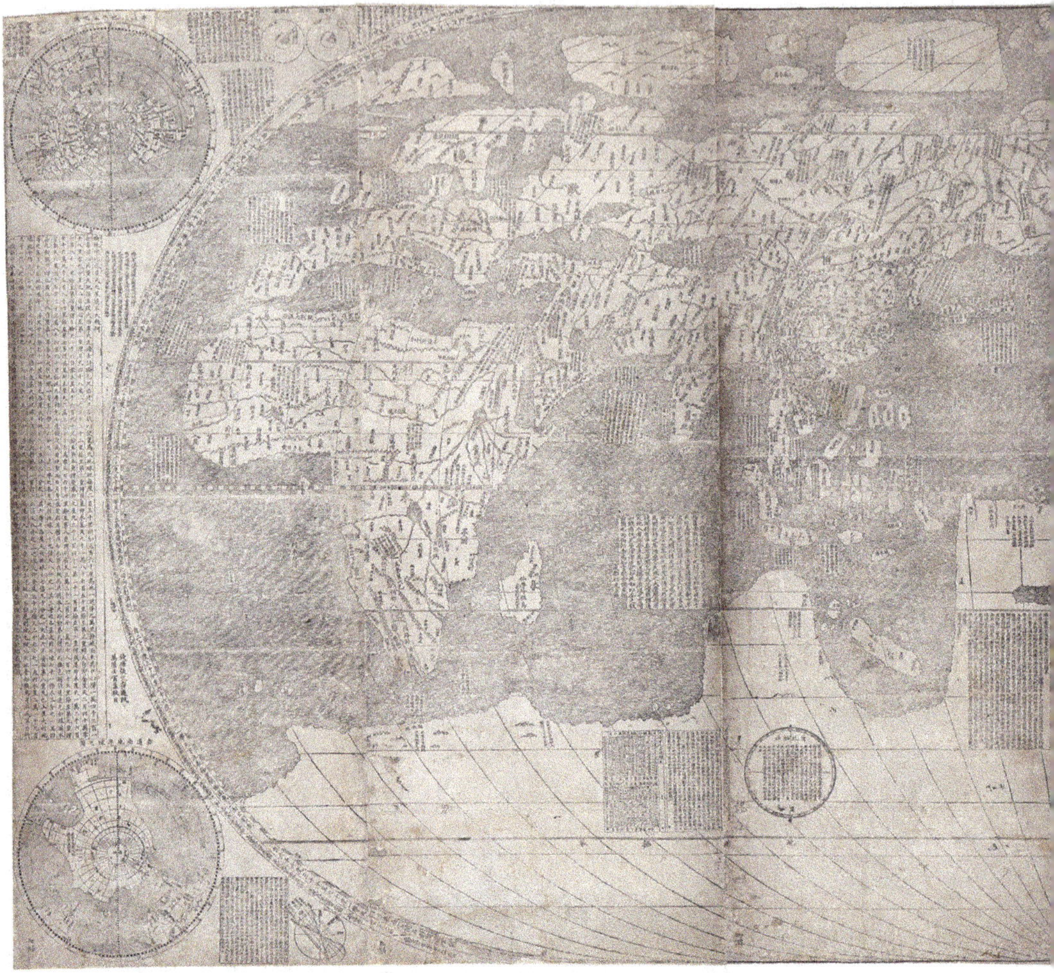

FIGURE 0.1 *Kunyu wanguo quantu* 坤宇萬國全圖 [A Complete Map of the Myriad Countries of the World], more commonly known as Matteo Ricci's 1602 World Map in Chinese. James Ford Bell Library, University of Minnesota.

on the copious amount of Chinese characters the map contains, both in the margins and within the body of the map. They may also have been attentive to the kind of paper on which it was printed and the manner of its printing. Viewers more accustomed to analyzing content might have remarked on the map's depiction of lines of latitude and longitude, its distinctive placement of the continents, and the prominent display of the Pacific Ocean. And they may have waxed eloquent on the geographical features of the map with which they had some history or familiarity. These two distinct approaches to viewing the map are, of course, not mutually exclusive; those drawn first to form would

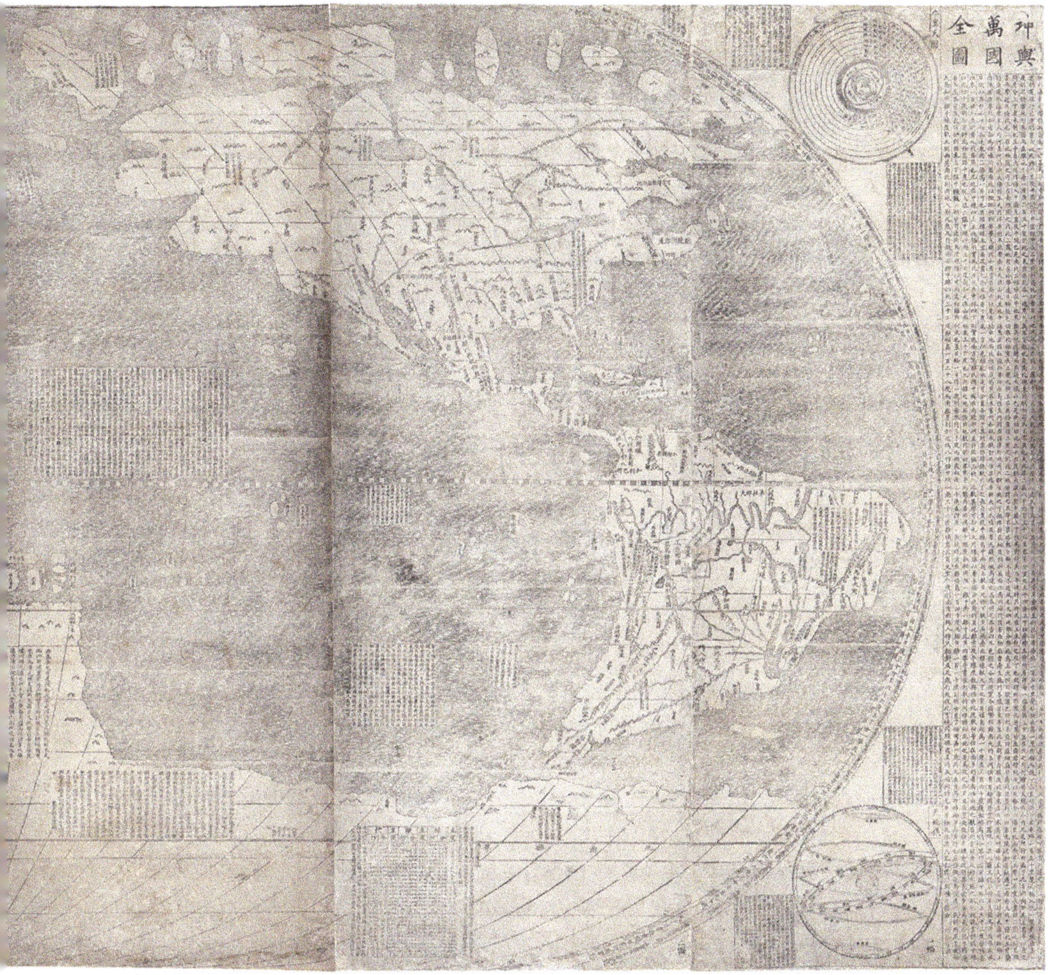

surely have given some attention to the content embodied within it, and those most attentive to content would undoubtedly have noticed the vertical scrolls on which the content is mounted.

And yet, this hypothetical example of different responses to the map on exhibit illustrates two distinct ways of orienting ourselves in relation to the depiction of cartographic knowledge. In the first case, viewers primarily consider the map as an object of material culture, among other similar objects, with which they reflexively compare it. Their vantage point is outside the map, onto which they project an "objective gaze" from a fixed vantage point—their own. As they take in the map, they tend to notice details and differences from other maps they have viewed, and by contrasting and comparing they build up their store of knowledge and expertise. They do so without being required to shift their own position in relation to the map. We can take this illustration even further if we understand their position not only literally according to their actual location in space, but also figuratively in relation to the body of cartographic familiarity they had already acquired, their level of education, cultural orientation, and national allegiance—as well as any number of other aspects of their identity that might influence what they are able to see and their interpretation of it.

In the second instance, while situated similarly in regard to the map, the viewers are more attuned conceptually to *where and how the map is simultaneously placing them*. To understand this approach, bring to mind a shopper at an unfamiliar mall—or a traveler in a complicated subway station—searching for a sign indicating "You are here" in order to orient themselves to the spatial layout in which they find themselves. This approach puts the viewer conceptually *inside* of the map. It demands an openness to thinking about one's own position in relation to someone else's conception of the same space.

Viewers who seek to locate themselves within the map may find their assumptions challenged, their standard points of reference defamiliarized, themselves de-centered. Confronted with radically new conceptions of space, some may reject what they see out of hand as "not making sense." Others may be open to taking in the new information as part of orienting themselves to what is presented on the map and begin to think about their place in the world somewhat differently. The viewers who remain figuratively outside the map may be less disoriented, but also more dismissive of content that conflicts with their familiar touchstones. They, too, may reject new content out of hand—or simply not see it—as when, to my horror, I discovered only recently that I had failed to notice that a map of China I have used occasionally in my classes over the course of several decades includes the controversial nine-dash line. How had I not seen it?

Whether we approach maps primarily by observing their form, or the content contained within that form, we bring with us our own world view, the sum of our geographical knowledge, our own epistemologies made up of the expectations we have formed based on the stories, scholarship, and images with which we are already familiar. What we see on the map might expand or alter that view, pose questions in our minds, or even prompt us to reject competing representations out of hand. If this is the case for us today, imagine the consternation that late Ming viewers of Matteo Ricci's World Map in Chinese may have felt. Responses included accusations of both deception and effrontery. Deception for portraying Ricci's own homeland as being much farther away than was thought possible—and as such minimizing the potential threat posed by his people—and effrontery for his use of the term "Great" (*da* 大) in regard to his homeland, for how could the stature of his homeland possibly be compared to that of the Great Ming (大明)?[3]

One thing a casual viewer would not have been able to discern from an initial look at Ricci's map is that Ricci was, among other purposes, using the map to communicate to his Chinese audience *who he was* and *where he had come from*. As elucidated cogently by Florin Stefan Morar, Ricci's map laid out the geographic parameters according to which he cast his identity in China as a person of the Great Western Ocean, 大西洋人, a rough equivalent for European.[4] The term "Western," or "Westerner," as a loose, but malleable, equivalent for Western Europe (which later also came to include parts of North America) has stayed with us in our casual use ever since (and some would argue much too long).[5]

The staying power of "Western" or "European" in regard to the identification of cartographic practices in contradistinction to a Chinese implied "Other" may have something to do with the fact that established mapping traditions tend to introduce change gradually in ways that affect either the content *or* the form of maps, but not both simultaneously—at least not in any significant measure. For example, content may be added or refined by adding more place names to an existing base map, or the margins of the map may be moved outward to show a larger expanse, possibly reflecting new knowledge. In this way content is added without changing the fundamental characteristics of the layout or basic form of the map, and therefore does not challenge the basic assumptions or orientation of the viewer. Occasionally formal aspects of the map might change as well, such as the different projections employed in world maps by Mercator and Ortelius. Or a map might appear as part of an atlas where it can be securely stored between the covers of a bound book, or take the format of a wall map to be hung or rolled up when not in use. These formal changes often

3 For the first incident see Laura Hostetler, *Qing Colonial Enterprise: Ethnography and Cartography in Early Modern China* (Chicago: University of Chicago Press, 2001), 54, and Kenneth Ch'en, "Matteo Ricci's Contribution to, and Influence on, Geographical Knowledge in China," *Journal of the American Oriental Society* 59, no. 3 (September 1939): 349. For the second, see Morar, Florin-Stefan, "The Westerner," 15. On the Ming as a Great State see Timothy Brook, *Great State: China and the World* (New York, NY: Harper, 2020).

4 Morar, Florin-Stefan, "The Westerner," 14.

5 For an insightful exploration of the ways that the usage of "Western" has morphed over time see Martin W. Lewis and Kären Wigen, *The Myth of Continents: A Critique of Metageography* (Berkeley: University of California Press, 1997), esp. Chapter Two: "Spatial Constructs of Orient and Occident, East and West."

happen without radical adjustments to the content, and so again are accepted as innovations, rather than as challenges to an existing tradition. If the form is familiar, changes in content are less disorienting. Likewise, when the content is familiar, changes in format can more easily be accommodated. This may be why scholars of Chinese cartography have long distinguished between "Western" or "European" maps of China, and "Chinese" cartography. Forms associated with cartographic practices that emerged and evolved in each of these regions have been understood to define the sum of their practice, i.e., to subsume content. These oppositional categories of "European" (or "Western") and "Chinese" as used in relation to early modern cartography are only beginning to be challenged.[6]

Matteo Ricci made major innovations to both form and content in his 1602 World Map in Chinese. Perhaps this is why it is impossible to categorize Ricci's map as either "European" or "Chinese." Note that we refer to his creation as Matteo Ricci's World Map in Chinese, not Ricci's Chinese map. Our language intimates that we see it as a translation of something non-Chinese rendered into that language, for it is not possible conceptually for us to understand Ricci, who was born in Macerata and educated in Europe, as having created something that was in fact Chinese. But at the same time his innovations to form, and to presentation of content, preclude the possibility that the map is a simple translation of something else into a Chinese version. What Ricci did was to give a world map created by Ortelius a new form, which in turn "created a new way of seeing, a new way of thinking." This is the work of the artist.[7]

Changes or innovations to cartography in China stemming from interactions between Chinese cartographers and Ricci's world map have been difficult to see because they did not affect the *form* of Chinese world maps in ways

6 Cordell D.K. Yee, "Traditional Chinese Cartography and the Myth of Westernization," in *The History of Cartography*, 2.2: 170–202 (Chicago: University of Chicago Press, 1994), and "A Cartography of Introspection: Chinese Maps as Other than European," *Asian Art* 5, no. 4 (1992): 28–47. For pushback against these categories see Laura Hostetler, "Global or Local? Exploring Connections Between Chinese and European Geographical Knowledge During the Early Modern Period," *East Asian Science, Technology, and Medicine* 26 (2007): 117–135 and "Contending Cartographic Claims: The Qing in Manchu, Chinese, and European Maps" in *The Imperial Map: Cartography and the Mastery of Empire* (Chicago: University of Chicago Press, 2009), 92–132; Mario Cams, *Companions in Geography: East-West Collaboration in the Mapping of Qing China (c. 1685–1735)* (Leiden: Brill, 2017); Mario Cams and Elke Papelitzky, *Remapping the World in East Asia*; Timothy Brook, *Completing the Map of the World.*

7 Jeanette Winterson, "Testimony Against Gertrude Stein," in Winterson, Jeanette, ed. *Art Objects: Essays on Ecstasy and Effrontery* (London: Jonathan Cape, 1995), 52.

easily recognizable to those outside the tradition.[8] However, Timothy Brook has identified at least one way in which they have been important and lasting. Through linguistic research on the titles of world maps, he has shown that the concept of the "complete map," *quan tu* 全圖, was introduced by Ricci, and that it would become a regular feature of world maps in China in the following centuries.[9] The term also appears regularly in other genres of mapping by at least the Kangxi reign period (1662–1722) when we see it used both in regard to complete maps of China as in A Complete Map of China Divided by Prefecture, (*Fenfu Zhongguo quantu* 分府中國全), or the Kangxi era Complete Map of All Surveyed, *Huangyu quanlan tu* 皇輿全覽圖 itself, in which the word "surveyed,"—as in "seen"—modifies what constitutes the completeness of the map, that completeness being equivalent neither to Qing China's borders nor the world itself.[10] *Quantu* also reappears in the 1767 *Da Qing wannian yitong dili quantu* 大慶萬年一統地理全圖 [A Complete Geographical Map of the Great Qing's Everlasting Unification], and its many subsequent early nineteenth-century versions (Figure 0.2). Interestingly, while this Qianlong era map centers China in a different way than Ricci's map, like his, it was constructed in a series of vertical panels.[11]

Whether focusing on form, content, context, or their intersection, it is clear that there is always more to learn about Matteo Ricci's World Map in Chinese. The essays in this volume bring together a variety of approaches stemming from different scholarly disciplines, as well as the voices and insights of authors of different national origins. Collectively the contributions cover a

8 Just as the Chinese origin of information about Asia on maps made in the European style is not evident from their form alone. Luo Hongxian's *Guang yu Tu* 廣輿圖 [Enlarged Territorial Atlas], formed the basis for early Jesuit maps of China. Brook, *Completing the Map of the World*, 104.

9 Superseding the earlier *huayi tu* 華夷圖, or "Chinese-and-barbarians maps." Brook, *Completing the map of the world*, 76.

10 For a detailed description of each of these maps and their similarities despite their differences in form, see Laura Hostetler, "Contending Cartographic Claims: The Qing in Manchu, Chinese, and European Maps" in *The Imperial Map: Cartography and the Mastery of Empire* (Chicago: University of Chicago Press, 2009), 93–132.

11 The map was reprinted in many editions and under a variety of names in the early nineteenth century. The version held by the Bibliothèque nationale de France (Rés. Ge A 1096, Cartes et Plans) may be viewed on their website, Gallica. This map, in its various versions, has recently received quite a bit of attention. See Richard A. Pegg, "China: The Realm," in *Cartographic Traditions in East Asian Maps* (Honolulu: University of Hawai'i Press, 2014), 18–27, and the essays by Timothy Brook, Laura Hostetler, and Matthew Mosca on the website of the Rockefeller Library of Brown University connected with the *Depicting Glory* site, under "Unit One: Mapping the Empire and the Mandate" at: https://storymaps.arcgis.com/stories/27dae33edbb141ba8af26c697f97fa29.

FIGURE 0.2 *Da Qing wannian yitong tianxia quan tu* 大清萬年一統天下全圖 [The Great Qing Dynasty's Complete Map of All under Heaven], 1811. Library of Congress.

大清萬年一統天下全圖

period of almost 500 years, from the establishment of the Jesuit order in 1540 to the present day. Using a geological metaphor, one can think of the different sections of the book in terms of strata. Part 1 begins the book by exploring the early modern historical context of Jesuit missions abroad, the networks of communication the order established, and publications about the China mission specifically. Part 2 shifts us from the broad context of Jesuit missions in China and their literary production to the production and dissemination of Ricci's map, and as well as other world maps in the style of Ricci, through that of Ferdinand Verbiest's *Kunyu quantu* 坤輿全圖 (1674). Part 3 addresses engagement with Ricci's map and its legacies in East Asia in light of the reception, circulation, and propagation of related maps in Korea, Japan, and China over time showing how in each case adoption or adaptation of the Ricci map and its derivatives was affected by domestic considerations. Within these sections, each author has been invited to write from the vantage point of their own expertise. Coverage is not meant to be comprehensive, rather the reader may find it useful to think of the essays within each part as magnified insets, as on a map, each giving a closer, more detailed, look at a specific aspect of the theme of that section of the volume.

In Part 1, "Jesuit Circuits of Communication and Publication," José Casanova provides a synthetic overview of the importance of Jesuit circuits of travel to establishing global networks of information gathering and exchange. Robert Danieluk takes a deep dive into the practices of written correspondence and reporting that were part of the structure of the missions. Paul Begheyn describes Jesuit descriptions of China published in the Netherlands up to the 1730s.

A more detailed overview of each of these chapters may prove useful to the reader. Casanova sets the scene with a chapter entitled "Jesuit Contributions to Global Connectivity and Global Consciousness in the Early Modern Era." In sketching out for us the multiply-inflected nature of the global connectivity of the order, he holds up a picture of global connection and cultural interaction apart from the extractive colonial and Orientalist practices with which globalization is more commonly associated. Alert to the personal spirituality developed through Ignatian practices that opened up members of the Jesuit order to personal discernment in their cross-cultural interactions, he describes how this freedom was always balanced by the rigorous expectations of the order as well as subject to the necessity of interacting with external structures (politics and power). Not only does he offer a synthetic overview, allowing us to understand the context that gave birth to the possibility of the Ricci map, in doing so he also provides a framework through which we can approach the future with a sense of possibility not often evident in more materialist analyses.

Casanova begins his essay with the dual assertions that "Global mobility was culturally encoded, as it were, into the make-up of the Jesuit order from its inception," and that "Jesuit ministries encompassed all kinds of activities and spheres that today may be considered 'secular' but that they regarded as an intrinsic part of their religious mission" (28). No division was seen between sacred and secular, the word of God and the fruits of God's creation. Following Ignatius, "Jesuits were to find God 'in all things'" (31). The pursuit of religious truth and scientific knowledge were one.

The Jesuits' global orientation and search for knowledge together account for their role as what Casanova terms "pioneer globalizers." Furthermore, the flow of information back and forth between Jesuits in the field and their network of colleges in Europe allowed them to become premier educators during a period when demand for education was rising and individual states had not yet stepped in to create institutions to fill the demand.

Casanova credits both external and internal institutional structures for the specific ways that Jesuits interacted with the world and propagated the knowledge that they accrued. External structures included Iberian colonial expansion (which facilitated travel); the Catholic revival; and the "culture of Renaissance Christian Humanism." Internally, the order was distinguished by pairing a hierarchical structure that required direct reporting with the cultivation of Ignatian spirituality in which individuals were encouraged to foster a spirituality grounded in personal inquiry. Together these internal and external structures helped to set the stage for a culture of encounter that characterized Jesuit missions abroad, each of which was shaped by the specific circumstances of its locality. Ricci's world map, and subsequently Verbiest's, grew naturally out of this context in which global travel had become essential to the Jesuit order and the Propagation of the Faith, and the pursuit of scientific knowledge was considered part of the search for truth.

Robert Danieluk's chapter, "From Manuscript to Print: At the Origins of Early Jesuit Missionary Strategies of Communication," develops Casanova's observation that "The Society of Jesus's character as a highly centralized and hierarchic transnational organisation with a highly flexible and mobile structure, was made possible by a well-developed system of communication within the entire global Jesuit network based on regular and periodic letter writing and reporting to the superiors, to the provincials, and to the Roman curia."[12] This system of communications formed the basis of the global network that allowed Ricci's map to come into existence.

Danieluk's thesis demonstrates how Jesuit attention to geography, both in the form of cartography and descriptive writings about other places, grew

12 Casanova, Chapter 1 in this volume, p. 32.

naturally out of the *formula scribendi*, the rules of Jesuit correspondence initially set down by Ignatius of Loyola himself and formally approved and modified during the second half of the sixteenth century. Addressing the "why," "what," and "how" of Jesuit epistolary writing, he shows that the tradition of letter writing was initiated with the dual goals of maintaining unity, or connection, between Jesuit missionaries as they spread out into the world and reported to superiors within the hierarchy. In their writing they made a distinction between material directed specifically to the addressee, and that of more general interest that could be shared more widely, including through publication. Such letters soon began to include quite a bit of information on local customs, climate, geography, flora, fauna, and other points of interest. It was not unusual for geographical descriptions to use coordinates of latitude to identify locations, or for the letters to be accompanied by hand-drawn maps. The Jesuits, especially once their letters began to be published, consciously wrote in a manner designed to "edify and bring instructive news" (69) rather than devoting attention only to matters that specifically concerned the order.

Danieluk concludes his contribution by introducing us to a little-known work on China and Tartary that was researched and compiled by the Polish Jesuit Tomasz Ignacy Szpot Dunin, S.J. (ca. 1645–1713) in Latin. Szpot's work "describes the geography, social organisation, customs, and religion" of China and Tartary, and in this respect accords with earlier injunctions on what would interest readers, and can be seen something of a precursor to Jean Baptiste Du Halde's better known *Description … de la Chine*.[13] The work includes not a few maps, several of which are reproduced for the first time in this volume.

Paul Begheyn's "Dutch Publications on the Jesuit Mission in China in the Seventeenth and Eighteenth Centuries" moves the focus from Jesuit epistolary writing to the various genres of publication to which the China mission contributed. Focusing specifically on the Dutch Republic, a hub for the publication of books and atlases during the early modern period, Begheyn provides a close chronological description of works produced—or based on information transmitted—by the China mission between 1615 and 1736, after which time this spate of publication ceased.

13 Du Halde, J.-B., *Description géographique historique, chronologique, politique, et physique de l'empire de la Chine Et de la Tartarie Chinoise: Enrichie des cartes générales et particulières de ces pays, de la carte générale & des cartes particulières du Thibet, & de la Corée, & ornée d'un grand nombre de figures & vignettes gravées en taille-douce* (Paris: Chez P.G. Le Mercier, 1735).

Begheyn's examination of these publications demonstrates that the Jesuits served as a conduit of geographical information not only from Europe to China, as with Ricci's map, but also from China to Europe, as seen in the publication of Martin Martini's *Atlas Sinensis* in the 1660s. Largely based on Luo Hongxian's *Guang yu tu* (a copy of which also reached Europe via the Jesuits), Martini used the geographical information that appeared in his *Atlas* to foster connections with the Dutch East India Company (VOC)—a relationship he cultivated in order to facilitate travel to China for the order's members. This series of publications culminated with the Dutch edition of Jean Baptiste Du Halde's *Description ... de la Chine*, a work well known for its inclusion of copperplate engravings of the surveys of the Qing empire that were commissioned by the Kangxi emperor in the 1710s and carried out with significant Jesuit involvement. These maps remained the state of the art for geographical knowledge of China in Europe for over a century.

In Part 2, "Jesuit World Maps and their Introduction to Asia," Qiong Zhang invokes the concept of the "contact zone" to explore both the intersection in the interests of Chinese and Jesuit scholars and the different sets of interests and constraints that led to very different interpretations of the origin and significance of the scholarship associated with Ricci's map. Aoyama Hiro'o undertakes a fine-grained study of the introduction of Ricci's World Maps into Edo Japan. Paola Demattè focuses on maps and other geographical representations of the world made by Europeans between the times of Ricci and Verbiest. And finally, Mark Mir concentrates exclusively on Verbiest's 1674 *Kunyu quantu* 坤輿全圖. As a group, these essays devote particular attention to the choices regarding types of historical, ethnographic, and religious references and scientific explanations included on the maps, as well as their sources and later the circulation and interpretation of their content.

It is worth exploring the arguments of each of these chapters in a bit more depth here. In her essay, "Parallels, Engagement, and Integration," Qiong Zhang approaches Matteo Ricci's cartographic work from the perspective of an "intertwined global early modernity." Drawing on Mary Louise Pratt's concept of the contact zone, Zhang demonstrates how in selecting what to include on his map Ricci carefully considered the intellectual milieu of his Ming audience—including a taste for exotica and an already sophisticated cartographic tradition. At the same time, geo-political imperatives influenced his choice of what to share about the countries of Europe. Identifying the European origins of known aggressive powers in maritime Southeast Asia, and faithfully portraying conflict within a Christian Europe, would have had the potential to undercut Chinese trust in the dual scientific and religious outlooks that Ricci was so keen to put forward. For these reasons, he crafted a delicate balance between "what his audience wished to find and what he wanted them to know."[14]

The repercussions of Ricci's choices would be felt until at least the beginning of the nineteenth century. Zhang argues specifically that in the century following the production of Ricci's initial world map, the proclivity he had shown toward satisfying the Ming interest in exotica was heightened, likely at the expense of the more fundamental geographical information he was seeking to convey. Correspondingly, a lack of clarity on European countries, particularly with regard to the details of their differences and rivalries—as well as their colonial activities in Asia—would have lasting repercussions. Gradually, nativist forms of Chinese cartography—both symbolic and measurement-based—would come to show influence from the view of the world that Ricci and his colleagues had presented. For example, as the new theory of the sphericity of the earth took hold, earlier descriptions of a flat earth and round heavens came to be understood metaphorically. Yet, Ricci's framing of some of the science he used as having already been practiced in China earlier strengthened nativist attitudes critical of the Jesuits presence and "their" science as well.

Aoyama Hiro'o, Professor Emeritus of the National Museum of Japanese History, provides us with a detailed overview of the editions, copies, and derivative works based on Matteo Ricci's 1602 world map housed in Japan. In a three-pronged approach Aoyama begins with a close reading of the six extant exemplars of the 1602 Ricci World Map, carefully examining the alterations and revisions that they underwent. He also rethinks the relationship of the famous 1645 *Bankoku Sōzu* 萬國総圖 [General Map of the Myriad Countries of the World] to the Ricci maps, demonstrating that the 1602 edition was most likely not a direct precursor, but rather that the map relied at least in part on Ricci's double hemisphere map, even if projective consistency was ignored in the adaptation.

Aoyama then turns to hand-copied exemplars, which vary in their degree of faithfulness to the originals. As we will see too in Lim's chapter on Korean reproductions of the Ricci map in Part 3, in Japan later renditions were sometimes enlisted in the service of national self-definition. The astronomy school of the Sendai region, a strong proponent of *kokugaku*, or "national learning," read Ricci's description of the country's temperate climate as resonating with its own association of bountiful harvests and the cultivation of rice with Japan's "unique spiritual culture." Furthermore, widely reprinted and distributed maps derived from the Ricci maps, including a version of the *Bankoku*

14 Zhang, Chapter 4 of this volume, p. 100.

Sōzu, the *Bankoku Sōkai Zu* 萬國総界圖 [Map of the General Contours of the Myriad Countries of the World], 1688 and 1708, and a late eighteenth-century map by Sekisui Nagakubo that was widely popularized, eventually brought the look of Western-style world maps into the Japanese popular consciousness, aided in part by the status that the Ricci map had attained as a classic artifact in Japan by that time.

In "*Ad maiorem Dei gloriam*: Jesuit Mapping in China by Giulio Aleni, Francesco Sambiasi, Niccolò Longobardi, Manuel Diaz and Others," Paola Demattè examines the corpus of geographical scholarship produced by Jesuits between the times of Ricci and Verbiest. Bringing an art historian's attention to pictorial and textual detail, sources, and particularities among the maps, geographical treatises, and globes produced by Aleni, Adam Schall von Bell, Francesco Sambiasi, Longobardi, and Manuel Diaz the younger, Demattè explores the nuances among them, ultimately concluding that during this roughly seventy-year span we see a "publishing continuum." This continuity, she argues, is located in the shared purpose of these geographical works: to introduce into China the "Christian aims of the mission" as well as to promote the "reformist ideals of Chinese converts and sympathizers."

Mark Stephen Mir's chapter on Ferdinand Verbiest's *Kunyu quantu* 坤輿全圖 [Complete Map of the Earth] (1674) provides a close reading of the map, including its textual supplements, in light of the development of both European and Chinese science. Drawing on his extensive bibliographic skills, Mir identifies an array of both European and Chinese sources used by Verbiest in the creation of his map, even tracing the derivation of many of the illustrations of animals. He also traces the reappearance of some of the textual components of the map in later Chinese publications. Full translations of a number of the sections, including those on the movement of air, the four elements, earthquakes, humankind, the South and North Poles, and the relationship of earth to the heavens, give the reader a first-hand glimpse at the kinds of information and logics contained in the rich cartouches of the Verbiest map.

Mir makes the case that scientists, in both Europe and China, were eager for more information about the lands and peoples of the world, as well as the workings of the cosmos. While steeped in different traditions, they shared many interests and held some common approaches to understanding the natural world. Although the continental scheme that Ricci's map had introduced was new to China, the amalgam of early modern philosophical and scientific theories expressed in Verbiest's reedition was "often in harmony with contemporary Chinese science," particularly in regard to the conception of space and time. As the 1669 contest at the Qing court regarding calendrical accuracy attests, the rulers of China accepted science as an arbiter of truth, or at least as a necessary accoutrement to monarchical legitimacy.

Collectively, the chapters of Part 3, "Nationally-Inflected Reverberations of Ricci's Map in Asia," span the eighteenth to twenty-first centuries, but each is concerned with a specific context and time period within that broad span. Lim Jongtae demonstrates that in Chosŏn Korea Ricci's map came to represent the achievements of the Ming dynasty (during which it was produced), and its preservation and reproduction served as a means for the Korean state to express admiration for the Ming, implicitly criticize the Qing, and position itself as the true heir to Ming values. Kären Wigen explains how in Japan, Buddhist, "Japanist," and Neo-Confucian cartographic traditions adopted the European five-continent schema introduced by Ricci, accepting Japan as part of "Asia" in ways that represented Japan as leading the region. Laura Hostetler argues that in China scientific mapping that relied on the coordinate system and the related concepts of latitude and longitude introduced by Ricci became the basis on which expansive maritime claims were marked out in unprecedented ways during the twentieth century. In other words, as coordinate mapping was introduced and subsequently accepted in each Korea, China, and Japan, these "scientific norms" were also inflected with meanings and uses distinctive to each region's political and cultural context.

More specifically, in "Representing an Ideal World Order of the Past," Lim Jongtae provides important context for helping us understand the significance of the eighteenth-century Korean reproductions of two seventeenth-century Jesuit world maps in Chinese: Ricci's 1608 map (reproduced in 1708), and Giulio Aleni's world map from the *Zhifang waiji* 職方外紀 [Record of Foreign Lands] (reproduced ca. 1790). Historically, Korea had produced practically-oriented maps of its own lands with great precision. Yet when it came to world maps, as a tributary of China, Korea accepted and reproduced maps emanating from its larger neighbor. Generally speaking, Korean royal interest in world maps coincided with periods when Korea was redefining its relationship to China. The Chosŏn dynasty, when these two maps were reproduced, was just such a period. Most salient for our purposes is that the Chosŏn dynasty was not happy with the Manchu rulers of China.

Lim suggests that the Chosŏn selection of these two world maps for production was rooted in their origin during the Ming, thus implying Chosŏn stood as that Chinese dynasty's legitimate cultural heir. Thus, perhaps counterintuitively to us today, Chosŏn engagement with Western calendrical science and cartographic practices was framed in terms of Ming loyalism. In Lim's words, "Jesuit maps appealed to the Chosŏn officials not so much for their technical potential for 'scientific' cartography as for their historical origin in Ming China" (239). For the same reason, it is no coincidence that the Verbiest map, completed only under the Qing, was not reproduced in Korea during the eighteenth century.[15]

Wigen, in "Entering Asia: The Repositioning of Japan," takes a close look at cartographic representations made within the Neo-Confucian, Buddhist, and "Japanist" traditions in Japan between Ricci's time and the late nineteenth century so as to demonstrate how each of these traditions gradually came to accept and even embrace Japan's position within the European continental scheme. Neo-Confucians tended to think of China as the center, which was increasingly depicted cartographically as the center of *Asia*. In Buddhist cartography, by contrast, the Indian subcontinent was central because of the religious significance of its sacred sites, particularly Mount Meru. For them it was this region, relative to which Japan and Korea occupied a marginal position, that became increasingly associated with the European concept of Asia. Proponents of a Greater Japan favored world maps based on Ricci's that showed Japan at the center, and further rotated the view a quarter turn, making Japan appear to be on top of Asia, as if in a leading position. Although each of these traditions had their own ways of adapting to—or adopting—the five-continent system, by the beginning of the Meiji period, it was widely accepted that Japan was part of Asia. By 1885 the assumption of Japan as part of Asia was so engrained it was polemically possible to conceptualize it "leaving" the continent.

In "China's Nine-Dash Line: Cartographic Science and the Adoption of New Map Languages in the Transition from Empire to Nation State," Hostetler historicizes the emergence of China's maritime claims in the South China Sea, setting them in the context of the emergence of the nation-state in the 1930s. The chapter further explores actions China took in the 2010s to concretize these ongoing cartographic claims through transformation of the maritime landscape itself. But an exploration of the long durée is also important; coordinate mapping as introduced to China by Ricci in 1602 was gradually so thoroughly adopted over the course of 400 years that it became the logical medium for the nascent Republic of China to use when inscribing its claims not only to extensive territories that had been ruled by the Qing, but to vast maritime expanses as well.

Until the middle of the nineteenth century a variety of mapping traditions co-existed in China, making the influence of the cartographic concepts Ricci introduced into China somewhat difficult to trace. By the early eighteenth

15 The Verbiest map was reproduced in Korea in 1860, perhaps in response to new Western threats in the region.

century, coordinate mapping based on both surveys and triangulation was clearly employed by the Qing court, yet the versions of these maps that circulated most widely were devoid of markings of longitude and latitude. Furthermore, until at least the late eighteenth century the court continued to sponsor more traditional style maps of *tianxia*, "all under heaven." However, in the aftermath of the Opium War, Chinese outside of the context of the court became engaged in geographical research and began to adopt the coordinate system, integrating it more broadly into Chinese mapping practices. By the early twentieth century, maps defining the Chinese nation in the context, and terms, of the global integration of space had gained a strong foothold, and were edging out more "traditional" maps.[16] By the mid-twentieth century, the use of coordinate mapping had become so engrained in China that it was used as the preferred medium to make territorial claims. Today this practice continues in the South China Sea, but with the inclusion of the controversial nine- (or ten-, or eleven-) dash line laying claim in unprecedented ways to strategically important maritime areas beyond the mainland.

The study of maps is dependent on both their physical preservation and our ability to have access to them. The Postlude, "Reflections on the Curation of Cartographic Knowledge," examines questions of collection-building and curation from the perspectives of two librarians whose institutions house copies of the 1602 Ricci and 1674 Verbiest world maps, respectively. Marguerite Ragnow, Curator at the James Ford Bell Library, addresses the role of the preservation and circulation of knowledge in the form of the written word. Situating maps in this broader context of both printed and manuscript documents, she highlights the contingent nature of what has come down to us and the centrality of preservation of the written word for the transmission of knowledge. Ralph Ehrenberg, who served as Chief of the Geography and Map Division of Library of Congress until his retirement in 2016, focuses on the contours and organization of the East Asian Map Collections at the Library of Congress, addressing both how the East Asian Map Collection was built over time, and the library's continuing commitment to open access.

In "Writing Technologies and Special Collections: Agents and Arbiters of Change through the Transmission of Knowledge," Ragnow engages the thorny question of what constitutes knowledge, and how knowledge is constituted. She demonstrates that both individuals and institutions play an important role in determining what records are conserved and passed on, thereby shaping what kinds of knowledge will be transmitted to future generations and in what

16 For more on this process internationally see Laura Hostetler, "Mapping, Registering, and Ordering: Time, Space, and Knowledge," in Peter Fibiger Bang, ed., *The Oxford World History of Empire* (Oxford: Oxford University Press, 2020), 1: 288–317.

formats. There are many steps in this process of knowledge creation: What information is included—and thus transmitted—by any author in a given work? On what basis do libraries and individual curators make decisions on what to collect and preserve and how to organize those materials? How might these materials be received and interpreted in another cultural context? All of these factors go into forming the scope of the collections of documents with which scholars are able to work today. She concludes by recognizing that transmission is never a straightforward process, not only because of the exigencies of the chain of transmittal, but also because we cannot know how these artifacts might be received and interpreted by later generations or in different cultural contexts.

In "East Asian Map Collections in the Library of Congress: A Unique Source for the Study of Cartography and East–West Cultural Exchange," Ehrenberg provides an overview of the development of the Library of Congress's Asian map collection, which includes Verbiest's 1674 *Kunyu quantu*. The earliest maps of East Asia were acquired through government exchanges, and purchases made by Americans living overseas. Arthur Hummel, Director of the Asia Division (then Orientalia) from 1928 to 1964, facilitated the acquisition of the Verbiest map as well as a significant collection of additional Chinese maps from the Ming and Qing periods. The information the chapter brings together on the context of collecting and policies of open access raise interesting questions when read in light of global events and US power during the period that the collections were being built. Further information on these Ming and Qing maps held in the Library of Congress, and details on obtaining digital access, can be found in the collaborative work *Reading Imperial Cartography: Ming-Qing Historical Maps in the Library of Congress*,[17] to which Ehrenberg refers.

We invite readers to approach these essays as a compilation of map details, or cutouts, each of which enlarges a specific view for the reader's consideration. The views are not comprehensive, nor are they uniform, for neither is the terrain. The texture, scale, and angle of the vantage points offered for consideration vary as well. Each chapter offers a richness of its own even as the framework of the book invites readers to draw the pieces into conversation with each other. In these times when so much is contested both globally and within our own national arenas—as well as within the academy—one contribution we offer is a place where perspectives and points of view from those

17 Lin Tianren 林天人 and Zhang Min 張敏, *Huangyu suolan: Meiguo Guohui tushuguan suo zang Ming Qing yutu* 皇輿搜覽：美國國會圖書館所藏明清輿圖 [*Reading Imperial Cartography: Ming-Qing Historical Maps in the Library of Congress*] (Washington, DC: Library of Congress in association with the Academia Sinica Digital Center, 2013).

who stand in different places, and therefore see the objects of our study from different angles, can be read together. We hope that readers of this volume, like seventeenth-century viewers of Ricci's World Map in Chinese, may find themselves engaged by new content and opened up to new perspectives on what they may have expected would be familiar territory.

Bibliography

Batchelor, Robert. "Introduction: Jesuit Cartography." *Journal of Jesuit Studies* 6, no. 1 (March 11, 2019): 1–13.

Brook, Timothy. "China's World Map Transformed: The Complete Map of All under Heaven as Unified by the Qing Great State for Ten Thousand Years." *Depicting Glory*, n.d. https://storymaps.arcgis.com/stories/27dae33edbb141ba8af26c697f97fa29.

Brook, Timothy. *Completing the Map of the World: Cartographic Interaction between China and Europe* [*Quantu: Zhongguo yu Ouzhou zhijian de dituxue hudong* 全圖：中國與歐洲之間的地圖學互動]. Institute of Modern History, Academia Sinica, Lecture Series, no. 5. Taibei: 2020.

Cams, Mario. *Companions in Geography: East-West Collaboration in the Mapping of Qing China (c. 1685–1735)*. Leiden: Brill, 2017.

Cams, Mario, and Elke Papelitzky. *Remapping the World in East Asia: Toward a Global History of the "Ricci Maps."* Honolulu: University of Hawai'i Press, forthcoming.

Chen, Kennth. "Matteo Ricci's Contribution to, and Influence on, Geographical Knowledge in China." *Journal of the American Oriental Society* 59, no. 3 (1939): 325–359.

Cheng Fangyi. "Pleasing the Emperor: Revisiting the Figured Chinese Manuscript of Matteo Ricci's Maps." *Journal of Jesuit Studies* 6, no. 1 (March 11, 2019): 31–43.

Du Halde, J.-B. *Description géographique, historique, chronologique, politique, et physique de l'empire de la Chine et de la Tartarie chinoise: enrichie des cartes générales et particulières de ces pays, de la carte générale & des cartes particulières du Thibet, & de la Corée, & ornée d'un grand nombre de figures & de vignettes graveees en taille-douce*. 4 vols. Paris: Chez P.G. Lemercier, 1735.

Fontana, Michela. *Matteo Ricci: A Jesuit in the Ming Court*. Lanham, MD: Rowman & Littlefield, 2011.

Hostetler, Laura. "Complete Map of All Under Heaven Unified by the Great Qing." *Depicting Glory*, n.d. https://storymaps.arcgis.com/stories/7ea2ac73b94744f6bad68b89ae36297e.

Hostetler, Laura. "Contending Cartographic Claims: The Qing in Manchu, Chinese, and European Maps." In *The Imperial Map: Cartography and the Mastery of Empire*, edited by James R. Akerman, 92–132. Chicago: University of Chicago Press, 2009.

Hostetler, Laura. "Global or Local? Exploring Connections between Chinese and European Geographical Knowledge During the Early Modern Period." *East Asian Science, Technology, and Medicine* 26, no. 1 (June 15, 2007): 117–135.

Hostetler, Laura. "Mapping, Registering, and Ordering: Time, Space, Knowledge." In *The Oxford World History of Empire*, vol. 1, *The Imperial Experience*, edited by Peter F. Bang, C.A. Bayly, and Walter Scheidel, 288–317. Oxford University Press, 2020.

Hostetler, Laura. *Qing Colonial Enterprise: Ethnography and Cartography in Early Modern China*. Chicago: University of Chicago Press, 2001.

Lewis, Martin W. and Kären Wigen. *The Myth of Continents: A Critique of Metageography*. Berkeley: University of California Press, 1997.

Lin Tianren 林天人 and Zhang Min 張敏. *Reading Imperial Cartography: Ming-Qing Historical Maps in the Library of Congress [Huangyu suolan: Meiguo guohui tushuguan sou cang Ming Qing yutu* 皇輿搜覽:美國國會圖書館所藏明清輿圖]. Washington, DC: Library of Congress in association with the Academia Sinica Digital Center, 2013.

Morar, Florin-Stefan. "China's Great West: Matteo Ricci's World Map and the Quandaries of European Identity in the Late Ming Dynasty." *Journal of Jesuit Studies* 6 (2019): 14–30.

Mosca, Matthew W. "Introduction to Da Qing Wanning Yitong Tianxia Quantu." *Depicting Glory*, n.d. https://storymaps.arcgis.com/stories/ae532065bd764f07b0c00bfb99e9713d.

Pegg, Richard A. "The Realm." In *Cartographic Traditions in East Asian Maps*, 18–27. Honolulu: University of Hawaii Press, 2014.

Ricci, Matteo and Nicolas Trigault. *China in the Sixteenth Century: The Journals of Matthew Ricci, 1583–1610*. New York: Random House, 1953.

Song Gang. "Re-Locating the 'Middle Kingdom': A Seventeenth-Century Chinese Adaptation of Matteo Ricci's World Map." In *Mapping Asia: Cartographic Encounters between East and West*, 185–206. Cham: Springer International Publishing, 2019.

Winterson, Jeanette. "Testimony Against Gertrude Stein." In *Art Objects: Essays on Ecstasy and Effrontery*, 45–60. London: Jonathan Cape, 1995.

Yee, Cordell D.K. "A Cartography of Introspection: Chinese Maps as Other than European." *Asian Art* 5, no. 4 (1992): 28–47.

Yee, Cordell D.K. "Traditional Chinese Cartography and the Myth of Westernization." In *The History of Cartography*, vol. 2, bk. 2 *Cartography in the Traditional East and Southeast Asian Societies*, edited by J.B. Harley and David Woodward, 170–202. Chicago: University of Chicago Press, 1994.

Zhang, Qiong. *Making the New World Their Own: Chinese Encounters with Jesuit Science in the Age of Discovery*. Leiden: Brill, 2015.

PART 1

Jesuit Circuits of Communication and Publication

CHAPTER 1

Jesuit Contributions to Global Connectivity and Global Consciousness in the Early Modern Era

José Casanova

In this essay I would like to present the World Maps of Matteo Ricci and Ferdinand Verbiest as paradigmatic illustrations of Jesuit contributions to global connectivity and global consciousness in the early modern era.[1] In fact, both maps do not stand alone, but are rather part and parcel of the many other relevant contributions by numerous Jesuit astronomers and cartographers, botanists and naturalists, linguists and ethnographers who travelled all over the globe and pioneered some of the most significant cultural exchanges between the cultures and peoples of the world during the so-called "First Globalization."[2] One should not forget, however, that those Jesuit scientists were first and foremost Christian missionaries. Without taking this basic fact into account one cannot understand how they could have become pioneer globalizers in "the Age of Discoveries."

A few decades after their official foundation in 1540, the Society of Jesus had "missions" literally all over the globe, not only in the lands of Christian (Catholic, Protestant, and Orthodox) Europe, but in the Americas, Africa, and Asia. Following Roland Robertson's broad definition of globalization as the increased awareness of the unity of the world as a whole, taking the globe as a focus for human activities, one could argue that the Jesuits have been the first organized group in history to think *and* to act globally. Moreover, they did this before there were institutionalized global structures that could sustain such global practices. In a sense, the Jesuits' global consciousness transcended existing patterns of global connectivity, but their own global practices constructed

1 This essay draws upon two of my publications: José Casanova, "Jesuits, Connectivity, and the Uneven Development of Global Consciousness Since the Sixteenth Century," in Roland Robertson and Didem Buhari-Gulmez, eds., *Global Culture: Consciousness and Connectivity* (Farnham: Ashgate, 2016), 109–126; and "The Jesuits Through the Prism of Globalization, Globalization Through a Jesuit Prism," in Thomas Banchoff and José Casanova, eds., *The Jesuits and Globalization. Historical Legacies and Contemporary Challenges* (Washington, DC: Georgetown University Press, 2016), 261–285.

2 Geoffrey C. Gunn, *First Globalization: The Eurasian Exchange, 1550–1800* (Lanham, MD: Rowman & Littlefield, 2003).

© JOSÉ CASANOVA, 2024 | DOI:10.1163/9789004684782_003
This is an open access chapter distributed under the terms of the CC BY-NC-ND 4.0 license.

novel forms of global connectivity that contributed, at times directly but mostly indirectly, to the formation of global structures. In the early modern phase of globalization, up through the mid-1700s, no other group contributed so much to global connectivity and, through their correspondence and cultural and political influence, to a global consciousness linking the four quadrants of the world. The Society of Jesus, one could argue, was the first global NGO *avant la lettre*.

I first examine the opportunity structures, external and internal to the Society of Jesus, that help us to understand better the socio-cultural context of Jesuit practices that made possible Ricci's and Verbiest's world maps. Secondly, I focus briefly on some of the cultural aspects of the production of those maps in order to understand what could be called the characteristic Jesuit global "way of proceeding."

1 External Opportunity Structures

The Iberian colonial expansion offered the most basic opportunity structure for the Jesuit global mission. It made possible the connection of the Old World and the "discovered" New World, linking the East and West "Indies," thus forming for the first time one truly global world in novel transatlantic and transpacific exchanges. In May 1582 Alonso Sánchez, a Spanish Jesuit arriving from the Philippines met in Canton with Francesco Pasio, who had just arrived in Goa with his fellow Italian Jesuit, Michele Ruggieri, to begin the China mission. According to a typical Jesuit account of that meeting, "they embraced each other with great joy and gladness, giving thanks to the Lord that one of them having come by the eastern route and the other by the western, they had in that embrace encompassed the globe, like true sons of Ignatius, in fulfilment of their Institute."[3]

The eastern route referred to the Portuguese maritime route to the "Indies," circumnavigating Africa to reach Goa and radiating from there onto East and Southeast Asia. The western route was the one used by the Spaniards to reach the West Indies, then crossing the land of Mexico or New Spain, and continuing from Acapulco sailing the Pacific to the Philippines, joining the East and Southeast Asian routes. The Portuguese *Padroado régio* and the Spanish *Patronato real* constituted indeed the dual foundation for the Jesuit global mission. But the very nature of the Ricci and Verbiest world maps, produced in

3 Quoted in H. de la Costa, S.J., *The Jesuits in the Philippines, 1581–1768* (Cambridge, MA: Harvard University Press, 1961), 42.

China for the Chinese, not for any Western power or Western public, shows that the Jesuit global mission was not simply an instrument of Western imperial projects. The mission had a dynamic and rationale of its own that eventually clashed with the Portuguese and Spanish royal patronage, leading to the expulsion of the Jesuits in the second half of the 18th century from the Portuguese and Spanish empires.

The global Jesuit mission can also not be understood simply as a function of the expansion of trade and commerce. It is true that Jesuit missionaries sailed around the world in the same ships as *conquistadores*, traders, migrants, and colonial administrators. The great Jesuit missionary of Brazil, Antônio Vieira, put the matter most succinctly in his *Historia do Futuro*: "If there were not merchants who go to seek for earthly treasures in the East and West Indies, who would transport thither the preachers who take heavenly treasures? The preachers take the Gospel and the merchants take the preachers."[4] Portuguese traders and colonial administrators could land in Nagasaki, but the Jesuits continued onto Kyoto without them. Portuguese had a foothold in Macao, but further to the Chinese mainland and to Beijing the Jesuits travelled on their own, without colonial support.

In fact, the Jesuits were neither the only nor the first global missionaries. They followed literally in the steps of other older Catholic orders, Franciscans, Dominicans, Augustinians, etc., who had preceded them in colonial Spanish America as well as in Portuguese India. In this respect, the Jesuit global mission was part and parcel of the Golden Age of global Catholic missions that flourished throughout the 16th–17th centuries, as part of the broad Catholic renewal that emerged in Italy and the Iberian Peninsula before the Protestant Reformation.

This Catholic renewal, much broader than the so-called "Counter Reformation," forms another important opportunity structure for the emergence of the Jesuits as global missionaries. As pointed out by the historian R. Po-Chia Hsia "the centuries of Catholic renewal formed the first period of global history," in that the Early Modern era was shaped by "the encounter between Catholic Europe and the non-Christian world."[5] This was the age when Catholicism became a world religion.

4 Quoted in C.R. Boxer, *The Portuguese Seaborne Empire, 1600–1800* (London, Hutchison, 1965), 65; and Luke Clossey, "Merchants, Migrants, Missionaries, and Globalization in the Early-Modern Pacific," *Journal of Global History* 1 (2006): 41–58.

5 R. Po-Chia Hsia, *The World of Catholic Renewal, 1540–1770* (Cambridge: Cambridge University Press, 2005), 7.

But global mission became the specific foundational mission or ministry of the Jesuits from its very inception in a way in which it had not been the case of the mendicant or other Catholic orders. As expressed explicitly in the *Formula of the Institute*, the 1539 foundational charter of the order, Jesuits took an oath, to travel to any part of the world where there was hope of God's greater service and the good of souls in order to minister to "the Turks or any other infidels, even those who live in the regions called the Indies, or ... any heretics whatever, or schismatics, or any of the faithful."[6]

Global mobility was culturally encoded, as it were, into the make-up of the Jesuit order from its inception. Indeed, no other group took the entire globe as eagerly as the focus of their activities. Jerónimo Nadal, one of Ignatius's closest collaborators and the man who, in the words of John O'Malley, "more than any individual ... instilled in the first two generations their *esprit de corps* and taught them what it meant to be a Jesuit," coined the famous phrase, "the world is our home."[7]

The primary mission and ultimate end of the Jesuit global enterprise was the universal salvation of "souls" *ad majorem Dei gloriam*. According to O'Malley, no other expression "occurs more frequently in Jesuit documentation—on practically every page—than 'to help souls'."[8] To reduce the Jesuit mission to something else is not only to miss what clearly motivated Ignatius and the Society he founded, but also to misunderstand the very source of the globalizing dynamic of the Jesuit enterprise.

But how does one explain the fact that Jesuit missionaries, whose primary mission was "to help souls," spent so much time in what we would call today secular activities from cartography to military fortifications, from astronomy to botany, from building clocks to building cannons, from theater production to opera and ballet? Without ever losing sight of their primary salvational mission, Jesuit ministries encompassed all kinds of activities and spheres that today may be considered "secular" but that they regarded as an intrinsic part of their religious mission.[9] The Jesuits were not only the first professional

6 George E. Ganss, trans., *The Constitutions of the Society of Jesus*, by Ignatius of Loyola (St. Louis: Institute of Jesuit Sources, 1970), 68. The *Constitutions* may also be accessed at: http://jesuits-eum.org/readings/formula-of-the-institute-1550.

7 John W. O'Malley, S.J., "To Travel to Any Part of the World: Jerónimo Nadal and the Jesuit Vocation," in *Saints or Devils Incarnate?: Studies in Jesuit History* (Leiden: Brill, 2013), 147–164; and *The First Jesuits* (Cambridge: Harvard University Press, 1993), 12.

8 O'Malley, *First Jesuits*, 18.

9 Simon Ditchfield, "What did natural history have to do with salvation? Jose de Acosta SJ (1540–1600) in the Americas," in *God's Bounty?: The Churches and the Natural World*, ed. Peter D. Clarke and Tony Claydon (Woodbridge, UK: Boydell Press, 2010), 144–168.

missionary order. They became also the first Catholic teaching order, in fact the first transnational professional organization of schoolmasters.

Teaching had not been envisioned originally as a particular Jesuit ministry. But the establishment of the first Jesuit school in Messina, Sicily, in 1548, was to have immense repercussions on the character and development of the Society. O'Malley does not hesitate to call it "a crucial event in the history of schooling within the Catholic Church and in Western civilization."[10] The Collegio Romano established by Ignatius in 1551, became the first Jesuit university, the training ground for some of the most important Jesuit missionary scientists, including Ricci and Verbiest, and the model of every subsequent Jesuit university. It was at the Collegio Romano that Ricci learned from the great Jesuit scientist Christopher Clavius the foundations of mathematics, geometry, astronomy, and geography, while also applying himself to the use of the astrolabe, clockmaking, quadrants, planetary tables and calendar computations.

O'Malley's analysis shows how the Jesuits played a crucial role in developing a model of humanist liberal arts education, institutionalized in the 1599 *Ratio studiorum*, that was to become globalized through their extensive network of colleges throughout the world.[11] The Jesuit college, particularly in its higher form offering tertiary education, was a combination of two competing systems of education, the university system with its scholastic and professional training paradigmatically represented by the University of Paris and the humanistic college that had emerged in 15th century Renaissance Italy. The first Jesuits were themselves products of both institutions. But as they appropriated them, they imbued them with distinctive characteristics, adding a solid training in mathematics and early modern science. Most significantly, what distinguished the Jesuit college was a unique global orientation shaped by what the historian of science Steven Harris has called "the Jesuit geography of knowledge."[12] In fact, it was the virtuous feedback between the global network of Jesuit colleges and the global network of Jesuit missions that made the Jesuits into pioneer globalizers.

10 John O'Malley, "How the First Jesuits Became Involved in Education," in *Saints or Devils*, 199.

11 Vincent J. Duminuco, ed., *The Jesuit Ratio Studiorum: 400th Anniversary Perspectives* (New York: Fordham University Press, 2000); and Luce Giard, ed., *Les Jésuites à la Renaissance: Système éducatif et production du savoir* (Paris: PUF, 1995).

12 Steven Harris, "Mapping Jesuit Science: The Role of Travel in the Geography of Knowledge," in John W. O'Malley, Gauvin A. Bailey, Steven J. Harris and T. Frank Kennedy, eds., *The Jesuits: Cultures, Sciences and the Arts, 1540–1773* (Toronto: University of Toronto Press, 1999), 212–240; and Mordeccai Feingold, ed., *Jesuit Science and the Republic of Letters* (Cambridge: Cambridge University Press, 2003).

If the charisma of Ignatius and his early companions may explain the rapid initial growth of the Society, the even more extraordinary growth of Jesuit colleges accounts for the steady stream of recruits thereafter. At the death of Ignatius in 1556, the order that was made up of ten companions in 1540 had grown to over 1,000 members and administered already 46 colleges. In 1579 the Society had 5,164 members and 144 colleges. By 1626 the respective numbers were 15,544 members and 444 colleges plus 56 seminaries. In 1749, ten years before the expulsion from Portugal and its empire, the order had over 22,000 members and over 800 colleges spread throughout Europe, Latin America and Asia.[13]

The Jesuit "college" was developed at a time when there was an obviously rising demand for education from the nobility and the emerging middle classes well before the state took over the control of schooling. There was a particular historical window of opportunity which the Jesuits exploited: they were the pioneer constructors of a model of educational institution, which they themselves reproduced isomorphically around the world. As pioneer cultural brokers and translators between North and South and East and West, the Jesuits were important actors in the construction of the early phases of some aspects of an emerging world society.

Their most important contribution may have been the advancement of what Roland Robertson has defined as two of the four reference points of the global field, namely, a) *individuals* or *selves*, which in Jesuit-Christian parlance were called *souls*; and b) *humankind* or global humanity.[14] As is in so many other things, the Jesuits were not so much the originators but the effective carriers of a wider culture of universal Christian humanism that emerged from the confluence of Aristotelian-Thomist scholastic philosophy and Renaissance humanism, and crystallized in the "Schools of Salamanca and Coimbra." In their encounters with the non-Christian "other" Jesuit global missions and colleges became the effective global disseminators of this culture of universal Christian humanism.[15] This was to have unforeseen consequences in the

---

13 See Table 1.1, "Growth of the Society of Jesus, 1556–1749," in Alden, *Making of an Enterprise*, 17.

14 Roland Robertson, *Globalization. Social Theory and Global Culture* (London: Sage, 1992), 25 and passim.

15 Cf. Anthony Padgen, *The Fall of Natural Man. The American Indian and the Origins of Comparative Ethnology* (Cambridge: Cambridge University Press, 1982); Peter Stamatov, *The Origins of Global Humanitarianism. Religion, Empires and Advocacy* (Cambridge: Cambridge University Press, 2013); Guillermo Wilde, ed., *Saberes de la Conversión. Jesuítas, Indígenas e Imperios Coloniales en las Fronteras de la Cristiandad* (Buenos Aires: Editorial Sb, 2011).

Jesuits' sustained professional dedication not only to the humanities, but to science, technology, and the arts. Following the Ignatian instruction, Jesuits were to find God "in all things."

Besides being pioneer global missionaries and pioneer global educators, the Jesuits became global practitioners of what the first Jesuit Pope Francis has called "the culture of the encounter" with peoples and cultures of the world.

Those three interrelated processes—the Iberian colonial expansion, the early modern Catholic revival, and the culture of Renaissance Christian Humanism—formed the external opportunity structures for the development of the Society of Jesus as a global enterprise. All three had been operative for well over half a century by the time the society was established in 1540. All three helped to shape the institutional development and the global expansion of the new society in the following decades, and in turn the Jesuits became primary global carriers of the three processes in the early modern phase of globalization.

2 Internal Institutional Structures

Equally important in shaping the Jesuit global mission, however, were some internal institutional and structural characteristics of the society which gave them certain advantages over religious as well as secular competitors. Three in particular are worth mentioning. There is, in the first place, the characteristic Ignatian spirituality and ethic of critical self-examination and discernment, personal self-abnegation and control, dedication to the service of others and the common good, and openness to the world so that one may find God "in all things." This was an ethic grounded in the "Spiritual Exercises," a peculiar spiritual manual and technique of the self, first tested by Ignatius on himself and then applied to every individual who joined the society.[16] It contributed to a peculiar organization of self-directed and relatively autonomous individuals who could be sent to any remote place in the world where they would tend to reproduce the peculiar Jesuit "way of proceeding."

Both Ricci and Verbiest were in a sense extraordinary, yet typical members of the Society. Both joined the order at a young age, underwent a lengthy spiritual and academic formation and training within various institutional settings of the Society, but were then able to reproduce and further this Jesuit ethos of

16 Michelle Molina, *To Overcome Oneself: The Jesuit Ethic and the Spirit of Global Expansion, 1520–1767* (Berkeley: University of California Press, 2013).

contemplative disciplined activism in the world in the most remote settings among foreign peoples and cultures for the rest of their lives.

Equally significant was the peculiar centralized hierarchic structure of the Society, written down in detail in the lengthy Constitutions. It was a structure that allowed both the extraordinary global mobility and the pattern of long-distance obedience and internalized self-control that distinguished the Jesuits from other Catholic religious orders. Any Jesuit anywhere in the world had to be ready to be sent *ad missiones* anywhere in the world. Members of the Society who came from "nations" that did not have colonial empires could be sent overseas by the Father General. This explains the large number of Italian, German, Austrian, Bohemian, Polish and Belgian Jesuits on overseas missions. It also gave those missions their distinctive "international" character. The Jesuit China mission in particular, although under the jurisdiction of the Portuguese *Padroado*, had a pronounced international character. Approximately a third of all Jesuit missionaries in China were Portuguese. Among them, there were prominent figures such as the linguist João Rodrigues, Gabriel de Magalhães, founder of the *Dongtang* or Eastern Church in Beijing, and Tomás Pereira, musician and close advisor of the Kangxi emperor. But along with them, one can point to the many prominent non-Portuguese Jesuits such as: the Italians Michele Ruggieri, Matteo Ricci, Giulio Aleni, Sabatino de Ursis, Ludovico Buglio, Martino Martini, and Claudio Grimaldi; the Germans Adam Schall von Bell, Johann Schreck and Ignaz Koegler; the Spaniard Diego de Pantoja; the Poles Jan Mikolaj Smogulecki and Michal Boym; the Flemish Nicholas Trigault and Ferdinand Verbiest; and the French Joachim Bouvet, Louis Le Comte and Jean-François Gerbillon. All of them played important roles in the inter-civilizational encounter between the sciences, arts and cultures of the Middle Kingdom and Europe in the early modern era.

The Society of Jesus's character as a highly centralized and hierarchic trans-national organization with a highly flexible and mobile structure, was made possible by a well-developed system of communication within the entire global Jesuit network based on regular and periodic letter writing and reporting to the superiors, to the provincials, and to the Roman curia. Carefully edited, much of this information was later redistributed as edifying letters to the more horizontal networks of global Jesuit missions and colleges and to the wider world through Jesuit publications.[17] Their autonomous structure as a universal papal

17 On the Jesuit global system of organization, information management and communication see, Markus Friedrich, "Circulating and Compiling the *Litterae Annuae*. Toward a History of the Jesuit System of Communication," *Archivum Historicum Societatis Iesu* 77 (2008): 3–39; "Government and Information Management in Early Modern Europe: The

order, bound by the forth vow to the universal Bishop of Rome, with ambiguous and overlapping loyalties to various religious and secular authorities and jurisdictions, well-fitted to accommodate the most diverse local contexts, also gave the Jesuits certain global structural advantages in the early modern phase of globalization which elicited much envy, dread, and competition from friend and foe alike. This ambiguous transnational autonomy was at the base of the anti-Jesuit "black legend" and ultimately it led to their expulsion from every Catholic kingdom and to the final suppression of the Society of Jesus by Pope Clement XIV in 1773.[18]

In sum, the Jesuits contributed to the development of global connectivity and global consciousness not only through their ubiquitous missions but through their prodigious production and global circulation of annual letters and edifying mission reports, scientific and ethnographic descriptions, mapping and cartographic exercises, through the construction of numerous scripts, lexicons and grammars of non-Western languages, through the translation of classical Greek and Latin texts into non-Western languages and the translation of non-Western classical texts into Latin, through the production of Catechisms in every possible vernacular, and through the global circulation of all kind of objects, from scientific instruments to printing presses and type scripts, from medicinal plants (Jesuit bark or quinine) to all kinds of sacred objects, icons, and paintings, church architectural styles, music, drama, and ballet.[19] Even the 1689 Sino-Russian Treaty of Nerchinsk, the

---

Case of the Society of Jesus (1540–1773)," *Journal of Early Modern History* 12 (2009): 1–25; and "Jesuit Organization and Legislation: Development and Implementation of a Normative Framework," in Ines G. Županov, eds., *The Oxford Handbook of the Jesuits* (Oxford: Oxford University Press, 2019).

18 Sabina Pavone, "The History of Anti-Jesuitism: National and Global Dimensions," in Banchoff and Casanova, eds., *Jesuits and Globalization*, 111–130; Pierre-Antoine Fabre et Catherine Maire, ed., *Les Antijésuites: Discours, figures et lieux de l'antijésuitisme à l'époque modern* (Rennes: Presses universitaires de Rennes, 2010); Peter Burke, "The Black Legend of the Jesuits: An Essay in the History of Social Stereotypes," in Simon Ditchfield, ed., *Christianity and Community in the West: Essays for John Bossy* (Aldershot: Ashgate, 2001); Susana Monreal, Sabina Pavone, and Guillermo Zermeno, eds., *Antijesuitismo y Filojesuitismo: Dos identidades ante la restauración* (Mexico, CF: Universidad Iberoamericana, 2014).

19 Cf. John O'Malley et al., *The Jesuits*; John W. O'Malley, Gauvin A. Bailey, Steven J. Harris, and T. Frank Kennedy, eds., *The Jesuits II: Cultures, Sciences and the Arts, 1540–1773* (Toronto: Toronto University Press, 2006); Gauvin A. Bailey, *Art on the Jesuit Missions in Asia and Latin America, 1542–1773* (Toronto: Toronto University Press, 1999); Charlotte de Castelnau, Marie L. Copete, Aliocha Maldavsky, and Ines Županov, eds., *Missions d'évangelisation et circulation des saviors. XVIe–XVIIIe siècle* (Madrid-Paris: Casa de Velasquez-EHESS, 2011); Joseph A. Gagliano and Charles E. Ronan, eds., *Jesuit Encounters in the New World: Jesuit*

first official diplomatic encounter setting the territorial borders between the expanding Qing and Russian empires, was written in Latin and mediated by the Jesuits Tomás Pereira and Jean-François Gerbillon. The treaty had been prepared by the role of Ferdinand Verbiest in the formation of Sino-Russian diplomatic relations.[20]

In the remainder of this chapter, I would like to reflect briefly upon the production of Ricci's and Verbiest world maps as a paradigmatic illustration of the kind of collaborative cultural exchange that exemplifies the Jesuit method of accommodation, the practice of inculturation and, most importantly, what could be termed in the words of the Jesuit Pope Francis, "the culture of the encounter." This dialogic culture of the encounter became a habitual part of Jesuit corporate culture in all their global enterprises. But, as indicated by Nicolas Standaert, each encounter was shaped as much by the particular context and by the power dynamics present in each encounter with the other, as it was by Jesuit corporate culture.[21]

As is well known, the Ricci map was the product of genuine non-hierarchic collaboration between European and Chinese cartography. It reflects a phase of globalization before Western hegemony, when less unequal exchanges between East and West were still possible, when the West did not function yet as a global center and the East as a global periphery. What is symbolically interesting about the Ricci world map is that neither the Middle Kingdom, nor Europe are at the center of the global map. In fact, both are decentered together within the "Western" hemisphere, which is filled by Afroeurasia, while "the new world" of the Americas and the Pacific fills the "Eastern" hemisphere, thus inverting East and West.

Most importantly, Ricci's and Verbiest's world maps stand within a continuum of geographical works authored and published by Jesuits in Chinese, from Ricci's 1584 *Shanhai yudi quantu* [A Complete Map of Mountains and Seas in the World] to Verbiest's 1674 *Kunyu quantu* [Complete Map of the Earth], which are symbolically paradigmatic of Jesuit missionary practices in

Chroniclers, Geographers, Educators, and Missionaries in the Americas, 1549–1767 (Rome: IHSI, 1997); and Wilde, ed., *Saberes de la Conversión*.

20 Cf. Joseph Sebes, *The Jesuits and the Sino-Russian Treaty of Nerchinsk (1689): The Diary of Thomas Pereira* (Rome: IHSI, 1962); Artur K. Wardega, S.J. and António Vasconcelos de Saldanha, eds., *In the Light and Shadow of an Emperor: Tomás Pereira, SJ (1645–1708), the Kangxi Emperor and the Jesuit Mission in China* (Newcastle upon Tyne: Scholars Publishing, 2012); and John W. Witek, S.J., ed., *Ferdinand Verbiest, S.J. (1623–1688). Jesuit Missionary, Scientist, Engineer and Diplomat* (Nettetal: Steyler Verlag, 1994), 273–325.

21 Nicolas Standaert, S.J., "Jesuit Corporate Culture as Shaped by the Chinese," in O'Malley et al. eds., *The Jesuits*, 352–363; and *Methodology in View of Contact between Cultures: The China Case in the 17th Century* (Hong Kong: The Chinese University of Hong Kong, 2002).

China. As Chen Minsun has pointed out, after Ricci's maps, Diego de Pantoja wrote some *Notes*, which were revised and enlarged by Giulio Aleni to produce his 1623 book on world geography, *Zhifang waiji* [Records of Regions beyond the Jurisdiction of the Imperial Geographer].[22] Fourteen years later, in 1637, Aleni published his *Xifang dawen* [Answers to Questions concerning the West], which offers a vivid account of Europe. In 1669, after the consolidation of the Qing dynasty, Ludovico Buglio and Gabriel de Magalhães prepared the *Xifang yao yi* [A Concise Account of the West], which Verbiest presented to the Chinese emperor. Three years later, in 1672, Verbiest published his *Kunyu tushuo* [Illustrated Explanations of the Earth], a global geographical work which, as he pointed out in the preface, built upon the works of his learned colleagues, Ricci, Aleni, Alfonso Vagnoni, and Sabatino de Ursis. Two years later, Verbiest published the *Kunyu quantu* to complement the *Kunyu tushuo*. Thus, as pointed out by Chen, "from 1584 to 1674, a total of at least seven geographical works in Chinese were produced by the Jesuit missionaries."[23]

In his analysis, Chen points out four common characteristics of the Jesuit geographical works: 1) "They were all produced or written in response to requests or demands" from Chinese scholar-officials or from the emperor; 2) In all of them the Jesuits express "a high esteem and praise towards China, its rulers and scholar-officials"; 3) "each Jesuit writer presented his work, not as the writing of a single author, but as the collective work of himself and his fellow Jesuits"; 4) "all Jesuit writers and editors were aware that they represented not only the Society of Jesus as missionaries in an alien country, but were also representatives of the European community, which had a different and highly developed civilization."[24] In other words, the Ricci and Verbiest maps were part of the collective corporate enterprise of the Jesuit China mission, but this mission itself was also embedded within a larger non-hierarchical civilizational encounter between Europe and China, when both could view each other as equal, before the onset of Western hegemony towards the 1800's.[25]

The Jesuit "catholic" missionary impulse had naturally, as a matter of course, the hegemonic purpose of universal conversion to the true Catholic faith. But what makes Jesuit global missionary practices particularly relevant is the

22 Chen Minsun, "Ferdinand Verbiest and the Geographical works by Jesuits in Chinese 1584–1674," in John W. Witek, S.J., ed., *Ferdinand Verbiest (1623–1688): Jesuit Missionary, Scientist, Engineer and Diplomat* (Nettetal: Steyler Verlag, 1994), 123–133.

23 Ibid., 127.

24 Ibid., 127–128.

25 David E. Mungello, *Curious Land. Jesuit Accommodation and the Origins of Sinology* (Honolulu: University of Hawaii Press, 1985); and *The Great Encounter of China and the West, 1500–1800* (Lanham, MD: Rowman & Littlefield, 2013).

fact that, under certain "circumstances," particularly when their mission was not embedded in structures of Western colonial power, their controversial method of "accommodation" took a form which we would call today "nativist inculturation." One should avoid, of course, anachronistic interpretations of early modern Jesuit practices from our contemporary global perspective of cultural and religious pluralism. Nevertheless, Valignano's method of accommodation, points to a formula of globalization which rejects unidirectional Westernization and opens itself to multicultural encounters and reciprocal learning processes.[26]

Significantly, Fr. M. Antoni J. Ucerler's analysis stresses that "cultural accommodation" was not simply a method invented by European Jesuits in order to become effective "missionaries," but was rather an initiative pressed upon them by their Japanese and Chinese interlocutors, particularly by Christian converts who often demanded that the Jesuit "fathers" engaged the local culture *on its own terms*. Notwithstanding their many limitations, Ucerler's conclusion is that "the rich cultural legacies of the early Jesuits in East Asia are worthy of renewed scrutiny in our own time ... as an original paradigm of inter-cultural engagement."[27]

Even in Spanish colonial America, where conquest, colonization, reduction of the indigenous peoples, and conversion to Christianity were so inextricably intertwined, José de Acosta already insisted that "hispanización" was not necessary to "preach the Gospel" to the Indians nor to "procure their salvation."[28] This was the rationale behind the simultaneous publication (i.e., translation from Latin) of the trilingual Lima Catechism (1583) in Spanish, *Quechua* and *Aymara*. Not Spanish, but *Guaraní* was the language spoken in the large Jesuit Paraguay mission and became the official language of independent Paraguay.

26 The literature on Valignano and the Jesuit method of *accommodation* is immense. Cf., Josef Franz Schütte, S.J., *Valignano's Mission Principles for Japan* 2 vols (St. Louis: The Institute of Jesuit Sources, 1980); Adolfo Tamburello, M. Antoni J. Ucerler, S.J. and Marisa Di Russo, ed., *Alessandro Valignano S.I.: Uomo del Rinascimento: Ponte tra Oriente e Occidente* (Rome: IHSI, 2008); M. Antoni J. Ucerler, S.J., *Christianity and Cultures: Japan & China in Comparison, 1543–1644* (Rome: Institutum Historicum Societatis Iesu, 2009); Bonnie B.C. Oh and Charles E. Ronan, ed., *East Meets West: The Jesuits in China 1582–1773* (Chicago: Loyola University Press, 1988); Jacques Gernet, *China and the Christian Impact, A Conflict of Cultures* (Cambridge: Cambridge University Press, 1985); Nicolas Standaert, *L'"autre" dans la mission: Leçons à partir de la Chine* (Bruxelles: Lexius, 2003).

27 M. Antoni J. Ucerler, S.J., "The Jesuits in East Asia in the Early Modern Age: A New 'Aeropagus' and the 'Re-invention' of Christianity," in Banchoff and Casanova, eds., *Jesuits and Globalization*, 27–48.

28 José de Acosta, *De Procuranda Indorum Salute o Predicación del Evangelio en las Indias* (1588) (Alicante: Biblioteca Virtual Miguel de Cervantes, 1999).

The practice of nativist inculturation, made explicit in the Jesuit method of accommodation, amounts to a formula of globalization of Christianity through the particularization of the universal, by going "local" or "native" through a process of reflexive inculturation and acculturation, which theologically amounts to a formula of ever renewed Christian "incarnation."

This is the famous and controversial formula of Jesuit cultural accommodation which led to the adoption of the Confucian "habitus" in China by Matteo Ricci, the Brahmin "habitus" in India by Roberto de Nobili, the Guarani "habitus" in the "Reducción de Paraguay," but also the, for us today, less commendable accommodating "habitus" of slave-owners in the Jesuit plantations in Brazil or Maryland. It was the differentiation of true universal *religion* and particular *culture*, as well as that between *civilization* and *idolatry*, first introduced by the Jesuits that allowed the various accommodating syntheses of supposedly Christian universalism and cultural particularism.[29] The fact that the method was so vehemently attacked by the other missionary orders and even by other Jesuits in India and China, before it exploded into the Chinese and Malabar Rites controversies in Rome and Paris indicates the extent to which it challenged Eurocentric notions of a uniform Roman Catholic globalization.[30]

What becomes clear is that, at least in some cases, what begins as a one-way mission of Christian evangelization that assumes the exclusivity of Christianity as "the true religion" and the superiority of Christian European culture turns into a mutual intercultural and inter-religious encounter that "under certain circumstances" transforms the missionary as much as the native. There are of course fundamental differences between the Jesuit missions in Goa or Macao which remained embedded within the Portuguese colonial establishments and Nobili's mission to Madurai or Ricci's mission to Beijing. It seems indeed that the more the Jesuit missionaries were on their own and in the peripheries, without the support and protection of the Iberian colonial powers, the more favorable became the circumstances for an open-ended non-hierarchic interaction and a genuine dialogue.

In Ibero-America, by contrast, the colonial circumstance and the assumed superiority not just of Christianity but of European civilization and culture practically precluded such a non-hierarchic interaction and open dialogue

29 Joan Pau Rubiés, "The Concept of Cultural Dialogue and the Jesuit Method of Accommodation: Between Idolatry and Civilization," *Archivum Historicum Societati Iesu* LXXIV fasc. 147 (2005): 237–280.

30 On the internal Jesuit disputes concerning missionary methods in India see Ines G. Županov, *Disputed Mission. Jesuit Experiments and Brahmanical Knowledge in Seventeenth-century India* (New Delhi: Oxford University Press, 1999).

with indigenous religions and cultures.[31] Yet, even in Ibero-America there were fundamental differences between the Jesuit missions at the center of the Viceroyalties of Peru or New Spain and their missions to the indigenous frontiers. Maldavsky's analysis, moreover, reiterates one of the central points of Ucerler's analysis, namely, that the "method of accommodation" was not an invention of European Jesuits applied to different non-European contexts, but rather it emerged from pragmatic interactions.[32]

What is striking is not the fact that Jesuit missionaries in most respects behaved no differently from other Catholic missionaries, but the fact that "under certain circumstances" the Jesuit "way of proceeding" became peculiarly different, attracting in the process much controversial attention from all quarters, friends and enemies. Particularly in the encounter with the multifaceted religions of Asia the old catch-all category of "pagan," "heathen" or "infidel" began to collapse and a new plural system of what later would be called "world religions" began to emerge.[33]

Without in any way attempting to settle the contested debate concerning the role of the colonial encounters and the later emergence of academic Orientalism in the European universities, it is undeniable that the Jesuits served as pioneer interlocutors in the religious, cultural, scientific and artistic encounter between East and West and between Old and New World. Particularly, pioneer Jesuits in Japan, China, Tibet, Vietnam, and India played an important role in transmitting and mediating the first knowledge about the foundational texts, religions, cultures and civilizations of the "Orient," which would later develop into full-fledged academic "orientalism."[34]

---

31 For a fascinating account of an incipient yet repressed possibility of such interreligious dialogue with Inca religion, see Sabine Hyland, *The Jesuit and the Incas: The Extraordinary Life of Padre Blas Valera, S.J.* (Ann Arbor: University of Michigan Press, 2003).

32 Aliocha Maldavsky, "Jesuits in Ibero-America: Missions in Colonial Societies," in Banchoff and Casanova, *Jesuits and Globalization*, 92–110.

33 Cf. Richard King, *Orientalism and Religion: Postcolonial Theory, India and 'The Mystic East'* (New York: Routledge, 1991); Wilfred Cantwell Smith, *The Meaning and End of Religion: A New Approach to the Religious Traditions of Mankind* (New York: Macmillan, 1963); Talal Asad, *Genealogies of Religion* (Baltimore: The Johns Hopkins University Press, 1993); Tomoko Masuzawa, *The Invention of World Religions: Or How European Universalism Was Preserved in the Language of Pluralism* (Chicago: The University of Chicago Press, 2005); Peter Beyer, *Religions in Global Society* (New York: Routledge, 2006); Peter van der Veer, *The Modern Spirit of Asia. The Spiritual and the Secular in China and India* (Princeton: Princeton University Press, 2014); and José Casanova, *Global Religious and Secular Dynamics: The Modern System of Classification* (Leiden: Brill, 2019).

34 The majority of the names, over two thirds, in the list of early modern European Orientalists provided by Urs App are Jesuits. See Urs App, *The Birth of Orientalism* (Philadelphia: University of Pennsylvania Press, 2010), 481.

It is instructive to look at the parallel tracts as well as the reciprocal influences shaping the Jesuit intercultural encounters in the West and East Indies. This communication across oceans indicates the extent to which Jesuits in Europe, the Americas, and Asia were involved in a complex global conversation. José de Acosta's developmental theory of Amerindian religions, as well as his comparative reflections on Amerindian cultures and the religions and cultures of Asia, presented in *De Procuranda Indorum Salute* (1588) and in *Historia Natural y Moral de las Indias* (1590), offers one of the best illustrations. Acosta's work, which went through numerous editions and translations in various European languages, marks the point of departure of modern comparative ethnology and anticipates many of the later Euro-centric stadial theories of human development, both being forms of imagining global humanity.[35]

In fact, despite their Christo-centric assumptions and their frequent recourse to divine and satanic devices as explanatory keys to all forms of cultural and religious diversity, the Jesuit early modern imaginary of global humanity and their dia-praxis of cultural "accommodation" and local inculturation appears less Euro-centric, less racist, and less unilinear than later imaginaries associated with the cosmopolitan Enlightenment or with the nineteenth and twentieth centuries' *mission civilisatrice* and imperial "White Man's Burden."

In their complex history, in their global consciousness as well as in their global practices, the Jesuits seem to reflect best cultural theories of globalization that emphasize simultaneous and seemingly contradictory dynamics of homogenization and heterogenization, the dialectics of the universal and the particular, the interweaving of the global and the local, as well as disjuncture and difference in a global cultural economy formed by overlapping ecumenes.[36]

The practical experiment in Christian inculturation that Jesuits, following Valignano's instructions, were willing to probe in Japan, in China, and in the Madurai mission ultimately failed for a combination of geo-political, civilizational, and ecclesiastical reasons. But if one takes seriously the argument that processes of globalization are contingent historical processes, not functionally necessary processes or consequences of modernity, then the most important

35 José de Acosta, *Natural and Moral History of the Indies*. Edited by Jane Mangan (Durham, NC: Duke University Press, 2002); Claudio M. Burgaleta, S.J., *José de Acosta, S.J. (1540–1600)* (Chicago: Loyola Press, 1999); and Fermín del Pino Díaz, "Humanismo renacentista y orígenes de la etnología: a propósito del P. Acosta, paradigma del humanismo antropológico jesuita," in *Humanismo y visión del otro en la España moderna: cuatro estudios*, B. Ares, ed. (Madrid: CSIC, 1992).

36 Robertson, *Globalization*, and Arjun Appadurai, *Modernity at Large: Cultural Dimensions of Globalization* (Minneapolis: University of Minnesota Press, 1996).

lesson from the Jesuit global story is that different historical processes, that is, different outcomes in the Jesuit Christian encounter in Japan, China, or India could have led to a different age of globalization.[37]

As we are entering a new decentered global age after Western hegemony the Jesuit global story of dialogic inculturation and deep inter-civilizational encounters still contains valuable lessons for us. Most of the issues they grappled with and their attempts to find viable resolutions to the tensions between universality and particularity, and between the global and the local are still with us.

Bibliography

Acosta, José de. *De Procuranda Indorum Salute o Predicación del Evangelio en las Indias* (1588). Alicante: Biblioteca Virtual Miguel de Cervantes, 1999.

Acosta, José de. *Natural and Moral History of the Indies*. Edited by Jane E. Mangan. Translated by Frances Lopez-Morillas. Annotated edition. Durham, NC: Duke University Press Books, 2002.

Alden, Dauril. *The Making of an Enterprise: The Society of Jesus in Portugal, Its Empire, and Beyond: 1540–1750*. Stanford: Stanford University Press, 1996.

App, Urs. *The Birth of Orientalism*. Philadelphia: University of Pennsylvania Press, 2010.

Appadurai, Arjun. *Modernity at Large: Cultural Dimensions of Globalization*. Minneapolis: University of Minnesota Press, 1996.

Asad, Talal. *Genealogies of Religion*. Baltimore: The Johns Hopkins University Press, 1993.

Beyer, Peter. *Religions in Global Society*. New York: Routledge, 2006.

Boxer, C.R. *The Portuguese Seaborne Empire, 1600–1800*. London: Hutchison, 1965.

Burgaleta, s.j., Claudio M. *José de Acosta, s.j. (1540–1600)*. Chicago: Loyola Press, 1999.

Burke, Peter. "The Black Legend of the Jesuits: An Essay in the History of Social Stereotypes." In *Christianity and Community in the West: Essays for John Bossy*, edited by Simon Ditchfield. Aldershot: Ashgate, 2001.

Casanova, José. *Global Religious and Secular Dynamics: The Modern System of Classification*. Leiden: Brill, 2019.

Casanova, José. "Jesuits, Connectivity, and the Uneven Development of Global Consciousness Since the Sixteenth Century." In *Global Culture: Consciousness and Connectivity*, edited by Roland Robertson and Didem Buhari-Gulmez, 109–126. Farnham: Ashgate, 2016.

---

37 José Casanova, "Locating Religion and Secularity in East Asia through Global Processes: Early Modern Jesuit Religious Encounters," *Religions* 9.11 (2018): 1–12.

Casanova, José. "The Jesuits Through the Prism of Globalization, Globalization Through a Jesuit Prism." In *The Jesuits and Globalization. Historical Legacies and Contemporary Challenges*, edited by Thomas Banchoff and José Casanova, 261–285. Washington, DC: Georgetown University Press, 2016.

Casanova, José. "Locating Religion and Secularity in East Asia through Global Processes: Early Modern Jesuit Religious Encounters." *Religions* 9, no. 11 (2018): 1–12.

Castelnau-l'Estoile, Charlotte de, Marie-Lucie Copete, Aliocha Maldavsky, and Ines G. Županov, eds. *Missions d'évangelisation et circulation des saviors. XVIe–XVIIIe siècle.* Madrid-Paris: Casa de Velasquez-EHESS, 2011.

Chen Minsun. "Ferdinand Verbiest and the Geographical works by Jesuits in Chinese 1584–1674." In *Ferdinand Verbiest, S.J. (1623–1688): Jesuit Missionary, Scientist, Engineer and Diplomat*, edited by John W. Witek, S.J., 123–133. Nettetal: Steyler Verlag, 1994.

Clossey, Luke. "Merchants, Migrants, Missionaries, and Globalization in the Early-Modern Pacific." *Journal of Global History* 1 (2006): 41–58.

Costa, S.J., H. de la., *The Jesuits in the Philippines, 1581–1768*. Cambridge, MA: Harvard University Press, 1961.

Ditchfield, Simon. "What did natural history have to do with salvation? Jose de Acosta SJ (1540–1600) in the Americas." In *God's Bounty?: The Churches and the Natural World*, edited by Peter D. Clarke and Tony Claydon, 144–168. Woodbridge, UK: Boydell Press, 2010.

Duminuco, Vincent J. *The Jesuit Ratio Studiorum: 400th Anniversary Perspectives*. 1st ed. New York: Fordham University Press, 2000.

Fabre, Pierre-Antoine, and Catherine Maire. *Les Antijésuite: Discours, figures et lieux de l'antijésuitisme à l'époque modern*. Rennes: Presses universitaires de Rennes, 2010.

Feingold, Mordeccai, ed. *Jesuit Science and the Republic of Letters*. Cambridge: Cambridge University Press, 2003.

Friedrich, Markus. "Circulating and Compiling the *Litterae Annuae*. Toward a History of the Jesuit System of Communication." *Archivum Historicum Societatis Iesu* 77 (2008): 3–39.

Friedrich, Markus. "Government and Information Management in Early Modern Europe: The Case of the Society of Jesus (1540–1773)." *Journal of Early Modern History* 12 (2009): 1–25.

Friedrich, Markus. "Jesuit Organization and Legislation: Development and Implementation of a Normative Framework." In *The Oxford Handbook of the Jesuits*, edited by Ines G. Županov. Oxford: Oxford University Press, 2019.

Gagliano, Joseph A., and Charles E. Ronan. *Jesuit Encounters in the New World: Jesuit Chroniclers, Geographers, Educators, and Missionaries in the Americas, 1549–1767*. Roma: Institutum Historicum S.I., 1997.

Ganss, George E., trans. *The Constitutions of the Society of Jesus*, by Ignatius of Loyola. St. Louis: Institute of Jesuit Sources, 1970.

Gauvin A. Bailey. *Art on the Jesuit Missions in Asia and Latin America, 1542–1773*. Toronto: Toronto University Press, 1999.

Gernet, Jacques. *China and the Christian Impact, A Conflict of Cultures*. Cambridge: Cambridge University Press, 1985.

Giard, Luce, ed. *Les Jésuites à la Renaissance: Système éducatif et production du savoir*. Paris: PUF, 1995.

Gunn, Geoffrey C. *First Globalization: The Eurasian Exchange, 1550–1800*. Lanham, MD: Rowman & Littlefield, 2003.

Harris, Steven. "Mapping Jesuit Science: The Role of Travel in the Geography of Knowledge." In *The Jesuits: Cultures, Sciences and the Arts, 1540–1773*, edited by John W. O'Malley, Gauvin A. Bailey, Steven J. Harris and T. Frank Kennedy, 212–240. Toronto: University of Toronto Press, 1999.

Hsia, R. Po-chia. *The World of Catholic Renewal, 1540–1770*. Cambridge: Cambridge University Press, 1998.

Hyland, Sabine. *The Jesuit and the Incas. The Extraordinary Life of Padre Blas Valera, S.J.* Ann Arbor: University of Michigan Press, 2003.

King, Richard. *Orientalism and Religion: Postcolonial Theory, India and 'The Mystic East'*. New York: Routledge, 1991.

Maldavsky, Aliocha. "Jesuits in Ibero-America: Missions in Colonial Societies," edited by Thomas Banchoff and José Casanova, 92–110. Washington, DC: Georgetown University Press, 2016.

Masuzawa, Tomoko. *The Invention of World Religions: Or How European Universalism Was Preserved in the Language of Pluralism*. Chicago: The University of Chicago Press, 2005.

Molina, Michelle. *To Overcome Oneself: The Jesuit Ethic and the Spirit of Global Expansion, 1520–1767*. Berkeley: University of California Press, 2013.

Monreal, Susana, Sabina Pavone, and Guillermo Zermeno, eds. *Antijesuitismo y Filojesuitismo. Dos identidades ante la restauración*. Mexico, CF: Universidad Iberoamericana, 2014.

Mungello, David E. *Curious Land. Jesuit Accommodation and the Origins of Sinology*. Honolulu: University of Hawai'i Press, 1985.

Mungello, David E. *The Great Encounter of China and the West, 1500–1800*. Lanham, MD: Rowman & Littlefield, 2013.

Oh, Bonnie B.C. and Charles E. Ronan. *East Meets West: The Jesuits in China 1582–1773*. Chicago: Loyola University Press, 1988.

O'Malley, John W. *The First Jesuits*. Cambridge, MA: Harvard University Press, 1993.

O'Malley, John W., Gauvin A. Bailey, Steven J. Harris, and T. Frank Kennedy, eds. *The Jesuits: Cultures, Sciences and the Arts, 1540–1773*. Toronto: University of Toronto Press, 1999.

O'Malley, John W., Gauvin A. Bailey, Steven J. Harris, and T. Frank Kennedy, eds. *The Jesuits II: Cultures, Sciences and the Arts, 1540–1773*. Toronto: Toronto University Press, 2006.

O'Malley, John W. "To Travel to Any Part of the World: Jerónimo Nadal and the Jesuit Vocation." In *Saints or Devils Incarnate?: Studies in Jesuit History*, 147–164. Leiden: Brill, 2013.

Padgen, Anthony. *The Fall of Natural Man. The American Indian and the Origins of Comparative Ethnology*. Cambridge: Cambridge University Press, 1982.

Pavone, Sabina. "The History of Anti-Jesuitism: National and Global Dimensions." In *The Jesuits and Globalization: Historical Legacies and Contemporary Challenges*, edited by Thomas Banchoff and José Casanova, 111–130. Washington, DC: Georgetown University Press, 2016.

Pino Díaz, Fermín del. "Humanismo renacentista y orígenes de la etnología: a propósito del P. Acosta, paradigma del humanismo antropológico jesuita." In *Humanismo y visión del otro en la España moderna: cuatro estudios*, edited by Berta Ares Queija. Madrid: Consejo Superior de Investigaciones Científicas, 1992.

Robertson, Roland. *Globalization: Social Theory and Global Culture*. London: Sage, 1992.

Rubiés, Joan Pau. "The Concept of Cultural Dialogue and the Jesuit Method of Accommodation: Between Idolatry and Civilization." *Archivum Historicum Societati Iesu* LXXIV fasc. 147 (2005): 237–280.

Schütte, S.J., Josef Franz. *Valignano's Mission Principles for Japan*, 2 vols. St. Louis: The Institute of Jesuit Sources, 1980.

Sebes, Joseph. *The Jesuits and the Sino-Russian Treaty of Nerchinsk (1689): The Diary of Thomas Pereira*. Rome: IHSI, 1962.

Smith, Wilfred Cantwell. *The Meaning and End of Religion; A New Approach to the Religious Traditions of Mankind*. New York: Macmillan, 1963.

Stamatov, Peter. *The Origins of Global Humanitarianism. Religion, Empires and Advocacy*. Cambridge: Cambridge University Press, 2013.

Standaert, Nicolas. *L'"autre" dans la mission. Leçons à partir de la Chine*. Bruxelles: Lexius, 2003.

Standaert, S.J., Nicolas. "Jesuit Corporate Culture as Shaped by the Chinese." In *The Jesuits*, edited by John W. O'Malley, Gauvin Alexander Bailey, Steven J. Harris, and T. Frank Kennedy, 352–363. Toronto: University of Toronto Press, 1999.

Standaert, S.J., Nicolas. *Methodology in View of Contact between Cultures: The China Case in the 17th Century*. Hong Kong: The Chinese University of Hong Kong, 2002.

Tamburello, Adolfo, M. Antoni J. Ucerler, S.J., and Marisa Di Russo. *Alessandro Valignano S.I.: Uomo del Rinasciment: Ponte tra Oriente e Occidente*. Rome: IHSI, 2008.

Ucerler, S.J., M. Antoni J. *Christianity and Cultures. Japan & China in Comparison, 1543–1644*. Rome: Institutum Historicum Societatis Iesu, 2009.

Ucerler, S.J., M. Antoni J. "The Jesuits in East Asia in the Early Modern Age: A New 'Aeropagus' and the 'Re-invention' of Christianity." In *The Jesuits and Globalization: Historical Legacies and Contemporary Challenges*, edited by Thomas Banchoff and José Casanova, 27–48. Washington, DC: Georgetown University Press, 2016.

Van der Veer, Peter. *The Modern Spirit of Asia. The Spiritual and the Secular in China and India*. Princeton: Princeton University Press, 2014.

Wardega, S.J., Artur K., and António Vasconcelos de Saldanha. *In the Light and Shadow of an Emperor: Tomás Pereira SJ (1645–1708), the Kangxi Emperor and the Jesuit Mission in China*. Newcastle upon Tyne: Scholars Publishing, 2012.

Wilde, Guillermo, ed. *Saberes de la Conversión. Jesuítas, Indígenas e Imperios Coloniales en las Fronteras de la Cristiandad*. Buenos Aires: Editorial Sb, 2011.

Witek, S.J., John W., ed. *Ferdinand Verbiest, S.J. (1623–1688). Jesuit Missionary, Scientist, Engineer and Diplomat*. Nettetal: Steyler Verlag, 1994, 273–325.

Županov, Ines G. *Disputed Mission. Jesuit Experiments and Brahmanical Knowledge in Seventeenth-century India*. New Delhi: Oxford University Press, 1999.

CHAPTER 2

From Manuscript to Print: At the Origins of Early Jesuit Missionary Strategies of Communication

Robert Danieluk, s.j.

The missionary commitment of the Society of Jesus from its very inception is a well-known aspect of the Order's history and needs no introduction. Even a short bibliographical survey makes it amply clear that the historiography of Jesuit missions is very rich indeed. The studies in that field have been composed not only by the Jesuits themselves, but also (in more recent decades) by other historians interested in the same issues.[1]

An important part of the Jesuit missionary literature is the *litterae annuae*—reports of the Society's members spread worldwide. Those written in the foreign missions were sent to Europe to their authors' religious superiors and confreres. Very soon they started to spread out and, being also read by friends and benefactors of the Order, in a short time became "bestsellers," translated into several languages and reprinted many times.[2] The main series of this literary genre are *Annuae Litterae Societatis Iesu* (thirty-six volumes concerning the years 1581–1614 and 1650–54, published in 1583–1658); *Relation de ce qui s'est passé en la Nouvelle France* (forty volumes concerning the years 1633–72, published in 1634–73); *Lettres édifiantes et curieuses* (thirty-four volumes published in 1703–76); and *Der neue Welt-Bott* (five volumes subdivided in forty parts, published in 1726–61).[3] These letters began as a way for Jesuits to keep

1 For the titles of the books and the articles about the Jesuit missions, see Sommervogel, Carlos, *Bibliothèque de la Compagnie de Jésus*, vol. 11: *Histoire*, ed. Pierre Bliard (Paris: Picard, 1932), col. 1227–1366. For the publications which appeared during the twentieth century, see Polgár, László, *Bibliographie sur l'histoire de la Compagnie de Jésus, 1901–1980*, vol. 1: *Toute la Compagnie* (Roma: Institutum Historicum Societatis Iesu, 1981), 408–427. For the publications which appeared after 1980, see the supplements published each year in *Archivum Historicum Societatis Iesu* [hereafter AHSI], the revue of the Jesuit Historical Institute. These two bibliographies complete the main bibliographical tools for the history of the missions: *Bibliotheca Missionum* (30 vols., published in 1916–74) and *Bibliographia Missionaria* (published since 1934, at present by the Pontifical University Urbanianum in Rome).

2 Polgár, László, *Bibliography of the History of the Society of Jesus* (Rome/St. Louis: Jesuit Historical Institute; St. Louis University, 1967), 130–137.

3 About the *Annuae*, see Annick Delfosse, "Les *Litterae annuae* de la Compagnie de Jésus entre compte rendu factuel et construction identitaire: l'exemple de Bruxelles," in *Quatre*

© ROBERT DANIELUK, 2024 | DOI:10.1163/9789004684782_004
This is an open access chapter distributed under the terms of the CC BY-NC-ND 4.0 license.

in touch with their superiors and with each other; came to provide a record of the missions and their activities; and would also serve as a way to acquaint a broader public with information regarding the characteristics and geography of the lands where Jesuits served. Eventually some of these letters were used in compiling histories and were often supplemented by maps, as in the case of Polish Jesuit Tomasz Ignacy Szpot Dunin (ca. 1645–1713) described below.

The purpose of this essay is to show where this Jesuit missionary literature comes from, point out its origins and its first developments from the sixteenth through the eighteenth century, and explain how the general policies and strategies of communication, formulated in Rome for the whole Society of Jesus, had an impact on the preparation and publication of not only the annual letters but also of other kinds of Jesuit writing.[4] To follow their passage from manuscript to print means answering the questions "*Why*, *what*, and *how* did the Jesuits write?" All three questions come from an instruction of Juan de Polanco from 1547, which shall be quoted below.[5] All three are related to the origins of the *litterae annuae*; all three concern the collective literary activity of the members of the Society of Jesus. Some further examples with respect to maps and geography will also be provided.

siècles de présence jésuite à Bruxelles, eds. Alain Deneef and Xavier Rousseaux, 215–233 (Bruxelles: Édition Prosopon / Leuven: KADOC, 2012). I would like to express my gratitude to Mrs. Delfosse, who shared with me her text before its publication). Annick Delfosse, "La correspondance jésuite: Communication, union et mémoire: Les enjeux de la *Formula scribendi*," *Revue d'Histoire ecclésiastique* 104 (2009): 71–114; Markus Friedrich, "Circulating and Compiling the Litterae Annuae: Towards a History of the Jesuit System of Communication," *AHSI* 77 (2008): 3–40; Jörg Zech, "Die *Litterae Annuae* der Jesuiten: Berichterstattung und Geschichtsschreibung in der alten Gesellschaft Jesu," *AHSI* 77 (2008): 41–61. About the *Relations*, see Léon Pouliot, *Étude sur les Relations des Jésuites de la Nouvelle-France (1632–1672)* (Paris: Desclée de Brouwer, 1940). About the *Lettres édifiants*, see André Rétif, "Brève histoire des Lettres édifiantes et curieuses," *Neue Zeitschrift für Missionswissenschaft* 7 (1951): 37–50. About the *Welt-Bott*, see Charles G. Herbermann, "Neue Welt-Bott, or the New World Messenger," *The Woodstock Letters* 44 (1915): 371–379.

4 This article is an expanded version of the paper that the author first delivered on September 24, 2010 at the symposium *Early Missionary Printing in Asia and the Americas* organized by the Ricci Institute for Chinese-Western Cultural History at the University of San Francisco.

5 Juan Alfonso de Polanco (1517–76, Jesuit from 1541), secretary of three first generals of the Society and the favourite candidate to become the fourth (if not for the papal opposition to the idea of a Spanish general), enormously influenced the elaborate system of communication in the Order and its archives (see *Diccionario Histórico de la Compañía de Jesús* [hereafter *DHCJ*] 4: 3168–3169). For his instruction concerning the written communication between the Jesuits, see *Monumenta Historica Societatis Iesu* [hereafter *MHSI*], *Epp. Ign.*, 1: 536–549.

1 *Scribere sequitur esse*, or *Why Did the Jesuits Write?*

The classical principle of *agere sequitur esse* could be of some help in researching the origins of Jesuit writing.[6] In fact, without an understanding of the beginnings of the Society of Jesus and its early development, it could hardly be possible to answer the question, "*Why* did the Jesuits write?"

The Order, approved by Pope Paul III in September 1540, started to spread around the world even before this formal sanction. In fact, Francis Xavier, one of the first companions of Ignatius of Loyola, the Society's founder, left for India in March of that year, leaving in Rome his written statement concerning the proposed Order to be founded.[7] Other companions of the Society were sent to missions throughout Italy and other countries as well. As they spread across the continent, concerns arose about how the Jesuits were to preserve their friendship, a bond which was started in Paris during their studies, and how they would receive instructions on how to run the missions entrusted to them by the Church. These were the questions they faced as they were becoming, formally speaking, Jesuits.

Ignatius, who became the first general superior of the Society of Jesus in April 1541, set about answering them. Realizing the importance of communication between the former "friends in the Lord" who were now becoming his subordinates, he started to elaborate tools that would assure their unity in spite of their geographical distance. The whole eighth part of the *Constitutions*, entitled "Helps toward Uniting the Dispersed Members with Their Head and among Themselves," offers a full panorama of his thoughts on the matter.[8] In it Ignatius proposed to his companions a special tool serving the end of the required unity: writing. Several of his letters from these first years of his

---

6 A similar idea was proposed by Fr. Verástegui, S.J., during the meeting of the Jesuit archivists in October 2001 in Rome in order to think about the Jesuit archives, starting from a reflection about communication within the Society. See Verástegui, Nicolás R., "The Importance of Archives in the History and Spirituality of the Society of Jesus according to 'Our Way of Proceeding'," in *"Scriptis tradere et fideliter conservare": Archives as "Places of Memory" within the Society of Jesus* (Rome: General Curia, 2003), 25–47.

7 MHSI, *Xavier*, 1: 23–27 (the original in ARSI, *Hist. Soc.*, 1a, f. 258^{rv}). Francis Xavier (1506–52), one of the first companions of Loyola. He arrived in India in 1542; besides his missionary work there until 1545 he visited Ceylon, Malacca, and the Moluccan Islands. In 1549 he arrived in Japan, and after two years there he wanted to enter China, but died en route. In 1619 he was beatified and in 1622 canonized, together with Ignatius (see *DHCJ* 3: 2140–2141).

8 *The Constitutions of the Society of Jesus and Their Complementary Norms* [abbreviated: *Constitutions*], trans. George E. Ganss, ed. John W. Padberg, 673–676 (Saint Louis: The Institute of Jesuit Sources, 1996). Unless specified differently, the *Constitutions* shall be quoted according to its numbers.

mandate provide testimony on this point. Thus, both his epistolary output and the *Constitutions* answer the question of *why* the Jesuits generally wrote as well as why they composed the *annuae* or drew maps.

Starting from the *Constitutions*, a first answer to the *why* question is found in these words:

> The more difficult it is for the members of this congregation to be united with their head and among themselves, since they are so spread out in diverse parts of the world among believers and unbelievers, the more should means be sought for that union. For the Society cannot be preserved or governed or, consequently, attain the aim it seeks for the greater glory of God unless its members are united among themselves and with their head.[9]

The meaning is clear: maintaining unity was not only important for their friendship, it was the *conditio sine qua non* for the survival of the Order and the guarantee of its apostolic efficiency. Writing soon became a privileged tool for the required unity. This is why Ignatius ordered his companions to send letters to him: "Another special help will be communication by letter between subjects and superiors, and their learning frequently about one another and hearing the news and reports which come from the various regions."[10] He wanted to hear news from friends and he needed to have reports from his subordinates in order to better instruct them in their missions. The examples of this dual purpose of Jesuit correspondence are easy to be found in Ignatius's epistolary production. What is important for the *annuae* here is that in the first of those prescriptions concerning the exchange of letters there are already seeds for a future project of annual and missionary reports, as shown in the following examples.

Even before the formal foundation of the Society, Ignatius must have given some instructions on writing to him. In fact, two weeks after leaving Rome, Francis Xavier wrote the following response to him from Bologna: "I shall write in detail, as you have urged me to do; and I shall follow your instruction with respect to the *hijuelas*."[11] Unfortunately, Loyola's letter from March 21, 1540 has been lost.

---

9 *Constitutions*, 655.

10 *Constitutions*, 673.

11 MHSI, *Xavier*, 1: 29–30. The English version in: *The Letters and Instructions of Francis Xavier* [abbreviated: *Letters and Instructions of Francis Xavier*]. Ed. Joseph Costelloe (St. Louis: The Institute of Jesuit Sources, 1992), 11.

In 1541, when Paschase Broët and Alfonso Salmerón were sent to Ireland with a special mission entrusted to them by Pope Paul III, Ignatius provided them with instructions.[12] His letter from the beginning of September of that year contains, besides many other words of advice, indications or rather intuitions which would later organize all the Jesuit correspondence.[13] Ignatius wished that what his companions write to him should be read not only by the Jesuits but also by friends and benefactors of the Order. This purpose was specified later, but it is interesting to see its origins in this letter from 1541. In fact, Ignatius not only communicates to his two companions that what they would write to him should be forwarded to other readers, but also orders them to send copies of parts of their letters to certain ecclesiastical dignitaries.[14]

The main purpose of the annual letters was to ensure communication between the Jesuits. Looking for the origins of this literary genre, we can consider Ignatius's letter to the Jesuits in Italy from June 1, 1542 an early example.[15] With its short update regarding the missionary achievements of the Society's members, it could be seen as an "annua" *ante litteram*, which, after all, does not seem to have been exceptional at all, since Ignatius must have written more such letters. We can presume this from what he noted on March 18, 1542 in a letter to Simão Rodrigues: "Since I have written fully elsewhere about the dispersed Society and the spiritual fruit which our Lord is deigning to produce through them, there is no more to say here."[16]

---

12 Paschase Broët (ca. 1500–62), French, one of the first companions of Ignatius. After the failure of this mission, he worked in Italy until 1552 when he became the first provincial in France (see DHCJ 1: 552). Alfonso Salmerón (1515–85), Spaniard, another of the group of the first companions of Ignatius (see DHCJ 4: 3474–3476).

13 MHSI, *Epp. Ign.*, 1: 174–181.

14 Ignatius mentions the following persons: Reginald Pole (1500–58, cardinal from 1536), at that time the governor of the province of the Patrimony of St. Peter (in the area of Rome) (see *Enciclopedia Cattolica*, 9, col. 1663–1665); Girolamo Aleandro (1480–1542, cardinal from 1538), one of the closest collaborators of Pope Paul III, member of the commission for the reform of the Roman Curia (see *Enciclopedia Cattolica*, 1, col. 741–472); Prospero Santacroce (1513–89, cardinal from 1564), at that time a Roman cleric and consistorial lawyer (see *Enciclopedia Cattolica*, 10, col. 1825); and Rodolfo Pio (1500–64, cardinal from 1536), protector of the Society (see *Enciclopedia Cattolica*, 9, col. 1490–1491).

15 MHSI, *Epp. Ign.*, 1: 201–205.

16 MHSI, *Epp. Ign.*, 1: 196 (English version quoted after *Ignatius of Loyola: Letters and Instructions* [abbreviated: *Letters and Instructions*], ed. Martin E. Palmer, John W. Padberg, and John L. McCarthy (St. Louis: The Institute of Jesuit Sources, 2006), 74). Simão Rodrigues (1510–79), one of the first Jesuits. Though sent with Francis Xavier to India, he remained in Portugal and became provincial there 1546–52 (see DHCJ 4: 3390–3392).

Another proof of Ignatius's direct involvement in the circulation of the news between the companions is his letter from December 10, 1542, addressed to Peter Faber but copied and sent to all Jesuits.[17] Its contents reveal Ignatius's busyness in compiling the reports that he received in order to forward them to the others:

> If the copies of others' letters that I send you appear to have some order and be free of useless matter, the reason is that with enormous loss of my own time, I extract the edifying parts, rearrange the wording, and edit out the irrelevant matter, so as to give some pleasure in our Lord to all of you and edification to those who hear them for the first time.[18]

In this way, he was himself at the origins of the future *annuae*. He asked the Jesuits to write to him every second week, and in return he promised to answer them every month and to forward to them the copies of other Jesuits' letters every third month.[19]

A few years later, in the *Constitutions*, he specified the rules for the functioning of the network for circulating news among the Jesuits.[20] The main letter was to be prepared in each community in both the language of that province and Latin, then sent to the provincial in two copies; the provincial was to send one copy to the general superior, adding his own comments if necessary, and also to provide multiple copies of them to send to all the houses of his province. If necessary, some local superiors or rectors could be allowed to send their letters directly to the general and to exchange them between their communities (however, the provincial should always receive a copy of it). The general was to ensure that the copy sent to him from each province would be recopied and sent to all other provincials, who were then to distribute it to all the communities of their province. The direct exchange of such news was also allowed by the *Constitutions* (however, the general should always receive a copy of each letter).

In 1564, Diego Laínez decided that the edifying letters should be sent every sixth months by the local superiors to the provincials, who were then obliged

17 MHSI, *Epp. Ign.*, 1: 236–239. Peter Faber (1506–46), one of the first companions of Ignatius. After the approval of the Society, he was sent to several missions in Spain, Portugal, and Germany, where he was in 1542 (see *DHCJ* 2: 1369–1370). He was canonized in 2013.

18 MHSI, *Epp. Ign.*, 1: 238 (English version quoted after *Letters and Instructions*, 92).

19 The frequency of Jesuit written communication in its dual genres (letters *ex officio* and edifying reports) changed several times before assuming its definitive configuration in the *Formula scribendi* of 1580.

20 *Constitutions*, 675.

to prepare eight copies of them and send them to the general and to the other provinces or regions where the Jesuits were working (Italy, Portugal, Spain, France, Germany, Brazil, and India).[21] In order to facilitate their distribution, Laínez proposed that the provincials ask the local superiors to also send them a summary of their letters; this could help the provincials with the "provincial" letter. The purpose of the *annuae* was slowly taking form.

In 1547, Ignatius made a decision which influenced not only his generalship but the future of the Society of Jesus: he appointed Juan de Polanco as his secretary. Soon after this nomination, a detailed instruction was sent to the entire Society explaining the rules of regular correspondence to be observed from that time on, as well as a special letter explaining the importance of what was asked. Both documents have the same date, July 27, 1547, and are signed by Polanco; in both, the *why* question of the Jesuit writing is answered exhaustively.[22]

Introducing himself in his new function and recommending himself to his fellow Jesuits, Polanco declares that the regular exchange of letters offers the possibility of improving the function of the Society and, in this way, better serving the people in accordance with the Order's mission. Comparing the care that the people involved in business took of their administration, Polanco confesses that in his opinion the Jesuits should feel ashamed of doing little to improve their written communication, even when they know that it helps the spiritual progress of themselves and others.

Polanco then enumerates and briefly comments on twenty reasons proving the importance of having a regular and well-organized correspondence. Some of them are spiritual while others are purely administrative and functional. The exchange of letters should stimulate the prayer of thanksgiving for what has already been done and for what was being done. This should increase the glory of God and stimulate the growth of the Christian hope and love of God among the Jesuits and also be a cause of edification for those who will read it, for instance the friends and benefactors of the Order or the candidates for the novitiate. Strengthening unity and fraternal charity among the Jesuits, as well as their religious vocation, the letters shall not only encourage and stimulate them to compete in their mission, but also help the Society's administration to make the right decisions, based on knowledge of the situation. Those who need advice shall receive it, and those who could esteem themselves as having done a lot shall be kept in humility by discovering from the letters that others

21 MHSI, *Lainez*, 8: 322. Diego Laínez (1512–65), Spaniard, one of the first Jesuits. In 1558 he succeeded Ignatius as general of the Society (see *DHCJ* 2: 1601–1605).

22 MHSI, *Epp. Ign.*, 1: 536–549 (English version: *Letters and Instructions*, pp. 183–194).

have done much more than they have. News about Jesuit achievements would also be of great help to the whole Church, as information of what was happening in one region might possibly be useful in other regions.

From the same letter we learn that at that time there were in Rome already three or four secretaries busy with copying and forwarding such letters:

> Here we have to satisfy all the places where the Society is scattered, by writing to each place about what is happening not just here but everywhere else. And considering that for the reasons given above three or four of us are here occupied with a good will in this work of writing as our main and almost exclusive activity.[23]

The importance given to the correspondence is visible in many other places in the twelve volumes of Ignatius's letters published in the *Monumenta Historica Societatis Iesu*.[24] They show how writing letters was a way of keeping the Jesuits united and apostolically efficacious, thus answering the question *why* the Jesuits wrote.

Another answer to the same question can be seen in the great care that the Society always took in the formation of its candidates and, more generally, of all its members. It is instructive to read what the *Constitutions* say concerning the formation of the younger Jesuits: the novitiate candidates should be asked, among other questions, also about their spiritual reading; there will be such a reading during the meals in the communities; profane books must not be kept in the house; the Jesuits in formation should receive some spiritual instruction, also concerning the aforementioned reading; and the houses should be provided with a library and each Jesuit should have necessary books, but it was forbidden to annotate them.[25] Many of these rules have of course changed, transforming and evolving along the centuries with decrees of the following general congregations from the middle of the sixteenth century up to the most recent times.[26]

One more answer to *why* the Jesuits wrote is obvious—they wrote as a part of their mission. It is not difficult to understand how the pen became a tool of the apostolate if we consider the context of the early modern period, marked

23 MHSI, *Epp. Ign.*, 1: 540 (English version quoted after *Letters and Instructions*, 187).

24 MHSI, *Epp. Ign.*, vols. 1–7 (Madrid, 1903–11, reprinted in 1964–68).

25 *Constitutions*, 46, 251–252, 268, 277, 372–373.

26 About these changes, see Danieluk, Robert, "*Ob communem fructum et consolationem*: La genèse et les enjeux de l'historiographie de la Compagnie de Jésus," *AHSI* 149 (2006), 37–41.

as it was by humanism and the Reformation. Gutenberg's movable type served as a catalyst of new ideas and a tool for polemics, and their missionary work in the "new world" and increasing involvement in education in Europe compelled the Jesuits to not only write but also print their works.

It is true that writing and publishing books intended as a separate ministry could hardly be found in the *Formula Instituti*, by which Pope Paul III had approved the Society; the document recommended that the members of the new Order "strive especially for the progress of souls in Christian life and doctrine and for the propagation of the faith by the ministry of the word, by spiritual exercises and works of charity, and specifically by the education of children and unlettered persons in Christianity."[27] The second version of the *Formula* (approved in 1550 by Pope Julius III), however, repeats the previous recommendations concerning the "ministry of the word" as well as indicating "any other ministration whatsoever of the word of God."[28] As John O'Malley has noted, this expression gave Jerónimo Nadal the opportunity to point out that writing and publishing books in order to fight heretics and help souls is also part of those "other ministries."[29] From such a perspective, writing seems since the 1550s to have been used as a tool for the Jesuit apostolate, and to have been permitted under the general superior's authority.

The *Constitutions* mention the writing and publishing of books in only a few places. If a Jesuit prepares a book, he is allowed to publish only after the approval of the Father General and his accurate censorship; any difference of doctrine would not be admitted. The professed Fathers who were preparing books were allowed to live in the colleges and benefit from their financial

---

27 *Constitutions*, pp. 3–4 (the English version, published in 1996 in Saint Louis, omits the ministry of hearing confessions—*Christi fidelium in Confessionibus audiendis spiritualem consolationem praecipue intendat*—as it is exposed in the Latin original; see *Constitutiones Societatis Iesu a Congergatione Generali XXXIV annotatae et Normae Complementariae ab eadem Congregatione approbatae* (Rome: Apud Curiam Praepositi Generalis Societatis Iesu, 1995), p. 4.

28 *Constitutions*, p. 4.

29 O'Malley, John W., *The First Jesuits* (Cambridge, MA/London: Harvard University Press, 1993), 114. O'Malley quotes three writings of Nadal: 1) *Secundus Dialogus* (MHSI, *Nadal*, 5: 665–666); 2) *Exhortatio 6a* (ibid., 841); and 3) *Scholia in Constitutiones S.I.*, ed. Manuel Ruiz Jurado (Granada: Facultad de Teologia, 1976), 188–189. Jerónimo Nadal (1507–80, Jesuit from 1545), was one of the first Jesuits and closest collaborators of Ignatius. He was charged to introduce the *Constitutions* to the European provinces of the Order. Realizing this mission, he contributed enormously in the diffusion and consolidation of the authentic Ignatian spirituality and the structures of the Society of Jesus (see *DHCJ* 3: 2793–2796).

resources (which was an exception from the general rule which obliged them to live out of the colleges and sustain themselves on alms).[30]

In spite of all these rules, it is not evident that the first generation of Jesuits wrote and published. Having regard for religious poverty, Alfonso Salmerón observed in 1547 that publishing books could occasionally be an obstacle in the works of charity, even without directly contradicting the exigency of simplicity and modesty which should characterize the life of the Jesuits.[31] Ignatius of Loyola was open, however, to the idea of writing as a possible ministry and of using Gutenberg's movable type in the Society's evangelical mission to newly discovered territories as well as in keeping European nations faithful to the Catholic Church.

Perhaps a better proof of this than the above-mentioned paragraphs of the *Constitutions* are certain of Ignatius's letters. On August 13, 1554, he wrote to Peter Canisius that it would be beneficial if the Jesuits wrote replies to Protestant attacks:

> Since the heretics write booklets and pamphlets aimed at discrediting Catholics, especially the Society, and shoring up their false dogmas, it would seem expedient that in such cases our men should pen short and well-written replies or tracts which can be brought out quickly and widely purchased, and in this way remedy the harm wrought by the pamphlets of the heretics and disseminate sound doctrine. These works should be modest but lively, exposing the evil behaviour and deceits of adversaries. Moreover, where needed, a number of these pamphlets could be gathered in a single volume. It is important, however, that they be written by learned men with a grounding in theology and adapted to the capacity of the multitude.[32]

30 *Constitutions*, 273, 558, 653.

31 See his letter from September 01, 1547, to Paul d'Achille, quoted by O'Malley, John W., *The First Jesuits* (Cambridge, MA/London: Harvard University Press, 1993), 114.

32 MHSI, *Epp. Ign.*, 12: 262 (English version quoted after *Letters and Instructions*, 506–507); letter mentioned by O'Malley, John W., *The First Jesuits* (Cambridge, MA/London: Harvard University Press, 1993), 114. Peter Canisius (1521–97, Jesuit from 1543), famous theologian and writer, author of the catechism which had up to 200 editions. In 1925 he was canonized and proclaimed Doctor of the Church (see *DHCJ* 1: 633–635).

The same idea appears again in his letter to Nadal from June 8, 1555: "Contradicting the heretics by writing against their books seems to be very convenient."[33]

The context of the religious controversies which characterized that time explains the polemical dimension of the first instructions given by Ignatius to his Jesuits. It is not surprising to see said Jesuits agree with the point of view expressed by their founder. Some examples should be enough to prove that many members of the Order quickly recognized the importance of writing and the usefulness of printing in their activities.

In April 1558, Peter Canisius informed Diego Laínez about the projects of their fellow Jesuit from Ingolstadt, Jean Couvillon, who was preparing the translation of the annual letters from the foreign missions and wanted to publish them as well as another book, a commentary on the Book of Psalms.[34] Canisius testified about the great interest and excitement for the *annuae* and was convinced that both books would be a success among German-speaking readers for whom "one writer is more worthy than ten teachers."[35]

In November 1582 a Spanish Jesuit, Alfonso de Pisa, in the city of Poznań in Poland wrote a memorial about the necessity of writing and publishing books as a tool in the fight with the Protestants, *De excudendis adversus haereticos libris*.[36] Written in order to be forwarded to the superiors, with the approval of

33 "El modo de contradezir a los erejes scribiendo contra sus libros etc., parece muy conueniente," MHSI, *Epp. Ign.*, 9: 116; letter mentioned by O'Malley, John W., *The First Jesuits* (Cambridge, MA/London: Harvard University Press, 1993), 114.

34 Jean Couvillon (sometimes Ioannes Cuvillonius) (ca. 1520–81, Jesuit from 1543), teacher of philosophy and theology in Coimbra, Rome, Lyon, and Ingolstadt. He participated in the Council of Trent (see Sommervogel, Carlos, *Bibliothèque de la Compagnie de Jésus*, vol. 2 (Bruxelles/Paris: Schepens/Picard, 1891), col. 1594.

35 *Le littere de India ci danno grande consolatione; et spero che il Padre Cuvillonio presto transferira anchora questa parte; et così farà un libro, che si pensa essere di grande importantia in settentrione, ut videant caeci transferri a se regnum Dei. Non ce è dubbio, che la opera riuscirebbe con gran guadagno de autorità appresso della gente thudesca, la qual fa più conto de un scrittore, che de dieci lettori* (see Braunsberger, Otto, *Beati Petri Canisii Societatis Iesu epistulae et acta*, vol. 2 (Freiburg i.B.: Herder, 1898), 259. The letter is mentioned in Duhr, Bernhard, *Geschichte der Jesuiten in den Ländern deutscher Zunge*, vol. 1: *Geschichte der Jesuiten in den Ländern deutscher Zunge im XVI. Jahrhundert* (Freiburg i.B.: Herdersche Verlagshandlung, 1907), 646. Apparently, these two projects had not been realized (the bibliographies give only the information about Couvillon's manuscripts; see Sommervogel, Carlos, *Bibliothèque de la Compagnie de Jésus*, vol. 2 (Bruxelles/Paris: Schepens/Picard, 1891), col. 1594.

36 See ARSI, *Germ.* 160, f. 293^r–294^v. The Pisanus memorial is quoted in Duhr, Bernhard, *Geschichte* 1: 646–647. It was published in MHSI, *Mon. paed.*, ed. Ladislaus Lukács, 7: 555–561. Alfonso de Pisa (Pisanus) (1528–98, Jesuit from 1552), after having worked as professor of philosophy and theology in Loreto, Rome, Ingolstadt, Dillingen, and Hall, in

his confessor to whom he submitted it first, this memorial was sent to Claudio Acquaviva, general of the Society of Jesus.[37] The opinion of de Pisa was clear: it was not only useful but necessary to write books against the Protestants.[38] To justify his point of view, he includes many reasons. For one, he quotes the *Constitutions*, which allows such a ministry.[39] For another, making an allusion to the parable from chapter twenty-five of the Gospel of Matthew, he says that it would be against God's will to spoil the literary talents of those in the Society who have them (actually, it should be the superiors' responsibility to encourage them to write). He further observes that publishing Catholic books in those regions would be the best way to restrain the influence of the Protestants and their books since there was a great need for and lack of Catholic books by comparison. He also underlines that the influence of the book is greater than that of teaching or preaching, for the book reaches more people and its influence lasts longer.[40]

The original of this memorial is kept in the Jesuit Roman Archives in the same volume with a contemporary and very similar memorial by the Polish Jesuit Jakub Wujek, who was asking his provincial, Giovanni Paolo Campano, whether it would be advantageous to answer to the polemic writings published against him (his *Postilla*, a selection of his sermons, first published in 1567, had been criticized in an edition of a Protestant *Postilla*).[41] He also indicated several reasons in favour of a continuing debate, among them that an answer was necessary to strengthen the Catholics since silence could be interpreted as a sign of weakness and a defeat in front of their adversaries. Wujek had received advice from important personalities in the country telling him to write and

---

1577–84 was in Poznań, where he taught theology and was one of the leading activists of the Catholic Counter-Reformation (see *DHCJ* 4: 3146–3148).

37 Claudio Acquaviva (1543–1615, Jesuit from 1567), general superior of the Society from 1581 (see *DHCJ* 2: 1614–1621).

38 *[N]on tam utile quam necessarium esse, ut in provinciis infectis haeresi, et nominatim in hac nostras, aliquis, aut aliqui, designarentur, qui lingua latina, aut vernacula vel utraque libris editis haereticos impugnarent*, MHSI, *Mon. paed.*, 7: 555.

39 *Constitutions*, 653.

40 *[P]lus prodest unus noster libellus, quam mille conciones: Pater qui in uno collegio per unam horam docet paucos auditores, perpetuo in centum locis docet omnis generis lectores edito libro*, MHSI, *Mon. paed.*, 7: 556.

41 ARSI, *Germ.* 160, f. 292rv (published in Sygański, Jan, "Korespondencya księdza Jakuba Wujka z Wągrówca z lat 1569–1596," *Roczniki Towarzystwa Przyjaciół Nauk Poznańskiego*, vol. 46 (Poznań, 1920), 40–43. Jakub Wujek (1541–97, Jesuit from 1565), famous writer and translator of the Bible into Polish (see *DHCJ* 4: 4052–4053). Giovanni Paolo Campano (1540–93, Jesuit from 1563), provincial of the Polish Province 1581–91 (see *DHCJ* 1: 615–616).

he declared to the provincial that he too was ready to write, even though he understood the polemics would never end.[42]

In October 1583, Canisius wrote from Freiburg to Acquaviva, asking him to designate some of the Jesuits to this ministry because it would be of great help for the Church, which was facing the rapid progress of the Reformation. According to Canisius, it was important to defend the Catholic faith not only by preaching, but also by writing (*non viva voce tantum, sed et calamo*). Such an activity would conform with the Institute of the Society of Jesus. He also pointed out that writing has the same merit as missionary work in the "new world."[43]

Another interesting deliberation concerning the apostolate of writing took place at the end of the sixteenth century in Germany. Paul Hoffaeus proposed to establish a special community of writers whose task would be the defence of the Catholic faith against Protestant attacks.[44] The project was sent to Francisco de Borja in several letters between March and May 1571.[45] Not having received any decisive answer, Hoffaeus insisted and, on June 23, 1572, wrote to Nadal, who was the general vicar of the Society at that time.[46] The question was then studied by de Borja's successor, Everard Mercurian, who asked three

42 *[S]cribendorum librorum nullus est finis, nec haeretici unquam cessabunt: sed neque nobis cessandum est, ubi maior agitur Dei gloria*, ARSI, *Germ.* 160, f. 292^{v}. His project was accepted and he published the new edition of his *Postilla* in 1582–84 (see Sommervogel, Carlos, *Bibliothèque de la Compagnie de Jésus*, vol. 8 (Bruxelles/Paris: Schepens/Picard, 1898), col. 1228–1229.

43 The letter is quoted in German in Duhr, Bernhard, *Geschichte*, 1: 646, as follows: *Diese Arbeit entspricht unzweifelhaft im höchsten Grade unserem Institut, und dieses Werk des Gehorsams und der Liebe hat nicht weniger Wert als die Arbeiten der Unsrigen bei der Bekehrung von Indien.* Duhr indicates its provenience (*Codex latinus monacensis*, Munich: Staatsbibliothek). In the Braunsberger edition of the Canisius letters, however, are only two indications concerning the letters by Canisius to Acquaviva written at that time from Freiburg: The first letter was written January 19, 1587, the second must have been written either at the end of 1587 or at the beginning of 1588 (*ineunte anno 1588*). Neither letter was found by Braunsberger, who deduced their existence from the general's answers to Canisius (see Braunsberger, Otto, *Beati Petri Canisii Societatis Iesu epistulae et acta*, (Freiburg i.B.: Herder, 1923), 8: 250–254; 261–262).

44 Paulus Hoffaeus (ca. 1530–1608, Jesuit from 1554), provincial of the Upper Germany Province 1569–81 (see *DHCJ* 2: 1932–1933).

45 The letters from March 11, 1571 (ARSI, *Germ.* 133, f. 109^{r}–110^{v}; published in: MHSI, *Mon. paed.*, 3: 498–99), April 07, 1571 (ARSI, *Germ.* 133, f. 136^{r}–137^{v}), and May 12, 1571 (ARSI, *Germ.* 133, f. 184^{r}–187^{v}). On the project of the house of writers and the relevant documents, see Duhr, Bernhard, *Geschichte*, 1: 648–650. Francisco de Borja (1510–72, Jesuit from 1546), third general of the Society of Jesus 1565–72; canonized in 1671 (see *DHCJ* 2: 1605–1611).

46 ARSI, *Germ.* 134, f. 271^{r}–272^{v}. In 1571, a similar request came to Rome, also from the Province Congregation Province of *Germania Inferioris* (see MHSI, *Mon. paed.*, 3: 54).

members of the German Province (Canisius, Peltan, and de Pisa) for their opinion about this project.[47] Their answers were diverse. Canisius supported and highly recommended the project.[48] Peltan was against it: according to him, there were more urgent tasks for the apostolate; writing could be done in free time after other tasks were completed.[49] De Pisa agreed with the idea of writing but objected to the project of forming a separate community of writers.[50]

Finally, a decision was made by Rome after the request was sent in 1573 by the Congregation of the Province of Upper Germany asking the general to appoint some fathers for writing. Rome mandated that the general should dedicate some Jesuits to such a ministry, but without creating a special community for them.[51] The project of a separate house of writers was not put into practice at that time, and, with the two Belgian exceptions of the Bollandistes and the *Musaeum Bellarminianum*, the *domus scriptorum* as a separate Jesuit community did not appear until the nineteenth century.[52]

In summary, the Jesuits wrote in order to keep themselves informed about other Jesuits, to facilitate the administration of the Society and its apostolic efficacy, and also to assure a good level of formation for themselves and for other readers of their missionary relations. Writing for them was a tool of the apostolate as well as a weapon in religious polemics and controversies. The *litterae annuae* in general and the missionaries' editions quoted above were just

47 Everard Mercurian (1514–80, Jesuit from 1548), fourth general of the Society of Jesus 1573–80 (see *DHCJ* 2: 1611–1614).

48 This letter of Canisius to Mercurian from September 01, 1574 (ARSI, *Germ.* 183, f. 237^{r}–238^{v}) was published in Otto Braunsberger, *Beati Petri Canisii Societatis Iesu epistulae et acta* (Freiburg i.B.: Herder, 1922), 7: 236–238, and MHSI, *Mon. paed.*, 3: 544–546.

49 This letter of Peltanus to Mercurian from May 1573 (ARSI, *Germ.* 153, f. 377^{r}–378^{v}) was published in MHSI, *Mon. paed.*, 4: 486–487.

50 This letter of Pisanus to Mercurian from August 25, 1574 (ARSI, *Germ.* 135—II, f. 403^{rv}) was published in MHSI, *Mon. paed.*, 4: 542–544.

51 See ARSI, *Congr.* 20b, f. 241^{r}, and *Congr.* 42, f. 30^{r}; published in MHSI, *Mon. paed.*, 4: 237–238.

52 Both initiatives started in the time between the end of the sixteenth century and the beginning of the seventeenth century in Belgium. The purpose of the Bollandistes was to critically study the lives of saints, while the *Musaeum Bellarminianum* was a group of writers whose goal was to write first against the Protestants and then against the Jansenists. In the eighteenth century they changed their name to *Musaeum Historicum* with the mission of publishing documents about the history of their country (see *DHCJ* 1: 472–475; De Backer, Augustin et Aloïs; Sommervogel, Carlos, *Bibliothèque des écrivains de la Compagnie de Jésus*, vol. 1 (Liège/Paris: Chez l'Auteur, 1869), col. 1296–1297; Poncelet, Alfred, *La Compagnie de Jésus en Belgique: Aperçu historique à l'occasion du 75e anniversaire de l'érection de la Province Belge (3 décembre 1832–3 décembre 1907)* (Bruxelles, s.d.), 34–36.

one part of all Jesuit literary activities. The study of geography and cartography was another.

2 *Una carta principal la cual se pudiese mostrar a cualquier persona*, or *What Did the Jesuits Write?*

Having understood why the Jesuits wrote makes it easier to answer the question "*What* did they write?" Since it is obvious that the results of their literary activity must have depended on their motivation to write, their handwritten and published works will reflect their reasons for writing. Following the logic of the passage from manuscript to print, the first part of the answer to the *what* question of Jesuit writing should focus on different typologies of their handwritten documents, while the second should indicate the variety of literary genres represented in the oeuvres of the members of the Order. Answering the second part of this question is not difficult, since the bibliographies provide more than enough information for this essay.[53] In fact, as we can see in the index of the tenth volume of the monumental "Sommervogel" bibliography (without speaking about its predecessors), there is hardly any topic which has not become the subject of an article or a book written by a Jesuit.[54] Not only theology and philosophy in all their subdivisions but also law, the arts, and the sciences appear among the subjects studied by members of the Society. Catechisms and books on theology were certainly published more frequently than manuals on military arts or pharmacy, but all coexist in a testimony to the interdisciplinary interests of their authors. Maps—and geography in general—were not last on their list, as is clearly evident from the "Sommervogel," which dedicates one section of its index to cartographical endeavours.[55]

As for manuscripts, the Roman Archives of the Society of Jesus (ARSI) provide a good example of the typology of Jesuit handwritten material. Conserving the documentation of the central government of the Order from its beginning up to the present time, they could easily serve to answer our *what* question.

53 For more on Jesuit bibliographies, see Danieluk, Robert, *La "Bibliothèque" de Carlos Sommervogel: Le sommet de l'œuvre bibliographique de la Compagnie de Jésus (1890–1932)* (Rome: Institutum Historicum Societatis Iesu, 2006).

54 Carlos Sommervogel, *Bibliothèque de la Compagnie de Jésus*, vol. 10: *Tables de la première partie*, ed. Pierre Bliard (Paris: A. Picard, 1909).

55 Carlos Sommervogel, *Bibliothèque de la Compagnie de Jésus*, 10: 1385–1407.

Since their history and their content have been the subject of several studies,[56] the focus below shall be put only on what concerns the *annuae*.

Looking for their origins, it is necessary to consider once again the importance of written correspondence as indicated above. Indeed, the letters very quickly became the main tool for administration and communication in the Society of Jesus. That is why several detailed instructions were given on organizing the exchange of information according to the pyramidal structure of the Society's centralized government. The general residing in Rome received the letters and then sent them on to provincials as well as to local superiors, and even to individual Jesuits if necessary. Each Jesuit had the right to write directly to the general without passing through the intermediary superiors. However, normally a lower-level analogue of such communication also existed within each province. The provincial was in touch with his local superiors and individual Jesuits, being temporarily obliged to report to the general superior matters of particular importance and news from the province. Local superiors and individual Jesuits also were in contact with their provincial.

As a result of such a network of communication, ARSI, despite of all its vicissitudes, has maintained a rich and interesting collection of letters (received and sent) covering all periods of the Society's history, even if not always in a proportional and representative way (many letters were lost; from one period we have more than from others).[57] The same, after all, is also true for other Jesuit archives, or rather for other Jesuit documents, since the archives where they are kept today are most often not property of the Society.

The edifying letters and missionary relations were born as part of this system of communication. In fact, from the first years of the Society, besides the correspondence written by reason of office, there were also other letters, and one must distinguish between Jesuit correspondence's two complementary literary genres, as explained many times by the founder himself, by his

56 Of particular interest are Lamalle, Edmond, "L'archivio di un grande Ordine religioso: L'archivio generale della Compagnia di Gesù," *Archiva Ecclesiae*, 24/25 (1981–82): 89–120; Schurhammer, Georg, "Die Anfänge des römischen Archivs der Gesellschaft Jesu," *AHSI* 12 (1943): 89–118; Teschitel, Josef, "Das Generalarchiv der Gesellschaft Jesu in Rom," *Römische Historische Mitteilungen*, 4 (1960–61): 247–254; and idem., "L'organizzazione dell'archivio generale della Compagnia di Gesù," *Rassegna degli Archivi di Stato*, 22 (1962): 189–196.

57 On the importance and role of regular correspondence among the Jesuits, see Nicolás R. Verástegui, "The Importance of Archives in the History and Spirituality of the Society of Jesus According to 'Our Way of Proceeding'," *"Scriptis tradere et fideliter conservare": Archives as "Places of Memory" within the Society of Jesus* [also in Italian and Spanish] (Rome: General Curia, 2003): 25–47. For more on the correspondence preserved in ARSI, see Lamalle, Edmond, "L'archivio di un grande Ordine religioso: L'archivio generale della Compagnia di Gesù," *Archiva Ecclesiae* 24/25 (1981–82): 96–100.

collaborators, and by his successors.[58] Already in September 1541, Ignatius wanted Broët and Salmerón to send to him two kinds of letters. The first category would later be called "main letters," and was the direct ancestor of the *annuae*. Written in a separate sheet from the letters of business (i.e., the second category), it should, according to Ignatius, contain only the news that might edify its readers. Short reports of many things were recommended rather than a more detailed report of only few things:

> In the principal matters you should recount the course of events for greater edification; other news without exhortations, taking into account that the letter will be shown to shrewd persons who will want concise accounts of many things done rather than a lot of words about few if the subject does not require much.[59]

Unlike correspondence by reason of office, i.e., the second category, such a letter was destined to be shown to many people, as we can see in what Ignatius wanted to be communicated to Rodrigues in his name a year later:

> He should write us a letter that can be shown to anyone—great, middling, or small, good and bad—and which avoids speaking prejudicially of any individual. In other letters he can write whatever particulars he likes and judges profitable for what we all desire for the greater service of God our Lord.[60]

A key to understanding the distinction between these two kinds of letters is the already quoted instruction by Ignatius to Faber from 1542, containing the rules of a whole system of communication for the Society of Jesus. He repeats there what he had said earlier about the two letters which a Jesuit writing to the general should prepare. The first one, called the main letter (*carta principal*), could be shown to everybody: "Any member of the Society intending to write us ought to write out a main letter that can be shown to anybody."[61] The second one, the annexed letter (i.e., the business letter, also called *hijuela*), should contain the information destined only for its addressee. From then on, this distinction would become one of the points that would be returned in

58 Also, the *Diccionario Histórico de la Compañía de Jesús* considers Jesuit correspondence in this way: its article *correspondencia* is divided into *Litterae quadrimestres, semestres, annuae*, and *Litterae ex officio* (see DHCJ 1: 965–966).

59 MHSI, *Epp. Ign.*, 1: 177 (English version quoted after *Letters and Instructions*, 61).

60 MHSI, *Epp. Ign.*, 1: 235 (English version quoted after *Letters and Instructions*, 89).

61 MHSI, *Epp. Ign.*, 1: 236 (English version quoted after *Letters and Instructions*, 90).

the official admonitions that the generals wrote to the Jesuits concerning their correspondence.

The main letter, per Ignatius, should contain news that could edify the readers and present information about any Jesuit pastoral achievements which brought about spiritual good. Not having such activities to report, a missionary should write briefly about his health or any interesting conversations or meetings that he had had:

> In the main letter we should write what each man is doing by way of preaching, hearing confessions, [giving the] Exercises, and other spiritual work which God works through each man, in a way that will give the greatest edification to the hearers and readers. If the soil is barren and there is nothing to report, there should be a few words about health matters, a conversation with so-and-so, or the like.[62]

Other instructions concerning the main letter covered similar ground.

In 1547, Polanco, issuing further prescriptions, added that the writer should write about the challenges he faced, what he thought about the situation of his mission, how the local ecclesiastical and civil authorities had helped, and what the attitude of the local people was. The letters should contain information about the health of the Jesuits as well as the number of candidates entering into the novitiate and those who are preparing to enter, and the number of those who had left the Society or died. News should also be shared concerning the material situation of the community (house, clothes, and food) and the spiritual situation of all its members, including the difficulties and temptations that were faced, if writing about it could edify or help to resolve some problems.[63] In these prescriptions we can see not only the origins of the *annuae* but also of the Jesuit catalogues, i.e., lists of Jesuits from each province containing information about them.[64]

---

62 MHSI, *Epp. Ign.*, 1: 236 (English version quoted after *Letters and Instructions*, 91).

63 MHSI, *Epp. Ign.*, 1: 544–547.

64 There were two kinds of such lists: those compiled every year and called *catalogi breves* or *annuales*, containing only basic information about the members of each community of a particular province, and others, compiled every three years and called *catalogi triennales*, containing more detailed information about each Jesuit. These latter were organized in three parts: *catalogus primus* (more detailed biographical information, such as personal name, family name, origins, age, date of entrance to the novitiate, date of final vows, indication of ministries and offices held), *catalogus secundus* (information about the human qualities of each Jesuit, his health, character, experience, etc.), and *catalogus tertius* (short report of the financial situation of each community). The origins of these catalogues are similar to those of the correspondence: the first ideas and rules originated

All the rules concerning the Jesuit correspondence were put together in a practical manual called *Formula scribendi*.[65] It was approved by the second General Congregation in 1565 and modified and sent to the provinces in 1578 by Mercurian. It was officially published for the first time in 1580, codifying Jesuit communications policy for hundreds of years.[66] Polanco's letter from 1547 hinted that Ignatius wanted his prescriptions concerning the exchange of letters to be a part of a small manual explaining the way to proceed in different affairs.[67]

The *Formula* was the final result of a long process of building a whole system of Jesuit communication, which as attested above started shortly after the pontifical approval of the Society of Jesus in 1540 and the election of Ignatius as its first general superior in 1541. In its thirty-five points it resumes the previous prescriptions concerning both genres of the Jesuit correspondence (letters written by reason of office and edifying relations) and specifies what their content should be. For the next centuries, the *Formula* was the rule followed by Jesuits dispersed all over the world for their missionary activities. Most of the documents that survived the complicated vicissitudes of the Society's history were written according to what it prescribes.

One part of the *Formula* is dedicated to the *annuae*. It obliges the local superiors to be aware of what happens in their communities, to collect all that could edify (*quaeque ad Nostrorum consolationem ac proximorum aedificationem pertinent*), and to send it to the provincial at the end of each year.

---

with Ignatius and were then developed and received a fixed form which has lasted for centuries. Even if Polanco did not speak in 1547 about a catalogue, but rather about the letters, the *Constitutions* were more clear: "For fuller knowledge of everyone, every four months the provincial should be sent, from each house and college, a brief list in duplicate of all who are in that house, and of those who are now missing because of death or some other cause, from the time of the last list sent until the date of the present one, with a brief account of the qualities of these persons. In the same manner, every four months the provincial will send to the general the copies of the lists from each house and college" (see *Constitutions* 676). Such lists became the catalogues. Their double form was definitely fixed at the end of the sixteenth century by the *Formula scribendi*. From the beginning of the eighteenth century, the annual catalogues started to be printed. For more on Jesuit catalogues, see *DHCJ* 1: 967; Lamalle, Edmond "Les catalogues des provinces et des domiciles de la Compagnie de Jésus," *AHSI* 13 (1944): 77–101; MHSI, *Mon. Ang.*, 1: xxvii–xxxvii.

65 *Institutum Societatis Iesu*, vol. 3: *Regulae, Ratio studiorum, Ordinationes, Instructiones, Industriae, Exercitia, Directorium* (Florence: Ex Typographia A. ss. Conceptione, 1893), 41–45.

66 Some changes were introduced by Acquaviva at the end of the century, concerning mostly the annual letters.

67 MHSI, *Epp. Ign.*, 1: 541.

From what is sent to them, the provincials should prepare a Latin copy of the letter containing the news from all the communities of their province and send it to the general at the beginning of each year (a copy should be kept in case the official Latin version, concerning all the provinces of the Society and prepared in Rome, does not arrive). The *annuae* should contain: the number of active Jesuits (first altogether, then, specified for each house, the count of priests, teachers in colleges, scholastics, and brothers); the number of candidates and of Jesuits who died; news about the spiritual progress of the Jesuits (*agent primo de profectu Nostrorum in Domino*) and their ministries (preaching, teaching, giving the Spiritual Exercises, visiting hospitals and jails, etc.); the number of students and the situation of the colleges; any specific problems they might be encountering; and the help received from friends that the Jesuits experienced.

The letters coming from the "new world" excited the curiosity of friends and benefactors who wanted not only to hear news about the missionaries, but also to learn something interesting about the countries the Jesuits were in. That is why they were also told to include such information in their main letters.

Writing in the name of Ignatius on November 22, 1547 to Niccolò Lancilotto, rector of the college in Goa, Polanco pointed out that news from India was necessary to help Loyola in providing this mission with all it needed.[68] Thus, Lancilotto was told to send to Europe a messenger with five or six young Indians who would study in Rome and in Portugal. That messenger should also prepare a written relation about India, including information about its weather, food, customs, etc. (*de aëris temperie, victu, moribus et ingeniis locorum ac hominum*):

> he should also bring us in writing a full account of whatever would be worthwhile to know in providing for you and for India: climate, food customs, and the mentalities of persons and places, as well as whatever you think is needed for the worship of God and the help of souls there and throughout India.[69]

Specialists will have to determine whether *ingenium loci* can here be translated as "geography." In any case, other similar requests are easy to find in the epistolary exchanges from those years.

68 MHSI, *Epp. Ign.*, 1: 648–651 (English version: *Letters and Instructions*, 224–226). The letter is quoted in: Paul Nelles, "Jesuit Letters" in *The Oxford Handbook of the Jesuits* (New York: Oxford University Press, 2019), 58. Niccolò Lancilotto (?–1558), Italian, Jesuit from 1541, he was a missionary in India from 1543 (see *DHCJ* 3: 2276).

69 *Letters and Instructions*, 225.

On July 5, 1553, Polanco explained to Francis Xavier that in Rome there were several ecclesiastical dignitaries asking for news about the geography and climate of India. Thus, it would be good to satisfy their curiosity, for in this way they would continue to be friendly to the Society and its missions.[70] Among those dignitaries was "Cardinal Santa Croce," i.e., Marcello Cervini—the future Pope Marcello II:

> The cardinal of Santa Croce and others wish that in the letters coming here there should be some information with respect to the location where you are with regard to the sky, the climate, and the degrees, and so forth. I do not believe that everyone there is a cosmographer; but if anyone knows how to write something about this, he could do so in a few words; and if not in the same letter, in another separate missive, in order to satisfy these lords, who are very fond of the Society and of the things of India.[71]

A few months later, writing on February 24, 1554 to Gaspar Berze in India, Polanco asked him in the name of Ignatius the same news about the geography and climate, but also about the plants and animals of India, and for exactly the same reason:

> There are important persons in Rome who read the letters from the Indies with great edification and who frequently express a wish for geographical information about the countries where our members travel: how long the days are in summer and in winter, when summer begins, whether shadows fall to the left or to the right—in short, information should be furnished about anything else that is out of the ordinary, such as plants and animals that are unknown here, or of a larger size, and so on.[72]

A similar request was sent also on August 13, 1553, to Manuel Nóbrega in Brazil, who was asked to give information about the climate and the traditions of the country's inhabitants (including their buildings, their clothes, and their

---

70 MHSI, *Epp. Ign.*, 5: 164–165 (English version: *The Letters and Instructions of Francis Xavier*, 458–459). The same letter is also published in MHSI, *Doc. Ind.*, 3: 7–9.

71 *The Letters and Instructions of Francis Xavier*, 459. Marcello Cervini (1501–1555), cardinal in 1539; he was elected pope April 9, 1555 but died May 1st of the same year (see *DHCJ* 3: 2971).

72 MHSI, *Epp. Ign.*, 6: 358 (English version quoted after *Letters and Instructions*, 474). Gaspar Berze (1515–53, Jesuit from 1546), sent to India in 1548, from 1552 he was vice-provincial there (see *DHCJ* 1: 427).

food) as well as about the general situation.[73] In all these instructions we can see the origins of many of missionary reports then becoming so popular in Europe. Some of them have been published, while many others exist only in manuscript.

Among the answers we can find data useful to the fields of geography, meteorology, and even seismology. For example, on May 10, 1546, Francis Xavier wrote from the island of Ambon (Amboino, modern Indonesia) to the Jesuits in Europe mentioning, among other curiosities, earthquakes and volcanos:

> These islands have a temperate climate and large, dense forests. They have an abundance of rain. These islands of Maluco (Maluku) are so high and so difficult to traverse that in time of war they defend themselves by climbing to higher grounds and using these as their fortresses. They have no horses, nor would it be possible to ride on horseback across the islands. The earth frequently quakes, and also the sea, so much so that those who are sailing when the sea quakes gain the impression that they are striking on rocks. It is a frightful thing to feel the earth tremble, and even more so the sea. Many of these islands spew forth fire with such a roar that there is no barrage of artillery, however loud it may be, that makes so great noise; and the fire rushes forth with such violence from its place of origin that it carries huge rocks along with it.[74]

Eight years later and from the other extremity of the terrestrial globe, José de Anchieta wrote to Ignatius of Loyola from what later become the city of São Paulo, not only describing the progress of the mission in spite of the crude habits of the inhabitants (including cannibalistic practices), but also including some information about the geography of the country:

> This part of Brazil that we live in is at 24 degrees south. All this seacoast extends on 900 miles, starting from Pernambuco—the first place where

---

73 "[…] quanto á la región, dónde está, en qué clima, á quántos grados, qué vecindad tiene la tierra, cómo andan vestidos, qué comen etc.; qué casas tienen, y quántas, según se dize, y qué costumbres; quántos [cristianos] puede auer, quántos gentiles ó moros; y finalmente, como á otros por curiosidad se scrive muy particulares informaciones, así se scriuan á [Nuestro Padre], porque mejor sepa cómo se ha de proueer; y también satisfazerse ha á muchos señores principales, devotos, que querrían se scriuiesse algo de lo que he dicho." MHSI, *Epp. Ign.*, 5: 329–331 (the same letter was sent the same day also to Berze; see MHSI, *Doc. Ind.*, 3: 15–16). Manuel Nóbrega (1517–70), Jesuit from 1544, in Brazil from 1549; he became provincial in 1553 (see *DHCJ* 3: 2826–2827).

74 MHSI, *Xavier*, 1: 332–333 (English version quoted after: *The Letters and Instructions of Francis Xavier*, 142–143).

> Christians started to live, coming up to here and going further. It is inhabited by Indians who eat human flesh and enjoy it so much that they often make war and travel more than 300 miles because of it. If they capture four or five enemies, they come back and eat them entirely without throwing away even a nail, amid a large feast when they sing and drink a lot of wine made from roots. They consider this a major victory and reason for glory. As for the prisoners, they appreciate their fate because they esteem as cowards those who die and are buried, having to sustain the burden of earth that they find very heavy.[75]

Many other examples of such reports are available in both the *Monumenta Historica* series of published primary sources and in the Jesuit Roman Archives. They provide abundant information about *what* the Jesuits wrote about.

In summary, the Jesuits wrote a great deal and concerning an immense variety of topics as shown in bibliographies of the Order. Among the handwritten materials, the letters are of special interest for understanding the origins of Jesuit missionary relations, including geographical information and maps. The fundamental distinction between letters written by reason of office and those written to edify shows that the latter were direct predecessors of the series of *annuae* quoted at the beginning of this paper.

3 *No es todo para todos*, or *How Did the Jesuits Write?*

The Jesuit letters were written according to specific rules, as were all the rest of the writings produced by the members of the Society. In order to understand them well, it is necessary to know these rules. This will help us answer the question "*How* did the Jesuits write?" Understanding the rules also allows us to avoid misunderstandings or misinterpretations, e.g., attributing to the authors of those writings intentions they never had. The most important rules, which shaped this literary genre, were established at the beginning of the Society and never changed.

One of the rules was the already mentioned distinction between main letters and annexed letters, which was based on the pertinent principle expressed by Polanco in 1547 that "not everything is for everyone (*no es todo para todos*)."[76]

---

75 MHSI, *Mon. Bras.* 2: 96 (letter from September 1, 1554; the English translation is mine). José de Anchieta (1534–1597), Spaniard, Jesuit from 1551, missionary in Brazil from 1553 (see *DHCJ* 1: 156–158); he was canonized in 2014.

76 MHSI, *Epp. Ign.*, 1: 543 (English version quoted after *Letters and Instructions*, 189).

Some of his advice is very practical. He recommends that his audience look for and communicate to him the names of people who could help the Jesuits send the letters, e.g., merchants or ecclesiastical dignitaries having regular contacts with Rome. Other suggestions concern the organisation of the office: maintaining a special register with the dates of the letters received would help to keep them in order, while another book containing the material used for the edifying letters would be useful in case the letter was lost in transit and needed to be resent (for that reason he announced that the arrival of each letter to Rome should be confirmed; absent such confirmation, it would be resent).

Whoever wrote the main letter should, according to the advice, do it with requisite attention to not only its content but also its external form. It should be written on separate sheets in order to make it easy to divide it from the letter *ex officio*. The authors were asked to write in such a way that no suspicion of vainglory could be found. They should be prudent both in praising and in criticizing, the latter to be included in the letter only if really necessary. If there was a question of mentioning somebody from the civil or ecclesiastical authorities, it was recommended to add it in the annexed letter and in such a way that the concerned person could read that letter without problem. Even writing something that would offend somebody could be done in an implicit way such that only the addressee would understand what the author of the letter wanted to say: "Where certain details could not help giving offense and it was essential to mention them, a method might be employed to signify one thing by writing another."[77]

Polanco's instructions were based on Ignatius's directives, as we find them in many of Ignatius's own letters, even prior to 1547. Writing to Faber five years earlier, Ignatius revealed his own way of proceeding in correspondence—namely, first fully drafting and then carefully rewriting the letter to avoid inconveniences that might otherwise arise from possible misunderstandings or unintended offenses—and asked all Jesuits to follow his example. The reason for this was the importance he had assigned to the written word: "We must give even more thought to what we write than to what we say. Writing is permanent and gives lasting witness; we cannot mend or reinterpret it as easily as we can our speech."[78] His concern about the quality of the letters was confirmed by his experience, leading him to complain in the same letter about how, when asked by friends or benefactors to show the letters of the Jesuits, he could not do it since they were written in such a disorderly way that it would

77 MHSI, *Epp. Ign.*, 1: 548 (English version quoted after *Letters and Instructions*, 193).

78 MHSI, *Epp. Ign.*, 1: 237 (English version quoted after *Letters and Instructions*, 91).

not be edifying to show them; sometimes he managed to show just one part of the letter and to hide the rest.

One more example of what Ignatius expected from written communication between himself and his fellow Jesuits can be found in his letter to Nicolas Bobadilla from 1543.[79] Penned as a result of Bobadilla's objections to the above-mentioned rules, it repeated the principal points of what he had written some months earlier, and tried to persuade Bobadilla that it was important to follow the instructions, even if it took time.[80] The importance of the written word and of observing the difference between what is written and what is just said ("styles of speaking and writing are quite different")[81] could be seen, he argued, in the requests coming from those who wanted to see the letters sent by Jesuits they knew. If they were not allowed to read the letters, it would seem that they were not being treated as friends. If they were allowed to read them, but the letters were not written as Ignatius wanted them to be, then the readers would not be edified.

For these reasons Ignatius insisted that Bobadilla and all the Jesuits follow the instructions he reiterated in the letter, less he have to order it in virtue of obedience: "I wrote that unless you did this I would be forced, with a view to our common spiritual good and my own conscience, albeit much against my character, to command you under obedience to do so."[82] At the same time, however, he explained that his intention never was to push the Jesuits to change the way they expressed themselves in writing or to improve their literary skills at all costs; what he wanted was for everyone to do their best out of respect for the others and because of their resulting spiritual profit.

In 1554, Polanco reminded Berze that the author of the main letters should write about general matters which could edify and bring instructive news rather than focusing on specific things concerning the Jesuits; otherwise, it would be costly to prepare such letters for printing:

> It would be good for the writer, when preparing letters for showing to people outside the Society, to dwell less on matters specifically related to members of the Society and more on matters of general interest. Otherwise the letters cannot be printed here without considerable cutting.[83]

---

79 Nicolas Bobadilla (ca. 1509–90), Spaniard, one of the first companions of Ignatius (see *DHCJ* 1: 463–465).

80 MHSI, *Epp. Ign.*, 1: 277–282.

81 MHSI, *Epp. Ign.*, 1: 278 (English version quoted after *Letters and Instructions*, 94).

82 MHSI, *Epp. Ign.*, 1: 280 (English version quoted after *Letters and Instructions*, 96).

83 MHSI, *Epp. Ign.*, 6: 358 (English version quoted after *Letters and Instructions*, 474).

Another important word of advice concerning the correspondence was to send multiple copies of the same letter. Ignatius had already recommended it to Salmerón and Bröet in September 1541, suggesting they also send their letters through different countries such as Spain, Portugal, and Scotland. This later became an example of the different *viae* by which many Jesuit missionaries sent their letters to Rome.[84] When he realized that some letters sent to him by Rodrigues had never arrived in Rome, Ignatius repeated his order that the letters were always to be sent to him in a double copy, especially if they concerned an important matter.[85] These orders were repeated in other letters (Polanco suggested in 1547 that the letters concerning important matters could be sent even in three copies).[86]

According to the *Formula scribendi*, the authors of the edifying letters should put in it only important events (*ea tantum commemorentur, quae sunt alicuius momenti*) without writing any details of that which was common. Particular prudence was recommended for all that was connected to the ministry of confession and to other matters which required secrecy and discretion. All should be written to avoid any possible offense and in a way that would help future historians of the Society. Any exaggeration was to be avoided, including in that which concerned the style of writing, and the authors were told not to present as miracles those things which were not (*pro miraculis non ponantur, quae miracula non sunt*).[87]

As far as the frequency of edifying letters, the first dispositions of Ignatius were clear: he wanted news and he wanted it regularly. It is not clear, however, how often he wanted it. In 1541, he ordered Bröet and Salmerón to write often to him from diverse places they were passing through and, generally, once a month.[88] A year later, he specified in his letter to Faber that he wanted a letter from each Jesuit written to him every second week while promising to reply to the individual correspondents every month and to forward all news about the others every third month.[89] According to the prescriptions of Polanco from 1547, later confirmed by the *Constitutions*, edifying letters should be prepared and sent every fourth month.[90] Thus they became *litterae quadrimestres*. But

84 MHSI, *Epp. Ign.*, 1: 177.
85 MHSI, *Epp. Ign.*, 1: 233.
86 MHSI, *Epp. Ign.*, 1: 549.
87 *Institutum Societatis Iesu*, vol. 3: *Regulae, Ratio studiorum, Ordinationes, Instructiones, Industriae, Exercitia, Directorium* (Florence: Ex Typographia A. ss. Conceptione, 1893), 43–45.
88 MHSI, *Epp. Ign.*, 1: 176.
89 MHSI, *Epp. Ign.*, 1: 238.
90 MHSI, *Epp. Ign.*, 1: 548–549, and *Constitutions*, 675.

this title lasted only a few years, being replaced by *semesters* in 1564 and, a year later, were termed the *litterae annuae*, the name by which they are known today.[91]

To sum up, the edifying relations were prepared according to special rules and written so that they could be shown to and read by many readers, members of the Society as well as non-Jesuits. Their frequency, like their title, changed during their first decades and was finally established as annual. Their edifying purpose influenced both the letters' content and composition. This is why they can be profitably read by historians, though scholars need to be aware of the specifics of the genre.

4 Conclusion: Writing Jesuit History or Jesuit Writing of History?

The whole of Jesuit missionary literature as well as the *litterae annuae* have their origins in the decisions made by the founder of the Society of Jesus in consultation with his collaborators (especially Polanco); necessary rearrangements were then introduced by the immediate successors of Ignatius. Looking at the Order's official documents—letters written by superior generals, decrees of the general congregations, rules, and so-called "consuetudinaries" (customs books), one can easily see the differences between the first forty to fifty years of the existence of the Society and the following decades and centuries. Many of the decisions taken during the early period of the Order were modified once its structures became more institutionalized. Introducing changes was a necessary consequence of the Society's rapid growth. As for the policy of compiling and publishing missionary reports, the official document of the Order on the subject, the *Formula scribendi* of 1580, reiterated the importance of following established rules to which only a few changes were added afterwards.[92] These would be implemented in the so-called "new" or "post-restoration" (1814) Society, mostly in the twentieth century. To study them would be an obvious and interesting complement to the present essay.

The purpose of the *annuae* was to edify their readers. That is why they should be considered as a valuable but selective source for historical study. They are

91 The *quadrimestres* from the time of Ignatius and Laínez have been published (see MHSI, *Litt. Quad.*, vols. 1–7 (Madrid, 1894–1932)). More on their origins, finality, and content, can be found in MHSI, *Litt. Quad.*, 5: v–xx.

92 Some small changes had been introduced in the instructions issued by Claudio Acquaviva in 1599 and Muzio Vitelleschi in 1623; see Annick Delfosse, "La correspondance jésuite: Communication, union et mémoire: Les enjeux de la *Formula scribendi*," *Revue d'Histoire ecclésiastique* 104 (2009): 90–92.

just one part of a whole system of communication in the Society of Jesus, to which belong other literary genres such as the letters *ex officio*, the catalogues, and other biographical and historical materials produced by the Jesuits.

An interesting example of the latter is the work of the Polish Jesuit, Tomasz Ignacy Szpot Dunin. Sent to Rome in 1688, he spent the rest of his life there. His main endeavour was hearing confessions in St. Peter's Basilica. The fruits of his other activities include a history of the Jesuit mission in China from its beginning up to his own time. His unpublished manuscripts, preserved at the Jesuit Roman Archives, include the *Historia Sinarum* and two series of *Collectanea*, as well as shorter writings—all together, eight volumes and almost five thousand pages of Latin text written in clear and elegant calligraphy.[93]

His work belongs neither to the genre of *annuae* nor to that of the letters *ex officio*. It could hardly be catalogued as missionary reports (Szpot was never a missionary and never went to China) or as example of "edifying literature." Instead, it belongs to what we can call "material for writing history," i.e., that which the Jesuits wrote or collected as they contemplated composing a future history of the Society. Curiously, even though some members of the Order were in charge of systematically searching for and gathering all that might be of help to the future historian, it was rather rare that the same Jesuits would actually use that material to write real "histories." Szpot is one such rare example of that: he not only collected material available to him in the archives of the Roman *Casa Professa* but also used it in composing his *Historia*.

It is worth noticing that at the beginning of this work, Szpot describes the geography, social organisation, customs, and religion of China and several neighbouring regions that he refers to as *Tartaria* as if he were adhering to the earlier cited instructions to provide European readers with information about this area. The first pages of his *Historia* have as their title *Sinae et Tartariae brevis descriptio* (A Short Description of China and Tartary).[94] In fact, he provides the geographic coordinates and gives the latitude of China as being located between 22° and 44° North, taking as the borders of the Empire the Great Wall in the north and the South China Sea in the south. As for the eastern and western borders, he gives the Pacific Ocean for the former and Tibet for the latter.[95] With regard to *Tartaria*, or rather *Tartarias* (in the plural), he distinguishes

93 ARSI, *Jap. Sin.* 102–103 (*Historia*); *Jap. Sin.* 104–105; 109–111 (*Collectanea*). See more about Szpot and his work, in: Robert Danieluk, "Konfesjonał i pióro: Tomasz Ignacy Szpot Dunin, polski historiograf jezuickiej misji w Chinach," in *Iesuitae in Polonia - Poloni Iesuitae. Piśmiennictwo łacińskie czasów nowożytnych*, ed. Jarosław Nowaszczuk (Szczecin: Volumina, 2017), 75–108.

94 ARSI, *Jap. Sin.* 102, f. 1^r–13^v.

95 ARSI, *Jap. Sin.* 102, f. 1^r.

between *Tartaria Orientalis* and *Tartaria Occidentalis*, with the meridian at 155° East dividing one from the other; Szpot says that their external borders are marked by the Pacific Ocean in the east and the Caspian Sea and the River Ob in the west.[96]

Szpot also included several hand-drawn maps and pictures in the second part of his work, the *Collectanea*. There are ten altogether: five larger maps (more or less corresponding to a modern A4 format) and five smaller pages with maps and illustrations (more or less the equivalent of the modern A5 format).[97]

The large maps are all in colour (which is remarkably well preserved) and have descriptive titles that define their content: 1) *Imperium Sinicum*; 2) *Tartariae Imago*; 3) *Tabula itineris ex Moschovia in Chinam a Moschis facta*; 4) *Duplex iter terrestre in Chinam ex Persia et ex Mogor iuxta descriptionem P. Antonii Thomae missam anno 1690 ex Sina*; and 5) *Tabula geographica Orientis, iuxta autographum P. Antonii Thomae, Belgae e Societate Iesu, missum Pekino anno 1690, in qua demonstrantur etiam itinera in Chinam ex Moschovia, Persia et Mogor*. As these titles show, the interest of the author was either general (maps 1 and 2), or focused on the important issue of finding an alternative route from Europe to China for the missionaries (maps 3, 4 and 5).

On each map, Szpot has introduced elements of cartographical science such as the grid of parallels and meridians. He indicates the main cities, rivers and mountains, as well as the Latin names and the borders of the countries or regions, all according to the knowledge available to him. Whenever he lacked sufficient information, he would note that. For example, when he drew eastern Siberia (between Kamchatka and the Bering Strait), he wrote that "The Russians explored this tongue of land, but not up to its very end" (*Hanc Terrae lingulam lustrarunt Moschi, sed non ad finem*); and "It is not certain whether this tongue of land is adjacent to North America or whether there is still a bay there" (*Haec lingula an adjaceat Americae Septentrionali, an detur adhuc fretum* [*Anian?*] *incertum est*).[98] All of the smaller maps and illustrations are of the island of Sancian (Shangchuan), the place of the death and first sepulchre of Francis Xavier, which Szpot describes in his *Collectanea*. They are not reproduced here.

96 ARSI, *Jap. Sin.* 102, f. 10^r.

97 The bigger maps are in ARSI, *Jap. Sin.* 105-I, f. 1^r, 6^r, 98^r, 227^r; *Jap. Sin.* 110 (folio without number, at the beginning of the volume), while the smaller are in ARSI, *Jap. Sin.* 110, 477; *Jap. Sin.* 111, f. 293^r–294^r, 338^r.

98 ARSI, *Jap. Sin.* 105-I, f. 6^r; *Jap. Sin.* 110 (folio without number, at the beginning of the volume).

FIGURE 2.1 *Imperium Sinicum* [The Empire of China]. Archivum Romanum Societatis Iesu (ARSI), *Jap. Sin.* 105-I, f. 1r.

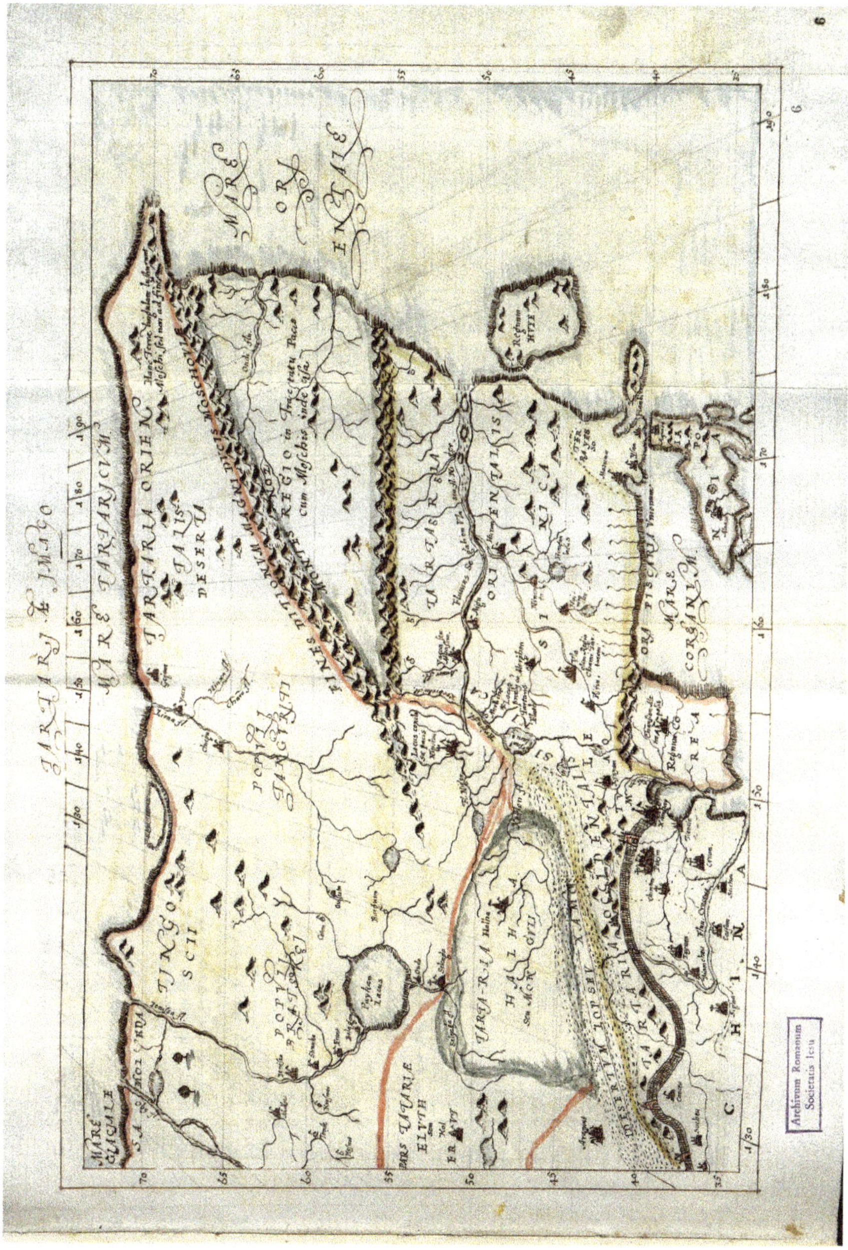

FIGURE 2.2 *Tartariae Imago* [Tartary]. ARSI, *Jap. Sin.* 105-I, f. 6r.

FIGURE 2.3 *Tabula itineris ex Moschovia in Chinam a Moschis facta* [Itinerary from Moscow to China]. Editor: The double dotted line indicates the route from the far western reaches shown on the map to the Great Wall of China, depicted in the lower right. ARSI, *Jap. Sin.* 105-I, f. 98r.

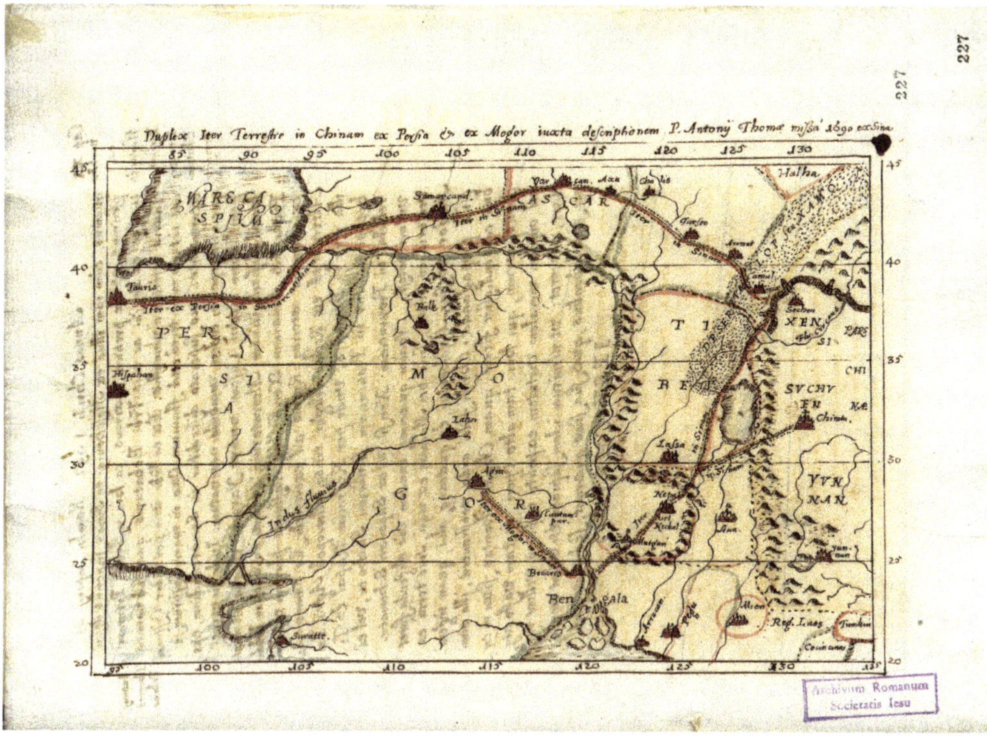

FIGURE 2.4 *Duplex iter terrestre in Chinam ex Persia et ex Mogor iuxta descriptionem P. Antonii Thomae missam anno 1690 ex Sina* [Land routes into China from Persia and from Mughal India, according to a description sent from China by Father Antoine Thomas in 1690]. Editor: Note that the northern route, from Persia, passes through Samarkand, Yarkand, and Turfan. The southern route begins in Agra and passes through Benaras, Nepal, and Lhasa. ARSI, *Jap. Sin.* 105-I, f. 227r.

Szpot is certainly not well known as a Jesuit cartographer, nor does his name appear among those who engaged in the study of geography.[99] Yet the above quoted material illustrates how an inquiry into the *annuae* should be accompanied by the study of other Jesuit sources. All of these materials were produced as elements of the same logical unity, reflecting the nature and purpose of the Order. Thus, they should be understood and studied as such. It is also worth noticing that, in the case of the *annuae*, to edify did not mean to manipulate or to attempt to hide the truth, especially from the readers external to the Society. These letters were constructed not only to provide interesting information

99 Chen, Hui-Hung, "Cartography in China," in *The Cambridge Encyclopedia of the Jesuits*, ed. Thomas Worcester (Cambridge: Cambridge University Press, 2017), 141–142.

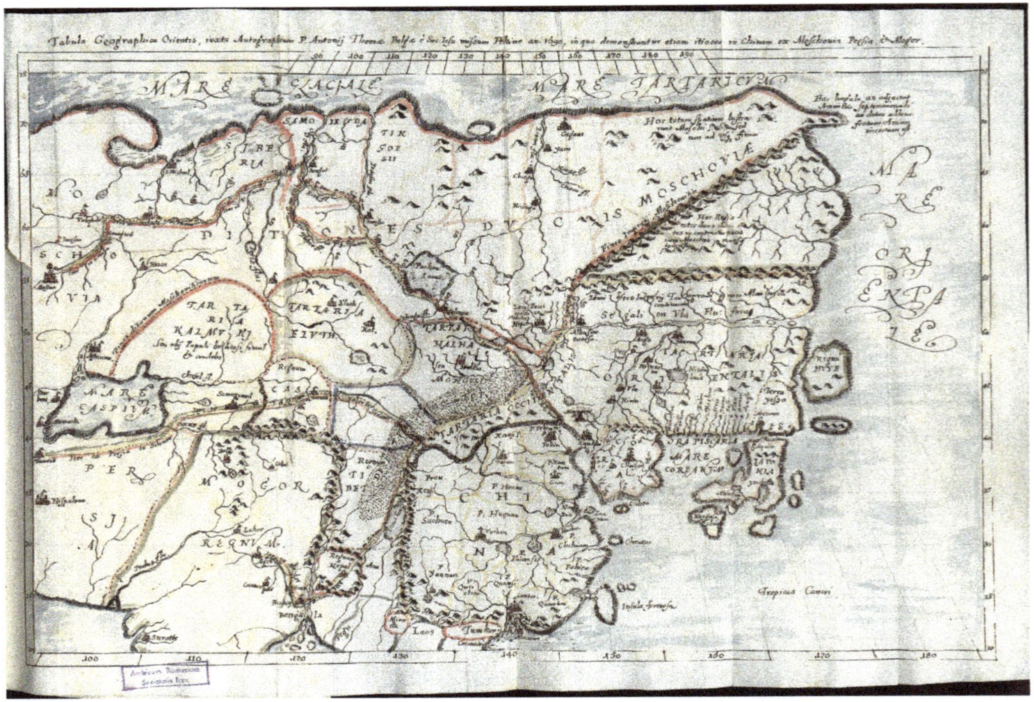

FIGURE 2.5 *Tabula geographica Orientis, iuxta autographum P. Antonii Thomae, Belgae e Societate Iesu, missum Pekino anno 1690, in qua demonstrantur etiam itinera in Chinam ex Moschovia, Persia et Mogor* [Geographical Map of the Orient according to the Manuscript sent from Beijing in 1690 by the Belgian Jesuit Father Antoine Thomas, showing also Routes to China from Muscovy, Persia, and Mughal India]. ARSI, *Jap. Sin.* 110 (folio without number, at the beginning of the volume).

from far countries and exciting news, but also and first of all to help readers spiritually and to encourage them to continue in their good work, as well as to strengthen the Jesuits and their religious communities and encourage prayers for each other.

Bibliography

Archival Sources

All Primary Manuscript Sources are from the Jesuit Roman Archives (*Archivum Romanum Societatis Iesu* [ARSI]).

Congregationes 20b, f. 241^{r}: Fr. General's answer to the postulates of the provincial congregation of the Province of Upper Germany (1573).

Congregationes 42, f. 30^{r}: Postulates of the provincial congregation of the Province of Upper Germany (1573).

Germaniae 133, f. 109^{r}–110^{v}: Letter by Paul Hoffaeus to Francisco de Borja (11 March 1571).

Germaniae 133, f. 136^{r}–137^{v}: Letter by Paul Hoffaeus to Francisco de Borja (7 April 1571).

Germaniae 133, f. 184^{r}–187^{v}: Letter by Paul Hoffaeus to Francisco de Borja (12 May 1571).

Germaniae 134, f. 271^{r}–272^{v}: Letter by Paul Hoffaeus to Jerónimo Nadal (23 June 1572).

Germaniae 135—II, f. 403^{rv}: Letter by Alfonso de Pisa to Everard Mercurian (25 August 1574).

Germaniae 153, f. 377^{r}–378^{v}: Letter by Theodor Peltan to Everard Mercurian (May 1573).

Germaniae 160, f. 292^{rv}: Letter by Jakub Wujek to Claudio Acquaviva (1582).

Germaniae 160, f. 293^{r}–294^{v}: Memorial by Alfonso de Pisa, *De excudendis adversus haereticos libris* (November 1582).

Germaniae 183, f. 237^{r}–238^{v}: Letter by Petrus Canisius to Everard Mercurian (1 September 1574).

Historia Societatis 1a, f. 258^{rv}: Statement of Francis Xavier from 15 March 1540 concerning the election of the general superior and the Constitutions of the Society of Jesus.

Published Materials

Braunsberger, Otto. *Beati Petri Canisii Societatis Iesu epistulae et acta*. Freiburgi Brisgoviae: Herder, 1922, 7: 236–238.

Canisius, Petrus, and Otto Braunsberger. *Beati Petri Canisii, Societatis Iesu, Epistulae et acta*. Friburgi Brisgoviae: Herder, 1896. http://archive.org/details/beatipetricanisi 05cani.

The Constitutions of the Society of Jesus and Their Complementary Norms: A Complete English Translation of the Official Latin Texts. Translated by George E. Ganss, and edited by John W. Padberg. St. Louis: The Institute of Jesuit Sources, 1996.

Correia-Afonso, John. *Jesuit Letters and Indian History, 1542–1773.* 2nd ed. Studies in Indian History and Culture of the Heras Institute, 20. Bombay: Oxford University Press, 1969.

Danieluk, Robert. *La "Bibliothèque" de Carlos Sommervogel: Le sommet de l'oeuvre bibliographique de la Compagnie de Jésus (1890–1932).* Rome: Institutum Historicum Societatis Iesu, 2006.

Danieluk, Robert. "*Ob communem fructum et consolationem*: La genèse et les enjeux de l'historiographie de la Compagnie de Jésus." *Archivum Historicum Societatis Iesu* 75 (2006): 29–62.

Danieluk, Robert. "Konfesjonał i pióro: Tomasz Ignacy Szpot Dunin, polski historiograf jezuickiej misji w Chinach," in *Iesuitae in Polonia – Poloni Iesuitae. Piśmiennictwo łacińskie czasów nowożytnych*, ed. Jarosław Nowaszczuk (Szczecin: Volumina, 2017), 75–108.

Delfosse, Annick. "La correspondance jésuite: communication, union et mémoire. Les enjeux de la *Formula scribendi*." *Revue d'Histoire ecclésiastique* 104 (2009): 71–114.

Delfosse, Annick. "Les *Litterae annuae* de la Compagnie de Jésus entre compte rendu factuel et construction identitaire: l'exemple de Bruxelles." In *Quatre siècles de présence jésuite à Bruxelles*, edited by Alain Deneef and Xavier Rousseaux, 215–233. Bruxelles: Édition Prosopon / Leuven: KADOC, 2012.

Diccionario Histórico de la Compañía de Jesús. Biográfico-temático. 4 vols. Roma, Institutum Historicum S.I. Madrid: Universidad Pontifica Comillas, 2001.

Friedrich, Markus. "Circulating and Compiling the *Litterae Annuae*: Towards a history of the Jesuit system of communication," *Archivum Historicum Societatis Iesu* 77 (2008): 3–40.

Friedrich, Markus. *Der lange Arm Roms? Globale Verwaltung und Kommunikation im Jesuitenorden 1540–1773.* Frankfurt: Campus Verlag, 2011.

Heiss, Gernot. "Die *litterae annuae* und die *historiae* der Jesuiten." In *Quellenkunde der Habsburgermonarchie (16. bis 18. Jahrhundert). Ein exemplarisches Handbuch*, edited by Josef Pausner, Martin Scheutz, and Thomas Winkelbauer, 662–674. Vienna: Böhlau, 2004.

Herbermann, Charles G. "Neue Welt-Bott, or the New World Messenger." *The Woodstock Letters* 44 (1915): 371–379.

Laborie, Jean-Claude. *Mangeurs d'homme et mangeurs d'âme. Une correspondance missionnaire au XVIe, la lettre jésuite du Brésil, 1549–1568.* Paris: Honoré Champion, 2003.

Lamalle, Edmond. "L'archivio di un grande Ordine religioso. L'archivio generale della Compagnia di Gesù." *Archiva Ecclesiae* 24–25 (1981–82): 89–120.

Pouliot, Léon. *Étude sur les Relations des Jésuites de la Nouvelle-France (1632–1672)*. Montréal/Paris: Desclée de Brouwer, 1940.

Rausch, Fred G. "Die gedruckten *litterae annuae SI*, 1581–1654. Ein meist übersehener Quellenschatz zur Jesuitengeschichte." *Jahrbuch für Volkskunde* 20 (1997): 195–210.

Rétif, André. "Brève histoire des Lettres édifiantes et curieuses." *Neue Zeitschrift für Missionswissenschaft* 7 (1951): 37–50.

Verástegui, Nicolás R. *The Importance of Archives in the History and Spirituality of the Society of Jesus according to "Our way of proceeding."* In *"Scriptis tradere et fideliter conservare": Archives as "Places of Memory" within the Society of Jesus*, 25–47. Rome: General Curia, 2003.

Wicki, Josef. "Von den gelegentlichen Veröffentlichungen der Missionsbriefe aus Übersee zu den offiziellen *Litterae annuae* der Gesellschaft Jesu (1545–1583)." *Neue Zeitschrift für Missionswissenschaft* 32 (1976): 95–129.

Zech, Jörg. "Die *Litterae Annuae* der Jesuiten. Berichterstattung und Geschichtsschreibung in der alten Gesellschaft Jesu." *Archivum Historicum Societatis Iesu* 77 (2008): 41–61.

CHAPTER 3

Dutch Publications on the Jesuit Mission in China in the Seventeenth and Eighteenth Centuries

Paul Begheyn, S.J.

In 2014, after more than twenty years of preparation, for the first time ever an overview was published of all publications by Jesuits, printed in the Dutch Republic and its Generality Lands for the period from 1567 until 1773, when the Society of Jesus was suppressed by the Pope worldwide.[1] During this period a total of 1,270 books were published by 266 different Jesuits as authors, editors and translators. Only thirty-five of them were indeed active in these areas. These books were printed in ten different languages: Armenian, Dutch, English, French, German, Italian, Latin, Persian, Portuguese and Spanish. They were published by 369 printers, editors and publishers in 31 towns spread all over the country, of which Amsterdam was the most important, with 176 printers, followed by Leiden with 43 and The Hague with 39. Although the number of Jesuits in the Dutch Republic and its Generality Lands in the so-called Dutch Mission was never higher than ninety-five, their influence was much greater, thanks to the many pirated publications by non-Jesuits.

The interest in everything Chinese among Dutch people grew remarkably during the Golden Age of the seventeenth century, and so did the artistic and cultural impact of Asia on the Netherlands. In connection to this fact any publication on China was popular and sought after.[2] Between 1615 and 1736 a total of 31 Jesuit publications on China appeared in the Dutch Republic, in Dutch, French, German, Latin and Spanish. These publications, mainly on history and cartography, will now be presented in chronological order.

In late 1612 the Walloon Jesuit Nicolas Trigault (1577–1628) was appointed by Niccolò Longobardi (1565–1654), superior of the China mission, as the mission's procurator in Europe. He sailed from Macao on February 9, 1613, and arrived in Rome on October 11, 1614. During this trip he edited and translated

1 Paul Begheyn, *Jesuit Books in the Dutch Republic and its Generality Lands 1567–1773: A Bibliography* (Leiden: Brill, 2014).

2 Karina Corrigan, Jan van Campen, and Femke Diercks, eds., *Asia in Amsterdam. The Culture of Luxury in the Golden Age* (Peabody Essex Museum, 2015); Thijs Weststeijn and Menno Jonker, *Barbaren en Wijsgeren. Het beeld van China in de Gouden Eeuw* (Nijmegen: Vantilt, 2017).

© PAUL BEGHEYN, 2024 | DOI:10.1163/9789004684782_005
This is an open access chapter distributed under the terms of the CC BY-NC-ND 4.0 license.

the "China Journal" of Matteo Ricci (1552–1610) from Italian into Latin, and gave it the title *De Christiana expeditione apud Sinas* [The Christian expedition to China]. Three months after his arrival in Europe a pamphlet of only 8 pages was published in Dutch by the printer Anthonie Scheffer in 's-Hertogenbosch, whose brother Paulus (1578–1620) had been a member of the Jesuit college in the same city (founded 1609), the second Jesuit college in the Netherlands after that in Maastricht (1575). It was an excerpt—hardly noticed by scholars of the Jesuit China mission—of the edition of 646 pages that would appear later in the same year in Augsburg. Its title translates as follows: "True story of some remarkable facts of the famous kingdom of China, by a priest of the Society of Jesus, a Netherlander, having recently arrived from there."[3]

In 1639, eleven years after the death of Nicolas Trigault, his description of the Chinese empire, *Regni Chinensis descriptio. Ex varijs authoribus* [The description of the Chinese Empire. From different authors], was printed by the famous publishing house of Bonaventura and Abraham Elzevier in Leiden. The small sized book (11 × 6 cm) was based upon the journals of Matteo Ricci, published already at Augsburg in 1615 by Christoph Mang. The text of Trigault covers the first 287 of 365 pages. Then he added additional topographies (pp. 299–330): some small texts by other authors who had written on China—the Italian discoverer Marco Polo (1254–1324), the Portuguese Jesuit brother Bento de Goís (1562–1607),[4] and the English cleric Samuel Purchas (1577?–1626)—were included. The book contains some woodcuts and is dedicated to Adriaen Pauw (1585–1653), who had been Grand Pensionary of Holland from 1631–1636. The *Regni Chinensis descriptio* is one out of fifty-four so-called 'Republieken' [Republics], a series started by the Elzeviers in 1628, and very popular among book collectors.[5] The Elzeviers were a family of Dutch booksellers, publishers,

3 Dutch title: *Waerachtich verhael van eenige merckelycke saecken des vermaerts coninckrijcx van Syna, door eenen priester der Societeyt Jesu, Nederlander, onlancx van daer ghecomen.* Only two copies are known (The Hague: Koninklijke Bibliotheek, Pamflet 2173 and 224 D 4); Nicolas Standaert, "Nicolas Trigault S.J., Litterae Societatis Iesu e regno Sinarum (1615)," in Paul Begheyn, Bernard Deprez, Rob Faesen and Leo Kenis, eds., *Jesuit Books in the Low Countries 1540–1773: A Selection from the Maurits Sabbe Library* (Leuven: Maurits Sabbebibliotheek, Faculteit Godgeleerdheid, 2009), 26–29; Begheyn, *Jesuit Books in the Dutch Republic*, 33, 62 (entry 1615.2).

4 The journey of Bento de Goís from Lahore via Tartary to China was also included in a Leiden publication of 1706 by Pieter van der Aa, translated from Portuguese, only 8 pages. See Begheyn, *Jesuit Books in the Dutch Republic*, 255–256 (1706.3).

5 Paul Begheyn, "De Elzeviers en de jezuïeten," in B.P.M. Dongelmans, P.G. Hoftijzer, and O.S. Lankhorst, eds., *Boekverkopers van Europa. Het 17de-eeuwse Nederlandse uitgevershuis Elzevier* (Zutphen: Walburg, 2000) 62, 70; J.A. Gruys, "De reeks 'Republieken' van de Elzeviers

and printers from the sixteenth through the end of the eighteenth century. Not without reason they were called "the booksellers of Europe."

On November 7, 1653, the famous Italian Jesuit missionary Martino Martini arrived in Amsterdam from Macao. In his capacity as procurator of the Chinese mission he was sent to Europe with a threefold task: seeking papal approval regarding the Jesuit accommodation to Chinese rites, recruiting new missionaries, and publishing books about China. At his arrival an unknown eyewitness wrote down:

> He [Martini] brings with him two volumes of the history of China from the Deluge until the present time, and he will publish the first volume, from the Deluge until Christ, in Antwerp. If he detects that this will be appreciated, he will add a second and even a third volume. What he calls the *Atlas of the Kingdom of China* will contain a representation and a description of China, which is different from that available in Germany until now to be seen. I have not seen these books, because Father has left them on the ship.[6]

Nine days later Martini went to see the famous Amsterdam cartographer Joan Blaeu (1599–1673) and started negotiations about the publication of his *Novus Atlas Sinensis*, the first European atlas on China, which would be published within two years. Probably during these days Martini also made a deal with Blaeu about the printing of the German translation of his *History of the Tartarian War*, that informed the European reader of the Manchu invasion of China in 1644 and the instalment of the Qing dynasty. That book appeared in the same year 1654 with the *impressum* of Blaeu himself, followed by a reprint in 1655.[7] Other publishers jumped on the bandwagon, and followed with five more editions, three in Dutch (1654, 1655, 1660), and two in Latin (1655, 1661).[8] The Dutch translation inspired the most famous Dutch poet of the seventeenth century, Joost van den Vondel (1587–1679), to write a theatre piece *Zungchin*

---

en Joannes de Laet," in Dongelmans et al., *Boekverkopers*, 95, 100; Begheyn, *Jesuit Books in the Dutch Republic*, 42, 93 (1639.2).

6 Henri Bernard, "Les sources mongoles et chinoises de l'Atlas Martini (1655)," originally published in 1947, reprinted in: Roman Malek and Arnold Zingerle, eds., *Martino Martini S.J. (1614–1661) und die Chinamission im 17. Jahrhundert* (New York: Routledge, 2000), 234–235 refers to a document in Nijmegen, Archives of the Dutch Jesuit Province [ANSI], which is now untraceable; Paul Begheyn, "The contacts of Martino Martini S.J. with the Amsterdam printer Joan Blaeu," *Archivum Historicum Societatis Iesu* 81 (2012): 219–232.

7 Begheyn, *Jesuit Books in the Dutch Republic*, 42–43, 120 (1654.2), 122 (1655.5).

8 Begheyn, *Jesuit Books in the Dutch Republic*, 42–43, 120 (1654.3), 122 (1655.4, 1655.6), 135 (1660.3), 138 (1661.6).

of ondergang der Sineesche heerschappye [Emperor Chongzhen or the downfall of the Chinese supremacy] in 1667. With this play he wanted to honour the Jesuits and their mission efforts in China, and at the same time his home town Cologne and the Jesuit who was born there, Johann Adam Schall von Bell (1592–1666), and who later would become Director of the Bureau of Astronomy.[9] The memoirs of the latter, published a year before, could have been an incentive for Vondel.[10]

The six months between December 1653 and July 1654 had been decisive for the production and publication of Martini's *Atlas*. On one hand Martini was able to work out the practical and financial details with Joan Blaeu, on the other hand he started negotiations with the *Lords XVII*, the governing body of the Vereenigde Oostindische Compagnie [Dutch East India Company], who had paid his travel expenses. These *Lords XVII* saw their chance to penetrate the Chinese market with the help of the Jesuits, and to get first-hand information about that country through these missionaries. In return Martini wanted to gain financial and other advantages for the costly transport of Jesuit missionaries and their luggage between Europe and China on the Dutch fleet.[11]

Since the publishing house of Joan Blaeu was located in Amsterdam, and thus was part of the Dutch Jesuit mission under supervision of the Flemish Church authorities, Martini needed a twofold approval for the printing of his *Novus Atlas Sinensis*. On March 1, after the book had been revised by two Jesuits, the Flemish Jesuit Provincial Jan Baptist Engelgraeve (1601–1658) gave his *imprimatur*, as did Antonius Sanderus (1586–1664), *censor librorum* and publisher of illustrated historical works himself, on March 25. In May 1654 the Jesuit curia in Rome sent a letter to Martino Martini, in which it expressed its

9 Alfons Väth, *Johann Adam Schall von Bell S.J., Missionar in China, Kaiserlicher Astronom und Ratgeber am Hofe von Peking, 1592–1666* (Köln: Rheinisches Museum, 1933), 348–350; Gregory Blue, "Johann Adam Schall and the Jesuit mission in Vondel's *Zungchin*," in Roman Malek, ed., *Western Learning and Christianity in China: The contribution and impact of Johann Adam Schall von Bell, S.J. (1592–1666)* (Nettetal, Germany: Steyer Verlag 1998) 2: 951–982.

10 Joannes Adamus Schall von Bell, *Historica narratio de initio et progressu missionis Societatis Jesu apud Chinenses, ac praesertim in regia Pequinensi* (Vienna: Matthäus Cosmerovius, 1665).

11 Noël Golvers, "Een detailblik op China in de zeventiende eeuw bij de 350ste verjaardag van Martino Martini's *Novus Atlas Sinensis* (Amsterdam, 1655)," *Verbiest Koerier* 18 (March 2006): 6–8; Noël Golvers, "Martino Martini, S.J., his stay in the Jesuit college of Brussels (1654) and the production of his 'Novus Atlas Sinensis'," in A. Deneef and X. Rousseaux, eds., *Quatre siècles de présence jésuite à Bruxelles / Vier eeuwen Jezuïeten te Brussel* (Brussels: Prosopon / Leuven, KADOC, 2010), 127–139; Noël Golvers, "Michael Boym and Martino Martini: A contrastive portrait of two China missionaries and mapmakers," *Monumenta Serica* 59 (2011): 259–271.

desires that he soon would complete his negotiations about the printing of the *Atlas*, so that he would be free and ready to depart for Rome.[12] Thanks to the mediation of Prince Leopold Willem, archduke of Austria and governor of Belgium and Burgundy, a privilege for Joan Blaeu could be obtained in Vienna, 7 January 1655, which was registered at The Hague 20 March and 11 May 1655. The latter privilege was necessary for every printer who wanted to publish in the Dutch Republic.

The history of China, written by Martino Martini, was published in Latin by Joan Blaeu in 1659 under the title *Sinicae historiae decas prima, res a gentis origine ad Christum natum in extrema Asia, sive magno Sinarum imperio gestas complexa*.[13] Of a French translation of this book, published in Amsterdam in 1692,[14] no copy could be traced, unless it should be identified with a Parisian edition of that same year. Of the Amsterdam edition of *Novo descobrimento de Gram Cathayo ou Reinos de Tibet* by the Portuguese Jesuit António de Andrade (1580–1634) in 1667,[15] no copy could be traced either.

In the Dutch dedication of the *Novus Atlas* to the commanders of the Dutch East India Company, Joan Blaeu wrote in 1662: "Aboard your ships and under your protection this book was written by the reverend Martini from the Chinese texts (which he often read to us as if it were his mother tongue) into Latin, and thus translated at the service of the public and also of those who love scholarship and good arts."[16] Martini had stipulated that on the maps of China the Jesuit settlements would be indicated by the Jesuit logo. These IHS-signs were beautifully accentuated with gold paint by the famous Amsterdam *afsetter* (colorist) Dirk Jansz van Santen (1637–1708) in the *Atlas Maior* in the collection of the Amsterdam Catholic lawyer Laurens van der Hem (1621–1678). The latter had the original eleven volumes expanded into 46 volumes and four supplements, with, in total, more than two thousand maps, prints, and drawings

12 Goswin Nickel to Martino Martini in Amsterdam, Rome, 16 May 1654, in: ARSI, Fl. Belg. 6, f.599v.

13 Begheyn, *Jesuit Books in the Dutch Republic*, 133 (1659.8).

14 Begheyn, *Jesuit Books in the Dutch Republic*, 217 (1692.6).

15 Begheyn, *Jesuit Books in the Dutch Republic*, 153 (1667.1).

16 "Dit boeck is van de E. Martinius op uwe schepen en onder uwe bescherming uyt der Sinesen schriften (die hy ons als zijn moederlijcke tale menighmalen voorlas) in 't latijn beschreven, en alzoo ten dienste van 't gemeen en ook van diegene, die de Geleertheydt en goede konsten beminnen, overgebracht," quoted in: H.J. Allard, "Een groet uit China voor Vondel en de Amsterdamsche 'vrunden' (1662)," *Jaarboekje Alberdingk Thijm* 46 (1897): 159–160.

which he had collected over the years.[17] In 2003 this atlas was added to the UNESCO list of World Heritage.

The *Novus Atlas Sinensis* is the sixth out of eleven volumes of the majestic *Atlas Maior*, with 594 maps, published in 1662 by Joan Blaeu. The earlier version in two volumes, entitled *Atlas Novus*, was printed by his father Willem Jansz Blaeu in 1635.[18] The *Atlas* of 1662, including the volume on China, was immediately a great success, and appeared in a Dutch translation in the same year. Reprints and translations of the atlases in Spanish, French, and German would come from the presses of Joan Blaeu between 1658 and 1667.[19]

The German Jesuit, Athanasius Kircher (1602–1680), in Rome, was another person involved in spreading the knowledge on China, although he had never been in that country personally.[20] From 1665 he had all his writings published in Amsterdam, in order to avoid Roman censorship.[21] Among these Amsterdam publications was a voluminous book on China, that appeared in Latin (twice in 1667), Dutch (1668) and French (1670), printed by Joannes Janssonius van Waesberge and (the widow and heirs of) Elizeus Weyerstraat, and copied by Jacob van Meurs.[22] In 1667 Athanasius Kircher gave a copy of his *China illustrata* to Jacob Golius (1596–1667), medical doctor, mathematician, astronomer and later Orientalist at the university of Leiden, where he founded the observatory. Golius had learned the basics of the Chinese language from Martino Martini, who gave him some Chinese books and manuscripts for

17 Jan van der Waals a.o., *Een wereldreiziger op papier: de atlas van Laurens van der Hem (1621–1678)* (Amsterdam: Stichting Koninklijk Paleis, 1992); Truusje Goedings, *"Afsetters en meester-afsetters": De kunst van het kleuren 1480–1720* (Nijmegen: Vantilt 2015), 111–150, esp. 118, ill. 10.

18 Djoeke van Netten, *Koopman in kennis. De uitgever Willem Jansz Blaeu in de geleerde wereld (1571–1638)* (Zutphen: Walburg, 2014).

19 Begheyn, *Jesuit Books in the Dutch Republic*, 122–123 (1655.7; 1655.8), 129–130 (1658.2), 140 (1662.6), 142 (1663.3), 144 (1664.4), 154 (1667.9).

20 Paul Begheyn, "Athanasius Kircher sj (1602–1680) en zijn contacten in Nederland," in Joep van Gennip and Marie-Antoinette Th. Willemsen, eds., *Het geloof dat inzicht zoekt. Religieuzen en de wetenschap* (Hilversum, 2010), 84–93.

21 Daniel Stolzenberg, "Utility, edification, and superstition: Jesuit censorship and Athanasius Kircher's *Oedipus Aegyptiacus*," in John W. O'Malley, Gauvin Alexander Bailey, Steven J. Harris, and T. Frank Kennedy, eds., *The Jesuits II: Cultures, Sciences and the Arts 1540–1773* (Toronto: University of Toronto Press, 2006), 336–354. Harald Siebert, "Kircher and His Critics: Censorial Practice and Pragmatic Disregard in the Society of Jesus," in Paula Findlen, ed., *Athanasius Kircher: The Last Man Who Knew Everything* (New York: Routledge: 2004), 79–104; Daniel Stolzenberg, "Oedipus Censored: Censurae of Athanasius Kircher's Works in the Archivum Romanum Societatis Iesu," AHSI 73 (2004): 3–52.

22 Begheyn, *Jesuit Books in the Dutch Republic*, 154 (1667.6; 1667.7), 157 (1668.4), 160 (1670.3).

his library. In return Golius sent seeds of the rhubarb plant—a rarity from China—from the hortus botanicus in Leiden to Martini.[23]

Thirty years after Martino Martini had visited the Dutch Republic, in 1653, another Jesuit from China arrived in the beginning of October 1683 in Enkhuizen: the Fleming Philip Couplet (1623–1693). He had been to Amsterdam already in 1654 before his departure to China. Now he returned to that city with many books and manuscripts. He was received by Nicolaes Witsen (1641–1717), who not only was thirteen times mayor of Amsterdam, but also cartographer, writer, diplomat, governor of the East India Company, and the first great collector of Asian art in Northern Europe.[24] Couplet offered him a copy of the first edition of the atlas *Guangyu tu* by Luo Hongxian (1504–1564), published in China 1555–1558.[25] In this copy mayor Witsen wrote a long informative text in Dutch, which translates as follows:

> Father Philippus Couplet, Jesuit, who returning from China, where he had lived for twenty-four years, in order to report to Rome about the situation of Christianity there, has given me this Chinese map book, and he also showed another map book of China, half its size, from which Martino Martini made his atlas, like it could be seen from the autographic notes on the maps of Martini himself. He added to this that the map book, from which Martini had copied his notes, already had been made before three hundred years. All the descriptions that one can read in the atlas about China, could be seen in the Chinese language from word to word. These maps, the Reverend said, were rather new and much more perfect.[26]

---

23 Golius to Kircher, Amsterdam 11 juli 1665 (Rome: Archives of the Gregorian University 562, ff. 139r–v); Jorink, Eric. *Het "boeck der natuere": nederlandse geleerden en de wonderen van Gods Schepping, 1575–1715* (Leiden: Primavera, 2006); Otto Zwartjes, "Jacob Golius (1596–1667) and Martino Martini (1614–1661): The Vocabularium Hispanico-Sinense (Bodleian Library, MS Marsh 696) and the Study of Chinese in the Netherlands," in Zbigniew Wesołowski, ed., *Early European (1552–1814) Acquisition and Research on Chinese Languages. The Sixth Fu Jen University International Sinological Symposium: Symposium Papers* (Taipei: Monumenta Serica Sinological Research Center, Fu Jen University), 305–384.

24 Willemijn van Noord and Thijs Weststeijn, "The Global Trajectory of Nicolaas Witsen's Chinese Mirror," *The Rijksmuseum Bulletin* 63 (2015): 324–361.

25 Ellinoor Bergvelt and Renée Kistemaker, eds., *De wereld binnen handbereik. Nederlandse kunst- en rariteitenverzamelingen, 1585–1735* (Amsterdam: Amsterdams Historisch Museum, 1992), 183–184; Marion Peters, *De wijze koopman. Het wereldwijde onderzoek van Nicolaes Witsen (1641–1717), burgemeester en VOC-bewindhebber van Amsterdam* (Amsterdam: Bert Bakker, 2010), 226–228; Corrigan et al., eds., *Asia in Amsterdam*, 200–201.

26 The Hague, Museum Meermanno, M 115 B 1.—"Pater Philippus Couplet jesuit welck terugkomende uit Sina daer hij vierentwintig jaren hadde geweest, om tot Romen

It would take fourteen years before another book on China would appear in the Dutch Republic. The author was the French Jesuit Louis le Comte (1655–1728), mathematician of the King of France, who lived in China between 1688 and 1691, as member of the French Jesuit mission. His *Nouveaux mémoires sur l'état présent de la Chine*, published in Paris in 1696, appeared a year later in Amsterdam in French, and two years later in Amsterdam and The Hague, in French and Dutch.[27]

The work of Louis le Comte was continued by two other Jesuits, and was printed in French in The Hague in 1699, and in Dutch in Utrecht, in 1699 and 1710. The original work had appeared in 1697 in Paris. The authors were French Jesuits Joachim Bouvet (1656–1730), missionary in China, and Charles Le Gobien (1653–1708), procurator of the China mission in Paris. The work focused on the Kangxi emperor (r. 1662–1722), from the beginning of his reign until 1698, followed by a historical description of his empire, with special attention to Confucius.[28]

The last publication on China to appear in the Dutch Republic came from the press in The Hague in 1736, a year after its publication in Paris.[29] Its author was Jean-Baptiste du Halde (1674–1743), another French Jesuit, who specialized in Chinese history without ever having been in that country. Voltaire said of Du Halde's work: "Although it is developed out of Paris, and he has not known the Chinese, he gave on the basis of the memoirs of his colleagues, the widest and the best description the empire of China has had worldwide."[30]

As was mentioned in the beginning of this essay: between 1615 and 1736 31 publications on China appeared in the Dutch Republic, in Dutch, French, German, Latin and Spanish. The most original and important is the *Novus Atlas Sinensis* by Martino Martini, printed for the first time in 1655, in Latin

verslag te doen, van den toestant des christendoms aldaer, heeft mij dit sinees kaertboek geschonken daer bij getoont, een ander half so klijn kaertboexken mede van Sina, uit het welcke Martinus Martini sijne Atlas heeft gemaekt, gelijk als mede daer aen de eigenhandige aentekeninge op de kaerten van Martini selve waeren te sien, hij voegde daer bij dat het kaertboekjen waer uit Martini de sijne had nageschreven al voor drihondert jaeren was gemaekt geweest alle de beschrijvinge so men in den Atlas over Sina leest, van woort tot woort genoegsaem in het sinesche walre te sien. dese kaerten daerentegen seijde hij E. waerde dat vrij nieuw en veel volmaekter waeren. N. Witsen. 1684."

27 Begheyn, *Jesuit Books in the Dutch Republic*, 230 (1697.2), 235 (1698.3, 1698.4).

28 Shenwen Li, *Stratégies missionnaires des Jésuites Français en Nouvelle-France et en Chine au XVIIième siècle* (Paris: L'Harmattan 2001).

29 Begheyn, *Jesuit Books in the Dutch Republic*, 330 (1736.6).

30 Isabelle Landry-Deron, *La Preuve par la Chine: la Description de J.-B. Du Halde, jésuite, 1735* (Paris: Éditions de l'École des hautes études et sciences sociales, 2002).

and Dutch. It is not clear why the publications on China by Jesuits in the Dutch Republic ended in 1736.

Bibliography

Allard, H.J. "Een groet uit China voor Vondel en de Amsterdamsche 'vrunden' (1662)." *Jaarboekje Alberdingk Thijm* 46 (1897), 159–160.

Begheyn, Paul. "Athanasius Kircher S.J. (1602–1680) en zijn contacten in Nederland." In *Het geloof dat inzicht zoekt. Religieuzen en de wetenschap*, edited by Joep van Gennip and Marie-Antoinette Th. Willemsen, 84–93. Hilversum: Verloren, 2010.

Begheyn, Paul. "The contacts of Martino Martini S.J. with the Amsterdam printer Joan Blaeu." *Archivum Historicum Societatis Iesu* 81 (2012): 219–232.

Begheyn, S.J., Paul. "De Elzeviers en de jezuïeten." In *Boekverkopers van Europa. Het 17de-eeuwse Nederlandse uitgevershuis Elzevier*, Zutphen, edited by Berry P.M. Dongelmans, Paul Gerardus Hoftijzer, and Otto Stephanus Lankhorst, 59–76. Zutphen: Walburg, 2000.

Begheyn, S.J., Paul. *Jesuit Books in the Dutch Republic and Its Generality Lands 1567–1773: A Bibliography*. 1st ed. Leiden: Brill, 2014.

Bergvelt, Ellinoor, and Renée Kistemaker. *De wereld binnen handbereik. Nederlandse kunst- en rariteitenverzamelingen, 1585–1735*. Amsterdam, 1992.

Bernard, Henri. "Les sources mongoles et chinoises de l'Atlas Martini (1655)." In *Martino Martini S.J. (1614–1661) und die Chinamission im 17. Jahrhundert*, edited by Roman Malek and Arnold Zingerle, 234–235. Sankt Augustin: Institut Monumenta Serica, 2000.

Blue, Gregory. "Johann Adam Schall and the Jesuit mission in Vondel's *Zungchin*." In *Western learning and Christianity in China. The contribution and impact of Johann Adam Schall von Bell, S.J. (1592–1666)*, edited by Roman Malek, vol. 2, 951–982. Nettetal, 1998.

Corrigan, Karina H., Jan van Campen, Femke Diercks, and Janet C. Blyberg. *Asia in Amsterdam: The Culture of Luxury in the Golden Age*. Amsterdam: Rijksmuseum Amsterdam, 2015.

"Dit boeck is van de E. Martinius op uwe schepen en onder uwe bescherming uyt der Sinesen schriften (die hy ons als zijn moederlijcke tale menighmalen voorlas) in 't latijn beschreven, en alzoo ten dienste van 't gemeen en ook van diegene, die de Geleertheydt en goede konsten beminnen, overgebracht." Quoted in *Een groet uit China voor Vondel en de Amsterdamsche "vrunden" (1662)*, edited by H.J. Allard, 159–160. *Jaarboekje Alberdingk Thijm* 46 (1897).

Goedings, Truusje. *'Afsetters En Meester-Afsetters': De Kunst Van Het Kleuren 1480–1720*. Nijmegen: Vantilt, 2015.

"Golius to Kircher." Amsterdam 11 juli 1665. Rome: Archives of the Gregorian University 562, ff. 139r–v.

Golvers, Noël. "Een detailblik op China in de zeventiende eeuw bij de 350ste verjaardag van Martino Martini's *Novus Atlas Sinensis* (Amsterdam, 1655)." *Verbiest Koerier* 18 (maart 2006): 6–8.

Golvers, Noël. "Martino Martini, S.J., his stay in the Jesuit college of Brussels (1654) and the production of his 'Novus Atlas Sinensis'." In *Quatre siècles de présence jésuite à Bruxelles / Vier eeuwen Jezuïeten te Brussel*, edited by A. Deneef and X. Rousseaux, 127–139. Brussels: Prosopon / Leuven, KADOC, 2010.

Golvers, Noël. "Michael Boym and Martino Martini: A contrastive portrait of two China missionaries and mapmakers." *Monumenta Serica* 59 (2011): 259–271.

Goswin Nickel to Martino Martini in Amsterdam, Rome, 16 May 1654. In ARSI, Fl. Belg. 6, f.599v.

Gruys, J.A. "De reeks 'Republieken' van de Elzeviers en Joannes de Laet." In *Boekverkopers van Europa. Het 17de-eeuwse Nederlandse uitgevershuis Elzevier*, edited by Berry P.M. Dongelmans, Paul Gerardus Hoftijzer, and Otto Stephanus Lankhorst, 77–106. Zutphen: Walburg, 2000.

Jorink, Eric. "Het 'boeck der natuere': nederlandse geleerden en de wonderen van Gods Schepping, 1575–1715." Leiden: Primavera, 2006.

Landry-Deron, Isabelle. *La Preuve par la Chine: la Description de J.-B. Du Halde, jésuite, 1735*. Paris: Éditions de l'École des hautes études et sciences sociales, 2002.

Li, Shenwen. *Stratégies missionnaires des Jésuites Français en Nouvelle-France et en Chine au XVIIième siècle*. Paris: L'Harmattan, 2001.

Netten, Djoeke van. *Koopman in kennis. De uitgever Willem Jansz Blaeu in de geleerde wereld (1571–1638)*. Zutphen: Walburg, 2014.

Noord, Willemijn van, and Thijs Weststeijn. "The Global Trajectory of Nicolaas Witsen's Chinese Mirror." *The Rijksmuseum Bulletin* 63 (2015): 324–361.

Peters, Marion. *De Wijze Koopman: Het Wereldwijde Onderzoek Van Nicolaes Witsen (1641–1717), Burgemeester En Voc-Bewindhebber Van Amsterdam*. Amsterdam: Bakker, 2010.

Schall von Bell, Joannes Adamus. *Historica narratio de initio et progressu missionis Societatis Jesu apud Chinenses, ac praesertim in regia Pequinensi*. Vienna: Matthäus Cosmerovius, 1665.

Siebert, Harald. "Kircher and his critics. Censorial practice and pragmatic disregard in the Society of Jesus." In *Athanasius Kircher: The last man who knew everything*, edited by Paula Findlen, 79–104. New York: Routledge, 2004.

Standaert, Nicolas. "Nicolas Trigault S.J., Litterae Societatis Iesu e regno Sinarum (1615)." In *Jesuit Books in the Low Countries 1540–1773: A Selection from the Maurits Sabbe Library*, edited by Paul Begheyn, Bernard Deprez, Rob Faesen, and Leo Kenis, 26–29. Leuven: Peeters, 2009.

Stolzenberg, Daniel. "Oedipus censored: Censurae of Athanasius Kircher's works in the Archivum Romanum Societatis Iesu." *AHSI* 73 (2004): 3–52.

Stolzenberg, Daniel. "Utility, edification, and superstition: Jesuit censorship and Athanasius Kircher's *Oedipus Aegyptiacus*." In *The Jesuits II: Cultures, sciences and the arts 1540–1773*, edited by John W. O'Malley e.a., 336–354. Toronto: Toronto University Press, 2006.

Väth, Alfons. *Johann Adam Schall von Bell S.J., Missionar in China, Kaiserlicher Astronom und Ratgeber am Hofe von Peking, 1592–1666*. 2 vols. Koln: Veröffentlichungen Des Rheinischen Museums in Koln, 1933.

Waals, Jan van der, Roelof van Gelder, and Y. Rozenburg. *Een wereldreiziger op papier: de atlas van Laurens van der Hem (1621–1678)*. Amsterdam: Stichting Koninklijk Paleis, 1992.

Waerachtich verhael van eenige merckelycke saecken des vermaerts coninckrijcx van Syna, door eenen priester der Societeyt Jesu, Nederlander, onlancx van daer ghecomen. The Hague: Koninklijke Bibliotheek, Pamflet 2173 and 224 D 4.

Weststeijn, Thijs and Menno Jonker. *Barbaren en Wijsgeren. Het beeld van China in de Gouden Eeuw*. Nijmegen: Vantilt, 2017.

Zwartjes, Otto. "Jacob Golius (1596–1667) and Martino Martini (1614–1661): The Vocabularium Hispanico-Sinense (Bodleian Library, MS Marsh 696) and the Study of Chinese in the Netherlands." In *Early European (1552–1814) Acquisition and Research on Chinese Languages. The Sixth Fu Jen University International Sinological Symposium: Symposium Papers*, edited by Zbigniew Wesołowski, 305–384. Taipei: Monumenta Serica Sinological Research Center, Fu Jen University, 2011.

PART 2

Jesuit World Maps in Chinese, from Ricci to Verbiest

CHAPTER 4

Parallels, Engagement, and Integration: The Ricci Maps and Their Afterlives in Ming-Qing China as a Case Study of Intertwined Global Early Modernity

Qiong ZHANG

This chapter develops the concept of an "intertwined global early modernity" using the world maps of Matteo Ricci (1552–1610) and their afterlives in Ming-Qing China as a focal point of historical analysis. Previous scholars have begun to speak of "early modernity" not as a phenomenon unique to Western Europe during the period from around 1450 to 1800, but as a global one more or less synchronically experienced by many other societies around the world.[1] The notion of an "intertwined global early modernity" does not only entail the recognition of multiple "early modernities" existing in parallel to each other across different parts of the world, each in its own way embodying certain common patterns in its state building, economic activities, material and popular culture, technologies, and intellectual life. It also acknowledges how newly emerging global networks of trade, colonial expansion, religious missions, and other contexts of interaction during this period put the various locally- and regionally-rooted "early modernities" into particular kinds of interconnection and interactive interdependence.[2]

The global integration of physical space was perhaps the first and foremost transformation marking the dawn of early modernity that paved

1 For some representative works exploring the idea of multiple early modernities, see the articles in *Daedalus*, vol. 127.3, *Early Modernities* (Summer 1998), and David Porter, ed., *Comparative Early Modernities* (Palgrave-Macmillan 2012).

2 Some seminal studies of global early modernity are Sanjay Subrahmanyam, "Connected Histories: Notes Towards a Reconfiguration of Early Modern Eurasia," in *Modern Asian Studies* 31.3 (1997): 735–762; Laura Hostetler, *Qing Colonial Enterprise: Ethnography and Cartography in Early Modern China* (Chicago: University of Chicago Press, 2001); Jerry Bentley, "Early Modern Europe and the Early Modern World," in *Between the Middle Ages and Modernity: Individual and Community in the Early Modern World*, ed., Charles H. Parker and Jerry H. Bentley (Plymouth, UK: Rowman and Littlefield, 2007), 13–31; Charles H. Parker, *Global Interactions in the Early Modern Age, 1400–1800* (Cambridge, UK: Cambridge University Press, 2010); and Benjamin Elman, "A Jointly Regional-Global Approach to Rethinking Early Modern East Asian History," in *The 'Global' and the 'Local' in Early Modern and Modern East Asia*, ed. Benjamin A. Elman and Chao-Hui Jenny Liu (Leiden: Brill, 2017), 66–78.

© QIONG ZHANG, 2024 | DOI:10.1163/9789004684782_006
This is an open access chapter distributed under the terms of the CC BY-NC-ND 4.0 license.

the way for other kinds of encounters, or at least for European inclusion in long-established networks in the Indian Ocean World and across Asia. For the first time in human history, the entire inhabited surface of the earth was connected by sustained networks of transportation when—following a series of European geographical explorations across the New and Old World, especially Magellan's circumnavigation of 1519–1521—new navigational routes were charted that linked East Asia with Western Europe, westward through the Indian Ocean and round the Cape of Good Hope and eastward across the Pacific and the Atlantic. This made possible and necessary the much slower and more difficult conceptual integration of space as various non-Western societies adjusted their images of the world and their own places in it in order to grasp and cope with the new geopolitical and economic realities of the wider early modern world, at the same time that their indigenous maps and geographical knowledge traveled to the West and contributed to the making of early modern European understanding and cartographical representation of the earth. In late Ming (1368–1644) and Qing (1644–1911) China, this was the process whereby the dominant view of the universe in terms of a round heaven and square earth (*tianyuan difang* 天圓地方) and of the world as "All Under Heaven" (*tianxia* 天下, but occupied primarily by China), flanked by the Four Seas (*sihai* 四海) along its fringes, eventually retreated into the background. In its place arose a new consciousness that the earth is spherical and that China shares this globe with many other civilizations, some of which are as ancient and sophisticated as its own.

Matteo Ricci's Chinese world maps and their "afterlives"—"afterlives" here referring to new varieties of visual and textual representations of the world by Chinese scholars who were inspired by versions of Ricci's maps—were key monuments to this historic transformation. This paper dwells on the creation of the 1602 Ricci map at the micro level and on the patterns of Chinese engagement with elements of the Ricci maps over a relatively longer period, from the late sixteenth to the early nineteenth century, on the macro level. I will apply a model of contact zone analysis that I have previously adapted, in my study of the early modern Chinese encounter with the Jesuit notion of the globe, from its original usage by Mary Louis Pratt in the context of colonial Latin America.[3] Pratt defined the "contact zone" as the "social spaces where cultures meet, clash, and grapple with each other, often in contexts of highly asymmetrical relations of power, such as colonialism, slavery, or their aftermaths as they are

3 For further discussions on the original formulation of the "contact zone" by Mary Louis Pratt and my adaptation of it, see Qiong Zhang, *Making the New World Their Own* (Leiden: Brill, 2015), 15–21.

lived out in many parts of the world today."[4] While no such "asymmetrical relations of power" were present in the case of the Sino-European exchange in the late Ming and early Qing, the historical context of the Jesuit mission in which this exchange took place—above all, the Jesuits' material dependence on the operations of the Portuguese colonial empire, their physical vulnerability inside China as isolated foreigners, and their ultimate goal of converting the Chinese—provided a source of tension that complicated the dynamic of this exchange. Thus, the metaphor of contact zone, more than any other model for studying cross-cultural interactions—among them "trading zones," "cultural borderlands," and "frontiers"—aptly captures the complex cognitive and psychological conditions for knowledge making by individual Jesuits and their Chinese collaborators and other respondents. It takes into account the "co-presence, interaction and interlocking understanding" of two hitherto historically and geographically separate cultures brought together by these encounters,[5] which allowed for close partnership and spontaneous, fluid, and even harmonious exchange, resulting in a much-broadened intellectual horizon for all. But it also acknowledges an ever-present tension between the Jesuits and their Chinese respondents, including even their closest associates.[6] Such tension made it inevitable that mapmaking and discourses on world geography also became a contentious arena wherein individual authors on both sides of the encounter exercised their power and agency, often at the expense of their counterparts.

Zooming in on the creation of his 1602 map, especially some of its place names and legends, I will show how Ricci selectively enmeshed certain parallel tropes in the ethnographical and world geographical discourses of late Ming China and Renaissance Europe and carefully disengaged others, weaving an image of the world that best served his missionary and personal agenda. My discussion of the afterlives of the Ricci maps, on the other hand, aims to demonstrate that the long-term integration of Renaissance European and Ming-Qing Chinese world geography and cartography also manifested itself through intertwinement, partly by following the models that Ricci and other Jesuits had set. However, this integration took place largely in the form of Chinese cultural renewal, rhetorically in some areas and substantively in

4 Mary Louis Pratt, "Arts of the Contact Zone," in *Professing in the Contact Zone: Bringing Theory and Practice Together*, ed. Janice M. Wolff (Urbana, IL: National Council of Teachers of English, 2002), 4.

5 Mary Louise Pratt, *Imperial Eyes: Studies in Travel Writing and Transculturation* (London and New York: Routledge, 1992), 7.

6 See, for example, Yu Liu's recent work, *Harmonious Disagreement: Matteo Ricci and His Closest Chinese Friends* (New York: Peter Lang Publishing, 2015).

others, nurturing new brands of Sino-centrism that countered the Jesuit master narrative of European cultural and religious superiority.

1 A Map That Mirrors and Manages the Contact Zone

The culmination of Ricci's decades-long mapmaking efforts, *Kunyu wanguo quantu* 坤輿萬國全圖 [Complete Geographical Map of Ten Thousand Countries], henceforth "Ricci map of 1602," was arguably the most influential of Ricci's Chinese works published during his missionary career. It has been noted that Ricci was first drawn into mapmaking in 1584 by his Chinese patrons and literati friends, and each revised edition of his map was printed on their initiatives and at their costs. This testified to the fact that there was a tremendous demand for new knowledge about the outside world among the late Ming elite. This demand was driven by two crosscurrents in the late Ming Empire, one intellectual and the other geopolitical.

In the intellectual realm, the late Ming saw the transformation of the discourse of exotica (*zhiguai* 志怪) from a marginal pursuit into a central focus of Confucian scholarship. This was attended by a simultaneous decline in orthodox Confucian moral discourse, in which the investigation of things was supposed to be a means of extending moral knowledge. Now scholars increasingly turned their attention to the world of things with an eye towards examining and experiencing the intricacies of their materiality. Of special interest to them were strange and rare things, ranging from peculiar animals, trees, minerals, foods, and medicines to extraordinary artifacts such as antiques, rare manuscripts, and works of art. By extension, they were also fascinated by exotic lands, "aboriginal" peoples in frontier regions of the empire, and even abnormal meteorological phenomena such as typhoons. An obsession with authentic encounters with such strange lands, peoples, and phenomena kept them constantly on the move, on the "hunt." The rapid expansion of the late Ming economy, partly fueled by New World silver, enabled them to pursue their hobbies to their hearts' content, traveling to inspect, collect, investigate, and record their findings in person. Indeed, the late Ming intellectual scene closely paralleled that of Renaissance Europe, where "discoveries" of formerly unknown lands, peoples, fauna, and flora, real and imagined, generated an unprecedented demand for printed maps, travel narratives, and ethnographical writings, and gave birth to cabinets of curiosities as precursors to the modern museum.

In the geopolitical sphere, the Ming empire also shared another connection with Renaissance Europe: it experienced the early modern European expansion from the other side, as its witness and victim. The first group of Europeans to arrive on the scene were the Portuguese, known to the Chinese as Folangji (佛郎機 or Franks). After capturing Malacca and gaining a foothold in Southeast Asia in 1511, the Portuguese moved into the South China Sea seeking to open trade with the Ming empire. With the Chinese ban on private shipping still in place, they sent an embassy under Tomé Pires to the Ming court in 1517 hoping to establish an official channel for trade. But the embassy's arrival in Beijing was preceded by reports of the Folangji's violent conquest of Malacca, then a tributary state of the Ming, and the embassy's inadvertent implication in court intrigue during the transition from the Zhengde to the Jiajing emperor further doomed that mission. What followed was three decades of constant clashes on the Chinese southern coast, in which the Portuguese demonstrated the advantages of their small speedy boats and the lethal power of their cannons, which also came to be called "Folangji" by the Chinese. It was not until after 1557, when the Portuguese negotiated a colonial settlement in Macao, that the relationship settled into a pattern of "controlled conflict."[7] A new security threat also loomed on the southeastern coast after 1565, when another group of Folangji, the Spaniards, arrived in Luzon and began their colonial conquest in the Philippines. With the lifting of the "Sea Ban" by the Ming court in 1567, a busy traffic developed between Chinese southeastern ports and Manila as part of a new triangular trans-Pacific trade generated by the Manila Galleons, as Chinese merchants emptied their shiploads of silks, tea, porcelain, cloth, and other commodities designated for the American and European markets in exchange for New World silver brought in by the Spaniards. Pirate activities off the Chinese coast spiked due to the rising volume of trade and traffic, and tensions between the Ming and the Spanish colonial empire would eventually erupt into direct confrontation, leading to the Spanish massacre of nearly the entire Chinese community in Manila in 1603 among other violent incidents. In a way anticipatory of what was to come in the post-Opium War era, these maritime encounters with Western Europeans in the sixteenth and early seventeenth centuries spurred an urgent need to know about the maritime world beyond Chinese shores, leading to a spate of geographical and ethnographical writings on various "island barbarians" by Chinese scholars.[8]

7 This term is borrowed from John Wills Jr., "Introduction," in idem, ed., *China and the Maritime Europe, 1500–1800: Trade, Settlement, Diplomacy, and Missions* (Cambridge, UK: Cambridge University Press, 2010), 20.

8 For further discussions on these two developments, see Zhang, *Making the New World Their Own*, 38–45 and 268–304. On Chinese encounters with maritime Europeans in the Ming, see also John Wills Jr., "Maritime Europe and the Ming," in idem, ed., *China and Maritime Europe*, 24–77; James Fujitani, "The Ming Rejection of the Portuguese Embassy of 1517:

Given the Renaissance atlases and other textual resources he commanded, his personal experiences of travel across half of the globe, and his knowledge of contemporary Europe and European overseas expansion—especially his intimate insight into the workings of Portuguese Asia from his prior extended stays in Goa and Macao—Ricci would have been in a unique position to satisfy his Chinese audiences' intellectual and geopolitical curiosities with the most updated information. But that is not what happened, nor could it have; the circumstances of Ricci's mission thrust him into a particular interdependent relationship with his Chinese interlocutors, ultimately limiting his choices as he constructed his map out of all the resources at his disposal. His priority was to promote Catholicism as a civilizing force, and Western Christendom as a Confucian utopia; to truthfully share what he knew about the rivalry and war that were tearing contemporary Europe apart and the brutal reality attending European overseas expansion across the New and Old World would not only have punctured the myth of European superiority but also put his own life, and those of his fellow missionaries in China, in imminent danger.

Thus, Ricci crafted a map that delicately balanced what his audience wished to find out and what he wanted them to know. Adopting a Mollweide projection of the globe, with parallel longitudes and latitudes meticulously drawn and precisely marked, the map was a stunning novelty compared to Chinese maps cast as a square (see Figure 0.1). Inserted in the four corners of the map are diagrams of the Nine Heavens, an armillary sphere, and images of Northern and Southern Hemispheres, accompanied by short essays noting the causes of solar and lunar eclipses and other astronomical matters. In a long introductory text running vertically to the right of the map, Ricci presented the notion of the terraqueous globe (a unified sphere of earth and water), newly crystallized among Renaissance-scholastic natural philosophers, which he designates with the term he himself coined, *diqiu* 地球. He introduced the concepts of the equator, north and south poles, and latitudes and longitudes, with which he explained the formation of the tropical, temperate, and arctic zones; seasons; and the variations of the lengths of day and night. He also outlined the five continents: Europe (*ouluoba* 歐羅巴), Libya, or Africa (*liweiya* 利未亞), Asia (*yaxiya* 亞細亞), North America (*bei yamolijia* 北亞墨利加), South America

A Reassessment," *Journal of World History* 27.1 (2016): 87–102; Han Qi 韓琦 (of Nankai University, Tianjin), "Manila dafanchuan maoyi dui Ming wangchao de yingxiang" 馬尼拉大帆船貿易對明王朝的影響 [The Impact of the Manila Galleon Trade on the Ming Dynasty], *Shijie jindaishi yanjiu* 世界近現代史研究 [Studies in Modern World History], vol. 10 (2013): 41–71; and Jin Guoping 金國平, *Xili dongjian: Zhongpu zaoqi jiechu zhuixi* 西力東漸：中葡早期接觸追昔 [Eastward Expansion of Western Powers: Historical Memories of Early Sino-Portuguese Contacts] (Macao: Macao Foundation, 2000).

(*nan yamolijia* 南北亞墨利加), and Magellanica (*mowalanijia* 墨瓦蠟泥加), the mythical continent in early modern cartography that only partially coincides with Antarctica. In other words, his essays and some of the legends on the map taken together provided a basic outline of Renaissance European astronomy, cartography, world geography, and Aristotelian scholastic natural philosophy.

Following a strategy of accommodation that he was also applying in his explanation of Catholic doctrines, Ricci did not flatly criticize the Chinese for believing that the earth is a square; instead, he alluded to the "egg-yolk" metaphor used by the Huntian 渾天 cosmologists in late antiquity as proof that they already had the idea of a spherical earth. As previous studies have shown, the Huntian cosmologists had intended this metaphor to convey the relative spatial relationship (exteriority and interiority) between Heaven and earth, and the sphericity of Heaven only. Early Huntian texts remained mute about the shape of the earth, while some authors from the Yuan (1271–1368) and Ming explicitly envisioned the universe as a flat block of earth floating above water, both of which were encircled by a round Heaven.[9] Furthermore, Ricci commented that when the Chinese ancients were quoted as saying "Heaven is round and Earth is square," they had been referring metaphorically to the state of immobility, rather than literally to the squareness of the earth. Thus, Ricci made the notion of the globe *Chinese* by adapting and reinterpreting existing, authoritative Chinese cosmological concepts and metaphors.

Ricci was quick to point out the implications of this notion of the globe. He explains that on the earth, "both 'above' and 'below' and all four sides are inhabited by living creatures. Being a globe, there is originally no such fixed distinction as above or below, since all of it is surrounded by the heavens. Wherever you are, if you look up, you see the heavens. Within the six directions, where your feet stand counts as 'below', and where your head points to counts as 'above'." To drive this idea home, he cites as proof his personal observation while sailing around the Cape of Good Hope: there, he says, "I still had the heavens above my head and not beneath my feet."[10]

While gently nudging his readers to give up their conceit that China occupied the center of a square that was the earth, he also took care to present the world as it was seen from the perspective of his contemporary Chinese

---

9 Cordell Yee, "Taking the World's Measure: Chinese Maps between Observation and Text," in Harley and David Woodward, eds., *The History of Cartography*, Volume 2, Book 2: 120–122; Zhang, *Making the New World Their Own*, 57–63, 150–152, and 179–180.

10 Ricci, *Li Madou zhongwen zhuyi ji* 利瑪竇中文著譯集 [The Chinese Works and Translations of Matteo Ricci], ed. Zhu Weizheng 朱維錚 (Shanghai: Fudan University Press, 2001), 173–174.

viewers. For example, he shifted the projection of the globe to make the Ming empire appear near the center of the map. He also translated the names of various seas and oceans on earth with such Chinese names as "North Sea" (*beihai*, marked on the area of the Arctic Ocean north of the Bering Strait) and "South Sea" (*nanhai*, referring to *Mare Lantchidol*, the sea between Java and modern day Australia), Little and Great Western Ocean (the Indian Ocean and Atlantic Ocean respectively), and Little and Great Eastern Ocean (the Western and Eastern Pacific respectively), reinforcing this China-centered view of the world. What resulted is an image of the globe that mirrored and stretched the dominant Chinese "All Under Heaven" and "Four Seas" model of the world.[11]

Also remarkable is the extent to which Ricci incorporated Chinese cartographical and world geographical knowledge into his map. His representation of the Ming empire and its immediate surroundings was modeled largely on the one in *Guangyu tu* 廣輿圖 [Maps of Extended Territory] 1541, drawn by Luo Hongxian 羅洪先 (1504–1564, *jinshi* 1529). To determine the longitudinal and latitudinal ranges of the empire and the coordinates of major cities and other landmarks in China, Ricci used, as his points of reference, the actual longitude of Zhaoqing 肇慶 (124 degree east) that he obtained during an eclipse there in 1583 or 1584, the latitudes of cities that he personally observed during his travels, and the latitudes of some other sites that had been recorded in the History of the Yuan (*Yuanshi* 元史), which resulted from a large survey conducted in the Yuan dynasty in 1276. Ricci estimated the rest of the coordinates based on various road distances recorded in Chinese gazetteers.[12]

But by far his most elaborate efforts to offer what his readers expected to find came in the form of place names and legends. Ricci consulted at least two Chinese world geographical texts that derived their data from the Zheng He voyages of 1405–1433 and constructed many of his Chinese place names and captions about Southeast Asia, South Asia, and coastal West Asia (such as *hulumosi* 忽魯謨斯 Hormuz) based on these sources.[13] Another region where Ricci's map draws extensively on Chinese sources is the vast stretch of land

11 For further discussions on the issues of linguistic and cultural translation involved in Ricci and later Jesuits' creation of a Chinese hydrographic nomenclature, see Zhang, *Making the New World Their Own*, 204–218.

12 See Hong Weilian 洪煨蓮, "Kao Li Madou ditu" 考利瑪竇地圖 [An Inquiry into Matteo Ricci's Maps] *Yugong banyue kan* 禹貢半月刊 [Yugong Biweekly Journal] 5, no. 3–4 (1936): 6–7.

13 For a detailed comparison of Ricci's captions and his Chinese sources about these regions, see Zhao Yongfu 趙永復, "Li Madou kunyu wanguo quantu suo yinyong de zhongguo ziliao" 利瑪竇坤輿萬國全圖所引用的中國資料 [The Chinese Sources Incorporated into Matteo Ricci's "Complete Geographical Map of Ten Thousand Countries,"] *Lishi dili yanjiu* 歷史地理研究 [Research in Historical Geography] 1 (1986): 200–207.

FIGURE 4.1 Detail from Ricci's 1602 World Map showing Gouguo and Yecha Guo, or "Dog Country and Land of the *Yakṣas*." James Ford Bell Library, University of Minnesota.

to the northeast of the Ming empire. It is here that we find, nestled among the various tribal peoples more or less familiar to the Chinese, such fabulous creatures as the "Niuti tujue" 牛蹄突厥 [Ox-Hoofed Turks], who "have human bodies with the feet of oxen," and denizens of "Guiguo" 鬼國 [the Land of Disembodied Spirits], who "wander about by night and hide themselves by day. They flay deer and wear the hide as their clothing. Their ears, eyes, and noses are like those of ordinary people, but their mouths are on the tops of their heads. They feed on deer and snakes."[14] Ricci also marked Gouguo 狗國 [literally "the Dog Country"] prominently at the eastern tip of the Eurasian continent.[15] Opposite Gouguo, on a large island depicted north of Canada straddling the arctic zone, one also finds a place called Yecha Guo 夜叉國 [Land of the Yakṣas; 叉 here appears as 乂, as was often the case in Ming prints] (Figure 4.1).

14 Ricci, *Kunyu wanguo quantu*, in *Li Madou Zhongwen zhuyi ji*, 205.

15 For a recent study on "Gouguo," see Roderich Ptak, "Gouguo, the 'Land of Dogs,' on Ricci's World Map," *Monumenta Serica* 66, no. 1: 71–89.

The *yakṣas* were a category of natural spirits in Hindu mythology presumably possessing dual personalities; they may be harmless spirits of the woods and mountains, but they may also haunt these places in search of human victims. These beings found their way into Chinese literature after the Tang Dynasty (618–907) due to Buddhist influence. In Chinese Buddhist texts the *yakṣas* play the role of benevolent attendants of Vaiśravaṇa (*duowen tianwang* 多聞天王), the Guardian of the Northern Quarter, but in the Chinese popular imagination they were predominantly fearful beings which haunted the wilderness looking to devour human beings alive. A land of the *yakṣas* was mentioned in some texts from the Tang and Song dynasty, and judging from its location immediately north of *Liugui* 流鬼 [Vagrant Spirits] on Ricci's map, he was following the account that Ma Duanlin 馬端臨 offered in *Wenxian tongkao* 文獻通考 [Comprehensive Examination of Literature], ca. 1280: "traveling northward for a month from the land of Liugui, one reaches the (land of the) *yakṣas*, who have fangs like those of a swine and eat human beings."[16] To my knowledge, Ricci was the first author to give cartographical expression to these legendary places and creatures depicted in Chinese literature, thereby visually authenticating their presence in space.

Such incorporation of Chinese sources may be partly motivated by Ricci's genuine desire to fill in gaps in contemporary European world geographical knowledge, but it no doubt also served to buttress the credibility of his map among his audience.[17] Trust-building remained at the front and center of Ricci's consciousness throughout his missionary career in China. Ricci emphasized, through his various communications with his Chinese associates and readers, that Catholicism forbade the telling of lies and that he personally only spoke the truth in accordance with reason and facts—a professed principle that earned him deep respect and friendship among the Chinese. On his maps he frequently invoked this principle with captions such as the one on Magellanica, which says: "Few people have been to this southern region, so the indigenous people and things produced there are not known." This conveyed to his readers that he was not one to speak on any subject unless he had evidence to support himself. In fact, Ricci stated his principle explicitly in a

16 Ma Duanlin 馬端臨, *Wenxian tongkao* 文獻通考 [Comprehensive Examination of Literature], *juan* 347, in *Wenyuange Siku quanshu*, 616: 809.

17 In his journals, Ricci reflected on the effect of including such familiar signposts in his map: "seeing the names of many places in perfect accord with those given by their own ancient writers, they admitted that the chart really did represent the size and figure of the world." Ricci and Nicolas Trigault, *China in the Sixteenth Century: The Journal of Matthew Ricci: 1583–1610*, translated from Trigault's Latin text by Louis J. Gallagher, S.J. (New York: Random House, 1942/1953), 167.

note explaining how he sifted through his European and Chinese sources and selected items for inclusion on his map:

> On this Map, besides passing on the old information, I also make a good deal of revisions—in things, to wit, where my personal observations contradicted that which I had heard, I did not dare to write recklessly and without evidence, deceiving the world through a desire of saying something new. As for the strange tales told in the past and recorded on maps showing foreign countries—that there were people with three heads, with a single arm; people without thighs or without bellies, with eyes in the rear, with bodies joined; people who do not die; etc.—I have never heard or seen anything the like, and the Europeans, too, who are busy traveling everywhere by sea and land, never have reported such things. How could I dare to add such exaggerations, which would be insults to the Creator?[18]

However, judging from what he did include on his map, we cannot but suspect that Ricci may have bent his own stated criteria of truth to embrace some of the strangest things featured in the Chinese discourse of exotica in order to manufacture trust.

Ricci's depiction of the rest of the world likewise is teeming with natural wonders, "savage" peoples, and subhuman species which he culled from a vast body of historical and contemporary European sources. In Africa, for example, one finds Mount Atlas, "the highest mountain peak in the world [...] where the inhabitants sleep without dreams," and a cat whose sweat is scraped off and collected for use as fragrance by Europeans. In South America, one encounters the cannibals of Brazil, who are said to "not live in houses but in underground caves and [who] like to eat human flesh, although they eat only the flesh of men and not of women," and the people of Patagonia, "also known as the Land of Giants" (*changren guo* 長人國), who "did not exceed one *zhang* (approximately 10 feet 4 inches) in height and adorned their faces by painting them in

18 Translation cited with slight modification from Pasquale M. D'Elia, S.J., "Recent Discoveries and New Studies (1938–1960) on the World Map in Chinese of Father Matteo Ricci, S.J.," *Monumenta Serica* 20 (1961): 158. This note appears on Ricci's map of 1603 edition, titled *Liangyi xuanlan tu* 兩儀玄覽圖 [A Mysterious Visual Map of the Two Forms], but since this edition differs only in a few details from the 1602 edition, his principle stated here must be understood to apply to both editions. The copy of *Liangyi xuanlan tu* 兩儀玄覽圖 I consulted is the one held at the Soongsil Christian Museum in Seoul, South Korea.

various colors."[19] The continent of Europe is not exempt from this treatment. Mimicking his Chinese counterparts, who placed various exotic creatures beyond the borders of their own empire, Ricci carefully arranged them at the eastern and northern fringes of Europe far away from Western Christendom. He located the Land of the One-Eyed People (*yimu guo* 一目國) (without a caption) and the Land of Women (*nuren guo* 女人國) on the northwestern and southwestern coasts of the Caspian Sea respectively. On the northwestern frontier of Russia, near the shore of what is now known as the Pechora Sea, Ricci marked out a place called the "Land of Dwarves" (*airen guo* 矮人國), and in the accompanying caption, he relates:

> The men and women of this realm are only a little over 1 *chi* (about 12.6 inches) tall. At the age of five they have children, and at eight they are already old. Being constantly devoured by cranes, they live in caves in order to stay safe. Here they wait until the three summer months, when they come out and destroy the eggs of their enemies. They travel by riding on goats.[20]

Here, Ricci is cartographically following Gerardus Mercator (1512–1594) by locating the pygmies in the Arctic zone rather than in Africa, India, or north of the Black Sea, as ancient and medieval Western authors had done. However, while Mercator said that the pygmies were up to four feet tall, Ricci's pygmies, at about 12.6 inches tall, were still more minuscule than those described by Homer, Aristotle, or Pliny (in fact, less than half the size of Pliny's). By this modification, Ricci's pygmies were rendered more comparable to those little people (*xiaoren* 小人) portrayed in the steady stream of strange encounter stories found in Chinese literature running up to the time of Ricci's arrival.[21]

In other words, Ricci selected and adapted his sources, Chinese and Western, to craft a world map that maximized its appeal to the late Ming discourse of exotica. On his map, his curious readers would find a wealth of formerly unknown lands, natural spectacles, strange people, and uncanny customs that not only profoundly titillated their intellect but also were "certified" as true by

19 Ricci, "Kunyu wanguo quantu," in idem, *Li Madou zhongwen zhuyi ji*, 204.

20 See Ricci, "Kunyu wanguo quantu," in idem, *Li Madou zhongwen zhuyi ji*, 210. In Ming times, 1 *chi* amounted to about 32 cm, or 12.6 inches.

21 For an example of Western classical authors' discussion of the pygmies, see Pliny the Elder (23–79 CE), *The Natural History of Pliny*, translated by John Bostock and H.T. Riley (London: Henry G. Bohn, York Street, and Covent Garden, 1855–57), 2: 46, 132, and 501. For Mercator's legend, see G.R. Taylor, "A Letter Dated 1577 from Mercator to John Dee," *Imago Mundi* 13 (1956): 65. For the Chinese discourses on the little people, see Zhang, *Making the New World Their Own*, 73–74.

him. Within this ocean of the strange, Western Christendom stood as another beacon of civilization that appeared to rival if not surpass China in its splendor. Using terms familiar to his Confucian readers, Ricci describes Europe thus:

> The continent of Europe has over thirty countries. All adopt the political system of the ancient kings. *They adhere to no heterodox doctrines, and all follow the holy faith of the Lord of Heaven.* Their officials are divided into three ranks. The top rank governs matters of religion, the second, secular civil affairs, and the third, the military. Their soil yields five cereals, five metals, and a hundred fruits of the earth, and their wine is made of grape juice. Their craftsmanship is exquisite, and their knowledge of the celestial and human realms is thorough and all-embracing. Their customs are simple and honest, and the rules governing the five human relationships are observed in earnest. Their material riches are opulent. Both rulers and subjects are healthy and prosperous. All year round they are in communication with foreign countries. Merchants and traders travel around the world. Being 80,000 *li* away, there had been no contact between Europe and the Middle Kingdom until some 70 years ago.[22] [Emphasis added.]

While some of this has a basis in historical fact, the overall image of Europe emerging from this caption—especially its supreme order and harmony, features that characterized the Confucian utopia of Great Unity (*datong*)—was a stark contrast to the reality of Post-Reformation Europe, torn apart by religious strife and national rivalries. In a caption placed below Rome, Ricci further highlights the ultimate source of this supreme unity and harmony in Europe—the Pope and the Catholic Church: "In this place there is a pope who leads a celibate life. He lives in Rome to perform his duties as head of the Catholic religion. *All European countries revere him.* [Emphasis added]"[23] Ricci referred to the abode of the Pope as *cifang* (此方, this area or region), rather than *ciguo* (此國, "this state," or the Papal States). Indeed, he refrained from going any deeper into the actual geopolitics of contemporary Europe, for to do so would be to open a can of worms—one only needs to think about the wars and conflict between the Papal States and their Italian neighbors, between Catholic and Protestant Europe, and between Spain and the Netherlands, among a myriad of other ongoing conflicts. As far as readers of his map were concerned, Western Europe stood as a Confucian Utopia-incarnate in its political stability, spiritual purity, scholarly perfection, and material prosperity, and Catholicism is what made it so.

---

22 Ricci, "Kunyu wanguo quantu," in idem, *Li Madou zhongwen zhuyi ji*, 214.

23 Ricci, "Kunyu wanguo quantu," in idem, *Li Madou zhongwen zhuyi ji*, 210.

The same deliberate silence extends to Ricci's coverage of European overseas colonial expansion. In the Americas, except for the place names "New Spain" (Xin Yixibaniya 新以西把你亞, placed upon Mesoamerica), the "Sea of New Spain" (Xin Yixibani Hai 新以西把你海, on the Gulf of Mexico), and "Little Spain" (Xiao Yixibaniya 小以西把你亞, on Hispaniola), one finds no clue to the Spanish conquest and colonial rule, nor any allusion to the native organized states that once thrived there. Nor is there any clue to the Spanish colonial governance in the Philippines, or the Portuguese colonial presence in Goa, Malacca, and Macao.

Especially telling of Ricci's struggle to manage the Chinese perceptions of Europe through this map is his cartographical representation of Portugal. As Aoyama Hiro'o points out in chapter 6 of this volume,[24] recent studies comparing the six extant copies of Ricci's 1602 map with over twenty Japanese hand-drawn copies of this map have demonstrated that this map went through three revisions in short succession: versions 0, 1, 1a, and 1b. All extant authentic copies of this map belong to version 1b. A key revision made in version 1b involves replacing the then prevalent Chinese usage "Folangji" with the transliteration "Bo-er-du-wa-er" (波爾杜瓦爾) as the name for Portugal. Most of the extant Japanese hand-drawn copies of Ricci's 1602 map are now believed to be based on versions 1 and 1a the originals of which no longer exist. On these imitation Ricci maps "Folangji" is used as the name for Portugal, which appears both on the Iberian Peninsula and in the long text box near Madeira (Mudao 木島) telling the story of the Portuguese colonization of the island (Figure 4.2). There is also a textbox off the west coast of Portugal, which reads: "Folangji is a mistaken appellation given by the Huihui [Muslims]. Its original name is Bo-er-du-he-er (波爾杜曷爾)."[25]

On the six extant authentic copies of the 1602 map, on the other hand, "Bo-er-du-wa-er" is marked on the Iberian Peninsula and appears in the legend on Madeira, in place of "Folangji," while the above-cited textbox off the west coast of Portugal explaining the origin of the name "Folangji" was erased altogether. A major clue indicating that these extant copies of the Ricci map of 1602 are revised versions of the original Ricci maps on which most extant Japanese hand-drawn imitations are based is the vestige of the erased textbox that remains visible on them. To erase a portion of a carved text, the typical

24 Aoyama Hiro'o, "The Introduction of Ricci's World Maps into Edo Period Japan," section II.

25 See, for example, the copies held at the Hong Kong Maritime Museum 香港海事博物館 (https://www.google.com/culturalinstitute/beta/asset/kunyu-wanguo-quantu-complete-map-of-all-nations-on-earth-c-17th-century-left-half/dwEyNpMhRJtBcw) and the Tohoku University Library (http://www.i-repository.net/contents/tohoku/kano/ezu/kon/kon.html).

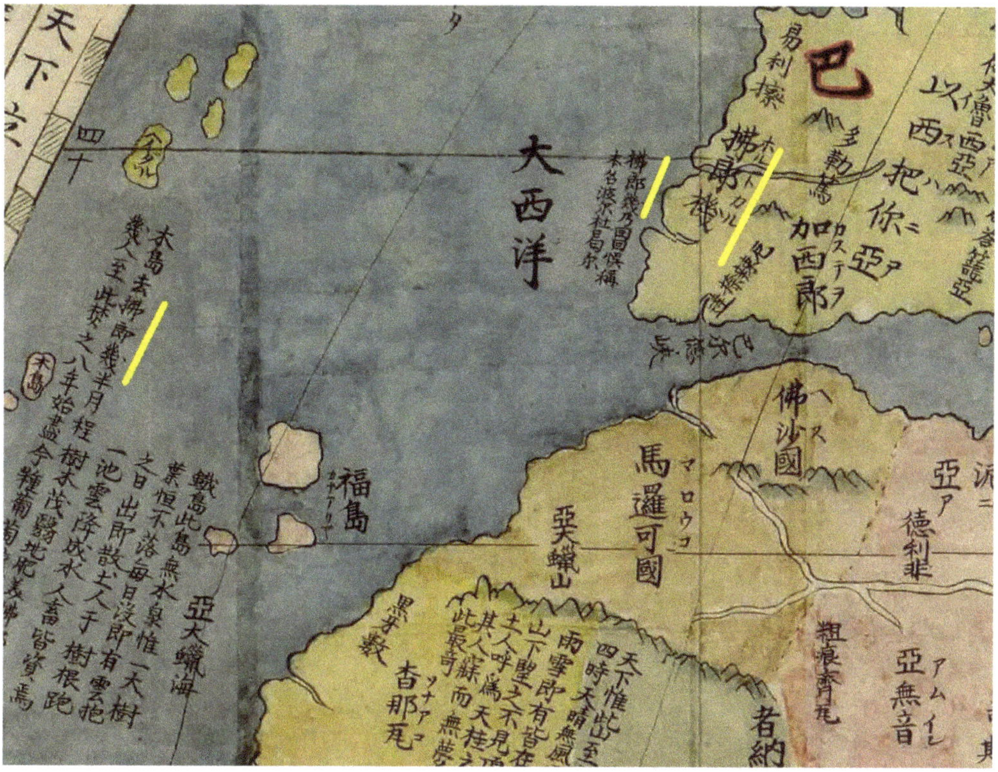

FIGURE 4.2 Detail of a Japanese hand-drawn copy of Ricci's 1602 map. Tohoku University, Japan. Yellow highlighting appears to the right of text referring to Portugal as Fulangji 拂朗機（幾）. (Here Ricci's original Chinese character 佛 (*fo*) in "Folangji" is substituted by the *kanji* 拂, which is pronounced "*fu*" in Chinese.)

practice at the time was to embed wood back into the carved characters to level the surface of the block. This technique is not perfect; it leaves vestiges of the original carvings which become more visible in the prints made later in the printing process, as some of the embedded wood pieces loosen and fall off after repeated use of the woodblock. Figure 4.3, excerpted from the Ricci map of 1602 held at the Bell Trust, shows a relatively light but still visible mark of the erased textbox off the west coast of Portugal. In comparison, the vestiges of the erased textbox are much more pronounced on the Ricci map held at Miyagi Prefectural Library in Japan, where five Chinese characters from the erased textbox are partially exposed.[26]

26 For further comparisons of different versions of Ricci's 1602 map, see Unno Kazutaka 海野一隆, "Ri matou <Kon'yo bankoku zenzu> no syohan" 利瑪竇《坤輿萬國全圖》の諸版 [The Various Editions of Matteo Ricci's *Kon'yo bankoku zenzu*], *Touhou*

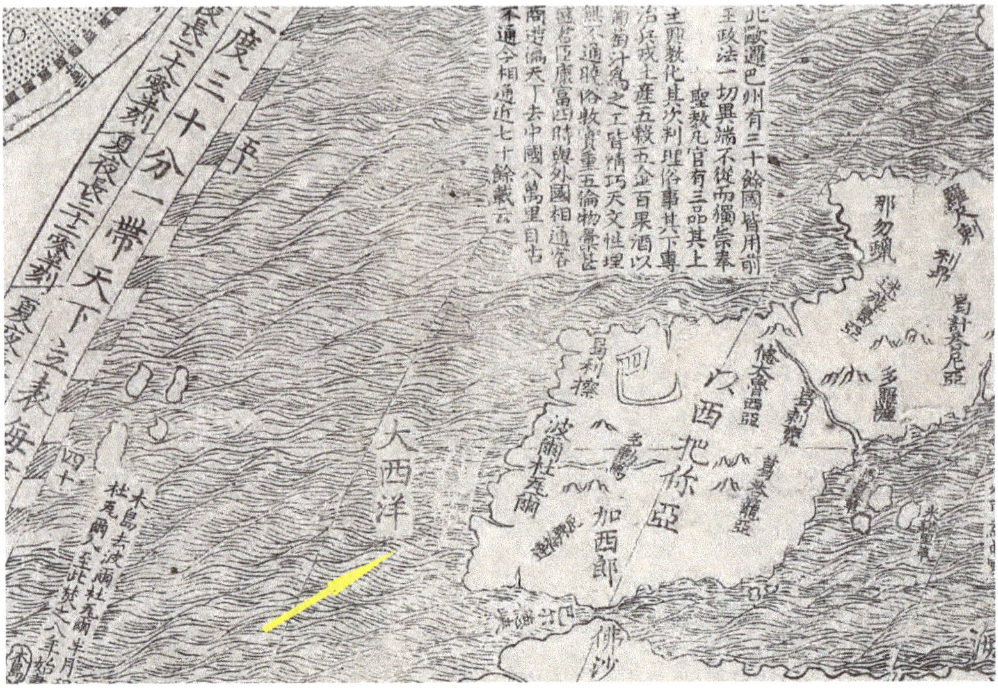

FIGURE 4.3 Detail from Ricci's 1602 World Map. The arrow indicates the area on the map showing what may be vestiges of an erased text box just west of Portugal. James Ford Bell Library, University of Minnesota.

We can only guess why Ricci had second thoughts and made these revisions apparently at much inconvenience to the woodblock carvers. But we may exclude the possibility that these changes were made due to purely technical considerations such as accuracy of translation, for if that were the case, Ricci would have also replaced the other two mentions of Folangji on the map located far away from Europe, one in the waterway between the Atlantic and the Indian Ocean and the other south of the Strait of Magellan. The latter caption reads:

> Mowalani 墨瓦蠟泥 [Magellan] is the name of a native of the country of Folangji (*folangji guo*). He passed through this strait some sixty years

---

gakuhou 東洋学報 [Journal of Oriental Studies] 87.1 (2005): 101–143. On p. 106, Unno reproduces an excerpt of the Ricci map held at the Miyagi Prefectural Library displaying some Chinese characters of that erased textbox. I wish to thank Liu Qing for translating this article for me.

> ago and arrived at this land. Therefore, gentlemen from Ouluoba 歐羅巴 [Europe] named the strait, the sea, and the land after him.[27]

In this legend, Ricci refers to Magellan's home country Portugal as the "country of Folangji." Supposing that Ricci would bother to undertake the revision at all, to achieve linguistic perfection, it is unlikely that he would stop short of revising these two other entries because of the additional work involved; it is even less likely (although not impossible) that he just forgot about these two other entries altogether. What is more plausible is that he purposely did not revise these legends. Given the omnipresence of the name "Folangji" in the late Ming public discourse on maritime security problems, it would have been strange that Ricci's map did not mention Folangji at all. In other words, these two legends featuring Folangji must have been Ricci's attempt to respond to such public interests. However, as previous studies have shown, while the Chinese remained mystified about the sudden emergence of Folangji, they did not know where it was based; many Chinese scholars and officials in the Ming thought it was a Southeast Asian power.[28] Based on his personal correspondence and journals, Ricci was well aware of the difficult beginnings of Sino-Portuguese relations, and indeed during his first years in Guangdong, he and his fellow Jesuits suffered violent attacks by local Chinese mobs who suspected them to be connected with the Folangji.[29] Thus it is only natural and understandable that he would not volunteer to enlighten his readers about the European identity of Folangji. One way or another, in his finalized version of the "1602 map" there was a clear disjunction between the two names he used for Portugal, "Bo-er-du-wa-er" (in Europe) and "Folangji" (in the Indian Ocean and the Pacific), which served to segregate the realpolitik of European colonialism operating in Asia from its home base in European Christendom.

I have analyzed the 1602 map as Ricci's map. A note is in order as to why I do not include Li Zhizao, a powerful Ming court official who sponsored the production of this map, as a co-author, as some previous scholars have done.[30]

---

27 Ricci, "Kunyu wanguo quantu," in idem, *Li Madou zhongwen zhuyi ji*, 204.

28 James Fujitani argued this point in his article, "Ming Rejection of the Portuguese Embassy of 1517." See also my discussion of Chinese writings focusing on Folangji in the mid- and late-sixteenth century in Zhang, *Making the New World Their Own*, 268–304.

29 For Ricci's reflections on Sino-Portuguese relations and Chinese views of Folangji, see Ricci and Trigault, *China in the Sixteenth Century*, 128. Regarding the violent attacks by local mobs that Ricci and his fellow missionaries suffered, see Ronnie Hsia, *A Jesuit in the Forbidden City* (Oxford and New York: Oxford University Press, 2010), 86–87, and 130–131.

30 See, for example, Roger Hart, *Imagined Civilizations: China, the West, and Their First Encounter* (Baltimore: Johns Hopkins University Press), 253.

While it is entirely possible that Li Zhizao assisted Ricci in the process of the latter's research and crafting of place names and captions for this map, we have no concrete evidence indicating that the above noted features of Ricci's map had been initiated or dictated by Li. Examinations of some of the extant Chinese sources which carry fragments and duplicates of the world maps Ricci created during his stay in Nanchang (1595–1598) suggest that these earlier maps already possessed the essential outlook of his 1602 map and that he was engaging the same types of Chinese sources back then as those he has been shown to have consulted in creating the 1602 map.[31] But Ricci did not meet Li Zhizao until 1601. Furthermore, in his preface to the 1602 Ricci map, Li himself puzzled over why the names of many countries recorded in Chinese dynastic histories as having paid tribute to the imperial court were not to be found on this Ricci map.[32] From these we may infer that if Li did participate in crafting this map's place names and captions, his role would likely have been minor and primarily limited to the technical domain of linguistic translation.

In her exploration of the "arts" of the contact zone in colonial Latin America, Mary Louis Pratt showed how members of subordinated or marginal groups selected and invented from materials transmitted by the dominant culture so as to craft a discourse of subversion.[33] Some notable parallels can be observed in Ricci's case seen above. As a foreign missionary still earning a marginal existence in a well-governed bureaucratic empire and yet seeking to convert an intellectual, political elite who presumed themselves to be at the pinnacle of the civilized world, Ricci chose to pursue his agenda by partially appropriating and reinventing the language and resources of his audience. His efforts were made possible by his keen awareness of the intellectual currents of the late Ming that resonated with those he had known back home, especially the fascination with curiosities, a new ideal of knowledge that prioritized first-hand observations and empirical evidence, and an emerging classical scholarship that consciously sought to reinterpret ancient classical texts to recover their true meaning. He selectively incorporated pre-existing Chinese knowledge of world geography and cartographical conventions, particularly as it pertained

31 See, in particular, Gong Yingyan 龔纓晏, "Xiancun zuizao de Li Madou shijie ditu yanjiu" 現存最早的利瑪竇世界地圖研究 [A Study of the Earliest Extant World Maps Made by Matteo Ricci,] *Lishi dili* 歷史地理 [Historical Geography] 2 (2018): 1–12, and Zhou Yunzhong 周運中, "Li Madou *Yutu zhi* yiwen kaoshi ji qita" 利瑪竇〈輿圖志〉佚文考釋及其他 [An Examination of Some Fragments from Matteo Ricci's *Yutu zhi* [World Map with Legends] and Other Relevant Issues,] *Ziran kexueshi yanjiu* 自然科學史研究 [Studies in the History of Natural Sciences] 4 (2010): 437–445.

32 Ricci, "Kunyu wanguo quantu," in idem, *Li Madou zhongwen zhuyi ji*, 180.

33 Pratt, "Arts of the Contact Zone," 36–37.

to the discourse of exotica, and veered from other concurrent Chinese discourses that could disadvantage him and his Jesuit confreres (such as those about the colonial activities of the Folangji and other maritime Europeans). The result was an ostensibly China-centered map of the world that, at its core, threatened to unseat Sinocentrism as a part of Chinese imperial ideology and as the dominant view of the world.

2 The Afterlives of Ricci Maps and Jesuit Cartography in Ming-Qing China

The same mode of selective appropriation and repurposing can be observed among Chinese scholars in the late Ming and the Qing who engaged the central ideas, techniques, and information conveyed by Ricci in his Chinese world maps. Since many of these scholars did not see copies of Ricci's maps *per se* but encountered these same elements of his maps carried over into the works of his Jesuit successors, especially Giulio Aleni (1582–1649) and Ferdinand Verbiest (1623–88), the following discussion will be focused on the afterlives of Ricci maps and of Jesuit cartography in general.

The most astounding development in the journeys of the Ricci maps and Jesuit cartography in Ming-Qing China is the broader framework of cultural renewal in which they took place. The early seventeenth century saw the (re)emergence of the "Chinese origination of Western learning" (*xixue zhongyuan* 西學中源) thesis. An antecedent of this thesis had already been invented by Chinese authors, especially the Daoists, in late antiquity and the medieval period to either justify their embrace of Buddhism or to downplay its originality. It claimed that Laozi, revered in historical times as the founder of Daoism, wandered late in his life beyond the northwestern border of China, where he either became known as the Buddha or had personally inspired the teachings of the Buddha. Ricci himself contributed to the reactivation of this narrative by his conscious and persistent efforts to seek confirmation from ancient Chinese authorities for the new ideas he preached on matters of religion and natural knowledge. Now the "West" was extended from the "Western region" previously known—that is, primarily Central and South Asia—to the "Far West," the homelands of the European Jesuits. Already in the prefaces to Ricci's maps written by his Chinese associates, one finds expressions of wonder at not only the maps' striking novelty but their uncanny resonance with the teachings of ancient Chinese authors. In the decades after Ricci's death, details were beginning to be fleshed out in the story that tells of how ancient Chinese authors had already known the earth to be spherical, how Chinese court astronomers and geographers of high antiquity painstakingly paced across the Eurasian continent to obtain its exact dimensions, and how during the chaos following the collapse of the feudal system of the Western Zhou (ca. 770 BCE), these astronomers and mathematicians drifted westward, bringing their secretly guarded knowledge with them. While their disciples in the West kept the knowledge alive and perfected it over the millennia, the Chinese had lost touch with its original core and fell behind. Formally endorsed by the Kangxi 康熙 emperor (1654–1722, r. 1662–1722) of the Qing dynasty, this narrative about the "homecoming" of Western learning became the orthodox doctrine. It was a mocking commentary on Ricci's method of accommodation in that while he (and his successors) tried to make a dent in Sinocentrism through propagating what they believed to be Western Christendom's superior system of learning, they had stoked the flames of the very attitude they hoped to expunge—an epistemological ethnocentrism which V.Y. Mudimbe has defined as the view that "there is nothing to be learned from 'them' unless it is already 'ours' or comes from 'us'."[34]

As the late historian Xiao Jiefu (蕭萐父, 1924–2008) pointed out, the "Chinese origination of Western learning" was not simply a face-saving device; by claiming that Jesuit science had roots in Chinese antiquity, Chinese scholars effectively made China its true home, paving the way for its organic integration into Chinese learning.[35] Under this banner scholars from the late Ming and Qing discovered simultaneously a new world vision in the works of the Jesuits and a similar one that lay hidden in their own ancient texts. A sense of mission to retrieve and restore the "lost learning" (*juexue* 絕學) from antiquity drove many of them to actively investigate and critique both their Jesuit sources and scholarship from the recent Chinese past, to identify weaknesses on both sides, and to venture novel solutions to them all.[36] Thus, some scholars, such as Jie Xuan 揭暄 (1613–1695), reinvented the notion of the globe by extricating it from the Aristotelian model of concentric crystalline spheres presented by Ricci and later Jesuits and developing instead a hypothesis of the formation and movement of the earth and celestial bodies that applied Chinese Qi-based cosmology.

Within the field of world geography, a complex process of integration began in the late Ming and continued to unfold in the following centuries. As I have

34 V.Y. Mudimbe, *The Invention of Africa: Gnosis, Philosophy, and the Order of Knowledge* (Indianapolis: Indiana University Press, 1988), 17.

35 Xiao Jiefu 蕭萐父, "The Rough Road to China's Philosophical Enlightenment," in idem, *Chuisha erji* 吹沙二集 [Blowing off Sand [to Sift Gold], Second Collection] (Chengdu: Bashu shushe, 1999), 553–599.

36 Zhang, *Making the New World Their Own*, 9–10.

argued previously, there were two sets of partially disjointed and partially contending discourses on world geography in pre-modern China. The geopolitical discourses championed the "Square-Earth-and-Four-Seas" model that had been articulated in classical texts formed around the middle of the first millennium BCE, such as the "Tribute of Yu" (Yugong 禹貢), a chapter in the Confucian classic *Book of Documents*. The embedding of this model in imperial ideology and state rituals since the Han dynasty (206 BCE–220 CE) guaranteed its preeminence. As a result, it often masqueraded as a description of the world and marginalized an alternative vision of a vast world with many lands and seas. First schematically sketched out by Zou Yan 鄒衍 (ca. 350–270 BCE) in his theory of the "Nine Great Continents," this alternative vision thrived in the empirical maritime discourses on the wider world that had, over the centuries, accrued a wealth of knowledge acquired through actual travel on land and sea by diplomats, soldiers, sailors, traders, pilgrims, and others, as well as through exchanges with foreign visitors to China.[37]

Ricci's depiction of the world, backed by his powerful personal stories and those of European explorers, reversed the relative position of these two sets of discourses in favor of the latter by rekindling Zou Yan's ancient theory while giving it a new context and substance. At the same time, a movement began among the classicists to discredit literal interpretations of the "Square Earth-and-Four Seas" model. In their exegesis of the *locus classicus* of this model in the "Tribute of Yu," evidential scholars such as Zhu Heling 朱鶴齡 (1606–1683), Gu Yanwu 顧炎武 (1613–1682), and Hu Wei 胡渭 (1633–1714) discovered that the "Four Seas" here had never meant the four physical "seas" that presumably surrounded the Nine Provinces—Yu's territory, and by extension China, as maintained by their predecessors such as Cheng Dachang 程大昌 (1123–1195) of the Song dynasty (Figure 4.4). Rather, they argued that it was a collective reference to foreign lands and peoples beyond the frontiers of the Chinese empire.[38] Indeed, Hu Wei, who did not explicitly engage any components of Jesuit cartography in his published works, created a map of Yu's Nine Provinces/China that appeared rather compatible with the Jesuit depictions of the world in terms of its allocation of land areas to China (Figure 4.5).

The harmonizing of these two pre-existing models of the world in Chinese thought, achieved by relegating the formerly dominant "Square-Earth-and-Four-Seas" model to a symbolic and metaphorical frame of reference and out of the realm of world geography *per se*, made possible another kind of

37 On the divergent discourses in pre-modern China, see Zhang, *Making the New World Their Own*, chapter 3, especially 108–147.

38 Zhang, *Making the New World Their Own*, 232–252.

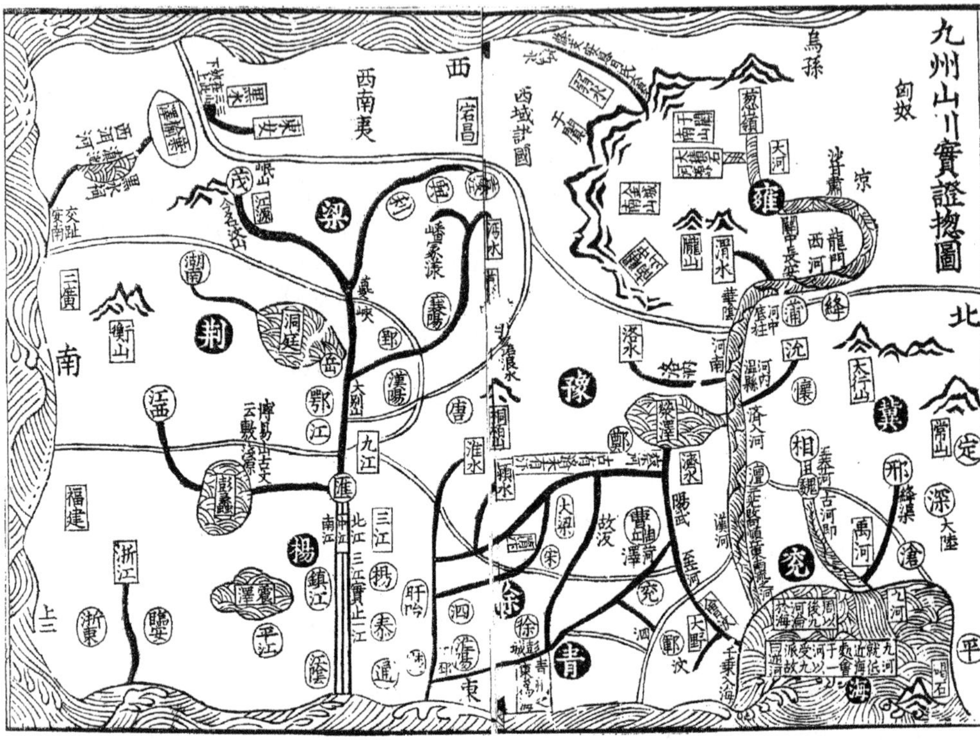

FIGURE 4.4 Cheng Dachang 程大昌, "Truthful and Verified General Illustration of the Mountains and Rivers in the Nine Provinces" [*Jiuzhou shanchuan shizheng zongtu* 九州山川實證總圖], Song dynasty. The land (with west facing up) is shown surrounded by the sea on three sides; China takes up most of the land area. Although the North Sea is not shown on the map, its presence is implied in the author's prefatory remarks. From: Cheng Dachang (1123–1195), *Yugong shanchuan di li tu* 禹貢山川地理圖 [Illustrations and Maps of Mountains and Rivers in the "Tribute of Yu"]. Beijing: Zhonghua shuju, 1985, beginning of *juan* 1, no page number.

integration, namely the synthesis of new data derived from Jesuit sources and data accumulated in Chinese historical records, ethnographic literature, and travel narratives. The seventeenth century saw the first Chinese attempts to construct a world geography that took account of most of the known inhabited regions of the earth. However, to the extent that these new world geographical syntheses depended much on Ricci and later Jesuits' presentation of the outside world, they tended to bear two distinct "birth marks," each in its own way reminding us of what Pratt aptly called the "risks of the contact zone."

One was the "Guinness World Record" approach that characterized much of Ming-Qing Chinese authors' treatment of foreign lands and peoples. This approach, strong in recording outstanding phenomena and weak in providing

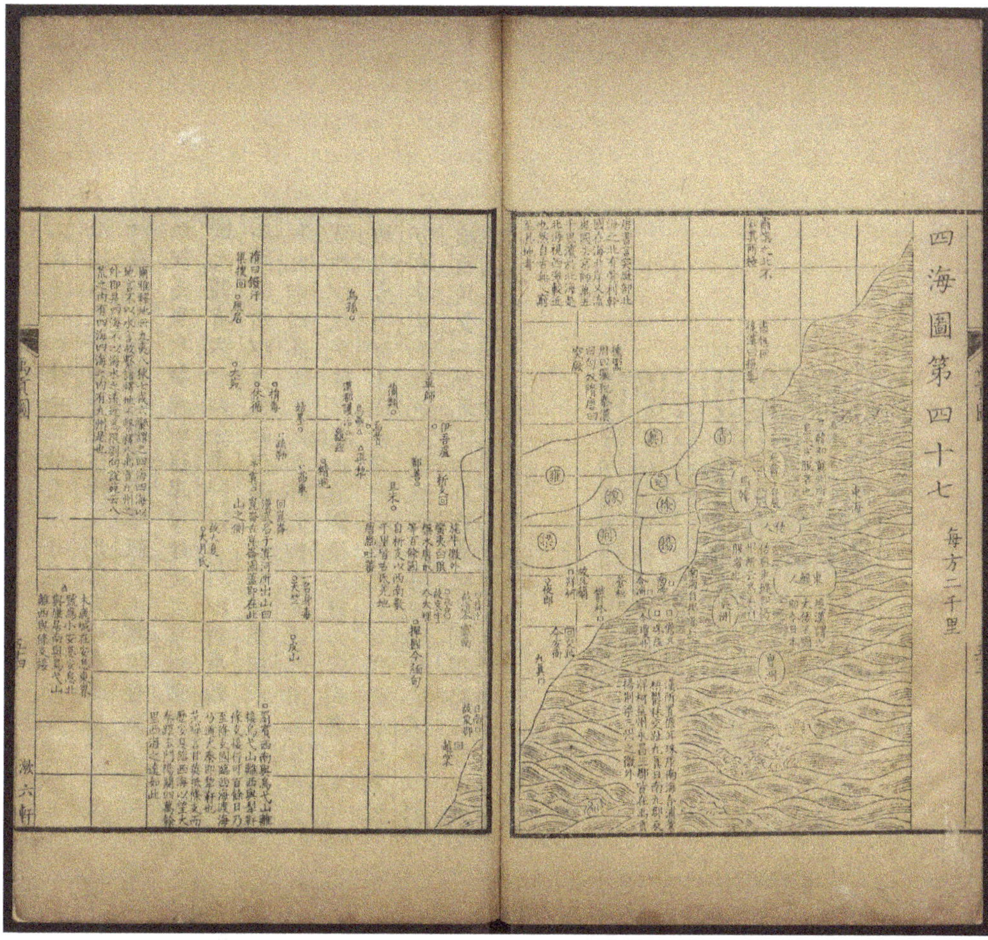

FIGURE 4.5 Hu Wei 胡渭, "Map of the Four Seas" [*Sihai tu* 四海圖], 1701/1705. Here the Nine Provinces constituting the Tribute of Yu appear close to the middle section of the map, each marked by an encircled character. From: Hu Wei Yugong zhuizhi 禹貢錐指 [Boring into *Yugong*], 1701/1705, *juan* 1, 53b–54a. Courtesy of Harvard College Library/Harvard-Yenching Library.

systematized geographical knowledge, was already prevalent in the late Ming discourse of exotica, but it certainly received a great boost through reinforcement by Ricci's maps and those of Aleni and Verbiest. This was especially pronounced in the works of Lu Ciyun 陸次雲 (fl. 1662). Lu compiled two texts related to this subject. The first, *Translated Accounts of Things within the Eight Corners of the World* (*Bahong yishi* 八紘譯史, preface 1683), covers foreign countries or polities the existence of which Lu deemed adequately proven. The book grouped these countries into four divisions according to their locations relative to China (i.e., the Northern, Southern, Eastern, and Western Divisions). For each country covered, he provides a few samples of its native tongue, rendered in Chinese transliteration; outstanding features of its natural landscape; and notable items among its local products and customs.

The second, *Accounts of the Wilderness within the Eight Corners of the World* (*Bahong huangshi* 八紘荒史, n.d.), pertains more to subjects in cosmography. It groups together various strange human or subhuman races that Lu considered to be not necessarily real but plausible, and thus worthy of recording. Among such creatures we find those of the *Ouluoba xi xiaoren guo* 歐羅巴西小人國 [Realm of the Pygmies in Western Europe], which is presented along with many other kinds of pygmies recorded in previous Chinese literature. The entry was derived from Aleni's *Zhifang waiji* 職方外紀 [Records of Lands beyond the Jurisdiction of the Imperial Geographer] 1623, which was in turn based on Ricci's legend of the pygmies in his 1602 world map seen above. Also among these entries were the ones on "Dog Country" and the Land of Yakṣas.[39]

A brief comparison of the afterlives of the "Land of Yakṣas" featured on Ricci maps in Qing China and Tokugawa Japan demonstrates how Ricci's overzealous embrace of the fabulous and fantastic, itself a reaction to the late Ming discourse of exotica, facilitated the circulation of related knowledge on his maps, but often at the expense of the deeper geographical concepts that he also communicated on these maps. As mentioned earlier, Ricci's place name "Land of Yakṣas" and its location north of the Land of Vagrant Spirits *Liugui*, echoing the descriptions of these two places in *Wenxian tongkao*, only served to authenticate the fearsome existence of the *yakṣas*, which had lingered in the Chinese cultural imagination for centuries. In the early Qing, new "travel narratives" were recorded in which alleged survivors of shipwrecks would tell of their gruesome encounters with giant man-eating *yakṣas*.[40] But the scenario in Tokugawa Japan, where Ricci's 1602 map was widely and diligently copied, was quite the opposite. On all the surviving copies of Japanese imitations and derivatives of this map that I have personally inspected, Ricci's "Land of

39 See Lu Ciyun 陸次雲, *Bahong yishi; Bahong huangshi* 八紘譯史； 八紘荒史 [Translated Accounts of Things within the Eight Corners of the World; Accounts of the Wilderness within the Eight Corners of the World], combined modern reprinted edition (Taipei: Guangwen shuju, 1969): Pygmies of Western Europe, 7; "Dog Country" and "Land of the Dog-Headed People," 3 and 22 respectively; Land of *Yakṣas*, 12–13.

40 For one of these accounts see the entry on "Island of Yakṣas" (*yecha dao*) recorded by Dong Han 董含 (1624–?), in his *Sangang shilue* 三岡識略 [Cursory Records from Sangang], *juan* 2. The copy consulted is the digital scan of the 1844 edition held at the State Library of Berlin, p. 76 v, at: http://digital.staatsbibliothek-berlin.de/werkansicht?PPN=PPN3346157571&PHYSID=PHYS_0076&DMDID=DMDLOG_0006 (accessed on 11/20/2017).

Yakṣas" (夜叉國) is consistently replaced with the "Land of Nocturnal Humans" (夜人國). The origin of this "mutation" remains murky and could be due to a mistaken copy of the character in the middle (叉 becoming 人) by the first copier(s); it was an easy mistake to make due to the simplified form—a common permutation in Ming Qing prints—of the character "叉" used on Ricci's original map, which resembles the uncommon character *yi* "乂" (Figure 4.1). However the change may have started, it was to have profound consequences: the new place name for this land, "Land of Nocturnal Humans," presumed the full humanity of its inhabitants; what was invoked by this name was instead the unique natural condition of the arctic zone that made these people spend a significant part of the year in total darkness. Taking a cue from Ricci's caption on the arctic zone in which he says that in the polar regions on earth day and night alternate only once every half year, Japanese cartographer Nagakubo Sekisui (長久保赤水, 1717–1801) created not just one but numerous "Lands of Nocturnal Humans" on his color world map modeled upon one of Ricci's Chinese world maps. He painted all of them dark and painted the zone near the South Pole on Magellanica similarly, marking it as "Land of the Night (夜國)" with a note indicating that nobody knows anything about this place (Figure 4.6).[41]

Meanwhile, in an illustrated digest of the captions on Ricci's 1602 Map compiled by Inagaki Shisen 稲垣子戩, dated 1802, we find that inhabitants of the Land of Vagrant Spirits and Land of the Yakṣas are grouped together. The caption that Ricci wrote on the island upon which he placed these two lands now appears as a description of their shared way of life: the prominence of ice in their environment; their lack of grains and other vegetation; and their resulting reliance on fish for diet, on fish oil for lighting, and on fish bones for constructing their houses and transportation vehicles; etc. All of these, and the image of a nocturnal human he drew, amounted to a reasonably close depiction of the Inuit as we know them today (Figure 4.7).[42]

In these Tokugawa discourses on the Land(s) of Nocturnal Humans, Ricci's ideas about climate zones and his explanations of the variations of daylight time and seasons inherent in his above-mentioned two captions came to

---

41 Nagakubo Sekisui, *Chikyu bankoku sankai yochi zenzu* 地球萬國山海輿地全圖 [Complete Map of Ten Thousand Countries, Mountains, and Seas on Earth], ca. 1785. Clip from Map held at the US Library of Congress digital collection: https://www.loc.gov/item/gm71005079/.

42 Inagaki Shisen 稲垣子戩, *Kon'yo zenzu setsu* 坤輿全圖説 [Legends on the Complete Terrestrial Map], prefaced 1802 (n.p.), copy held at Kyoto University Library, 26b (description of Vagrant Spirits and Land of Nocturnal Humans) and 27a (image of an inhabitant of the Land of the Night).

FIGURE 4.6 Detail from Nagakubo Sekisui's *Sankai yochi zenzu* 山海輿地全圖 [Complete Map of the Mountains and Seas on Earth], ca. 1785, Showing Multiple Lands of Nocturnal Humans (夜人國). Map held at the US Library of Congress digital collection: https://www.loc.gov/item/gm71005079/.

FIGURE 4.7
Image of an inhabitant of the Land of the Night from Inagaki Shisen's *Kon'yo zenzu setsu* 坤輿全圖説 [Legends on the Complete Terrestrial Map], prefaced 1802 (n.p.), 27a. Copy held at Kyoto University Library.

fruition. But in the Chinese context, they were somewhat derailed by the fabulous inner logic of the discourses on *yakṣas*.[43]

The second "birth mark" of early modern Chinese world geography is the state of misinformation and confusion in its coverage of contemporary Europe and overseas European expansion, a rather poignant fact given that its chief informants on these topics were none other than a group of well-educated Europeans. Previous scholars have pointed out the difficulty faced by

43 I discussed this development in my paper, "A Triangular Cartographic Encounter: The Strange Journeys of the Yakṣas in Ming-Qing China and Tokugawa Japan, via the Ricci Maps," presented at the international workshop, "Cartographic Operations: Art, Science and Politics in South East Asia," co-organized by Christine Anne Habbard of SUTD, and Jean-Marc Besse, CNRS, Paris, and held on 17 May 2018 at the National Library in Singapore.

Ming-Qing Chinese scholars in piecing together a credible picture of Europe.[44] Ricci's idealized depiction of Europe continued to guide later Jesuits in their narratives about Western Christendom. In his *Answers about the West* (Xifang dawen 西方答問, 1637/1642), for example, Aleni featured the following Q&A:

> Question: Are there bandits also in your country?
>
> Answer: They are very rare. According to the customs of my country, things lost on the road are not picked up. If something is lost, the finder hangs it on a wall to facilitate its recovery by the owner; or the finder keeps it for the time being and posts a notice on the church door. The owner then states what he has lost, and if the marks tally, the lost object is returned to him. If children or animals are lost, they are assembled in the market at specified hours and claimed by the loser after satisfactory proof has been given. … There are people who do not close their doors day or night; yet nothing is ever lost.[45]

Chinese critics of the Jesuits were quick to point out that these things sounded too good to be true, but they could do little besides admitting their own inability to falsify the Jesuit claims. Wei Jun 魏濬 (1553–1626?), among others, famously compared Ricci's depiction of Europe to "a painting of ghosts: nobody can tell whether or not the painter did a good job."[46] However, in Tokugawa Japan, a chance encounter with a European missionary of a different mindset would empower the neo-Confucian philosopher Arai Hakuseki 新井白石 (1657–1725) to respond another way. In the overview of Europe in his book on world geography, Hakuseki summarily presented the related captions on the 1602 Ricci Map, which he referred to as the "Ming map," and then commented:

44 See especially Pang Naiming 龐乃明, *Mingdai zhongguoren de ouzhou guan* 明代中國人的歐洲觀 [Ming Dynasty Chinese Views of Europe] (Tianjin: Tianjin renmin chubanshe, 2006); Timothy Brook, "Europaeology? On the Difficulty of Assembling a Knowledge of Europe in China," in M. Antoni J. Ucerler, s.j., ed., *Christianity and Culture: Japan and China in Comparison: 1543–1644* (Rome: Institutum Historicum Societatis Iesu, 2009), 269–293; and Zhang, *Making the New World Their Own*, 338–344.

45 Translation cited from John L. Mish, "Creating an Image of Europe for China: Aleni's *Hsi-Fang Ta-Wen* 西方答問. Introduction, Translation, and Notes," *Monumenta Serica* 23 (1964): 48.

46 Wei Jun 魏濬 (1553–1626?), "Lishuo huangtang huoshi" 利說荒唐惑世 [Ricci's Theory Is Absurd and Deludes the World], in Xu Changzhi 徐昌治, ed., *Shengchao poxie ji* 聖朝破邪集 [Collected Essays Exposing Heterodoxy, 1639], *juan* 3, reprinted in Xia Guiqi 夏瑰琦, ed., *Jidu jiao yu zhongguo wenhua congkan zhi yi* 基督教與中國文化叢刊之一 [Series in Christianity and Chinese Culture] (Hong Kong: Jiandao shenxue yuan, 1996), 1: 183.

"This is different from what I have heard. This continent [of Europe] has over a hundred countries and polities, large and small. Each country governs itself, and they could not be unified."[47]

The lingering confusion over the identity of the Folangji in Ming-Qing Chinese world geographical discourse is another example of the ambiguous Jesuit legacy. Xiong Mingyu 熊明遇 (1579–1649) and Xiong Renlin 熊人霖 (1604–1667), a father and son who both had extensive personal contacts with the Jesuits, received no help from the latter in their efforts to sort out who the Folangji were and where they were based. Xiong Mingyu, well known to modern scholars as an early enthusiast of Jesuit science, wrote a series of essays on various foreign groups active in Southeast Asian waters based on the information he gleaned from earlier Chinese literature and his own observations while serving as the military superintendent of Funing Circuit (福寧道) in northeastern coastal Fujian. In these essays Xiong referred to both the Portuguese stationed in Macao and points further southwest, and to the Spaniards in the Philippines as Folangji, a usage common among Ming authors, and considered both to be Southeast Asian powers. He also clearly separated them from Ouluoba *ren*, a term the Jesuits used for "Europeans," themselves included. His son Xiong Renlin inherited the same mistakes when incorporating these essays into his own large tome on world geography, *Diwei* 地緯 [Geography of the Earth], 1624.[48] Later Chinese geographers, through a hilarious misreading of a statement in Aleni's entry on France (translated as Fu-lang-cha 拂郎察) in his *Records of Lands beyond the Jurisdiction of the Imperial Geographer*, concluded that Folangji is France. The French, who did not arrive in the South China Sea until 1660, became the bearers of all the historical and fictional sins committed by the maritime Portuguese (and Spaniards) in many Chinese official and private narratives on world geography well into the mid-nineteenth century.

---

47 Arai Hakuseki 新井白石, *Sairan Igen* 采覧異言 [Gleanings of Sayings about Foreign Lands, 1713], reprinted edition of 1881, book 1, 1 *verso*. I consulted the digital scan of this text made available by Hathitrust.org at: http://hdl.handle.net/2027/keio.10812709705 (accessed 4/18/2017). The missionary was Giovanni Battista Sidotti (1668–1714), an Italian missionary who risked the Tokugawa ban to enter Japan and was taken prisoner. Hakuseki, serving as Shogunal Tutor, was sent to interrogate Sidotti over the course of many days.

48 See Xiong Mingyu, "Daoyi zhuan" 島夷傳 [Records of Foreign Islanders], in idem, *Luxue lou ji* 綠雪樓集 [Collections of Works Written in the Green Snow Villa], in *Siku jinhui shu congkan* 四庫禁毀書叢刊 [Books Marked as "To be Banned or Destroyed" by Compilers of the Imperial Library of Four Treasuries] (Beijing: Beijing chubanshe, 2000), 185: 176–187; Xiong Renlin, *Diwei* 地緯 [Geography of the Earth, 1624], in Xiong Zhixue 熊志學, ed., *Hanyu tong* 函宇通 [Penetrating [the Secrets of] the Universe, editor preface dated 1648], copy held at the United States Library of Congress, especially 79a–81a (on Folangji).

New phenomena of the contact zone also began to spring up in the field of cartography as early as the late sixteenth century. As Richard J. Smith has observed, there are two major approaches to mapmaking in premodern China which persisted alongside each other without being explicitly recognized as competing, or even different, traditions. One of these was based on relatively precise mathematical measurements, and the other was based primarily on cultural data.[49] Typical of the former were maps produced to serve administrative functions and for the purpose of military defense. They tended to represent distances and topographical features of the land with precision, following a grid-based system dating back to Pei Xiu 裴秀 (223–271) if not earlier. Maps with a strong ethnographic orientation, including those purported to depict the entire known world, tended to take the latter approach. Their visual representations were more symbolic and schematic than topographical, paying little attention to spatial coordination but instead relying heavily on textual annotations to communicate such information. For obvious reasons, the genre of symbolic mapping was the first to take account of the world geographical data introduced by Ricci and later Jesuits. In 1593, Liang Zhou 梁辀 pioneered a new genre of world map by tweaking an earlier map of the world drawn by the Ming cartographer Yu Shi 喻時, titled "Map of Advantageous Terrain, Chinese and Foreign, Past and Present" (*Huayi gujin xingsheng zhi tu* 華夷古今形勝之圖, 1555). He preserved the overall image of the world from Yu's map, showing a square-shaped earth taken up primarily by China with seas surrounding it, but expanded the map's exterior to include central and southern Asia (albeit much truncated) and some more distant foreign countries, such as Canada, found on an early version of Ricci's map.[50] In 1644 another cartographer, Cao Junyi 曹君義, created a more elaborate map of this genre using Aleni's world map instead of Ricci's as the source of its depictions of the outer

49 Richard J. Smith, *Chinese Maps* (New York: Oxford University Press, 1996), 34–35, and idem, *Mapping China and Managing the World: Culture, Cartography and Cosmology in Late Imperial Times* (New York: Routledge, 2013), 52–62.

50 Liang Zhou's map is titled *Qiankun wanguo quantu gujin renwu shiji* 乾坤萬國全圖古今人物事跡 [A Comprehensive Map of Heaven, Earth, the Myriad Countries, and Ancient and Modern Persons and Artifacts]. Its stated date, 1593, conflicts with the reference it makes to Ricci's 1602 map in the preface printed at the top of the map. It is possible that Liang's original map was created in 1593 and that it consulted an older version of Ricci's map, but the only known extant sample of Liang's map came from a post-1602 edition carrying an updated preface. In his talk at the Academia Sinica, Taipei, "Completing the Map of the World, 1644: Cartographic Exchange between China and Europe," on October 21, 2019, Timothy Brook suggests that this sample may be a post-1602 pirated edition made in Nanjing. See the summary of his lecture at http://mingching.sinica.edu.tw/en/Academic_Detail/808, accessed on 8/6/2020.

world (Figure 4.8). On this map, not only are all inhabited continents of the earth represented, albeit with North and South America being cleaved into two islands and squeezed into the northeastern and southeastern corners respectively, but parallel longitudinal lines are also shown running through the entire map except for the area that is the Ming Empire.

The map is still mostly occupied by China, yet it gives some symbolic reckoning to the immensity of the real world, just as the artist Saul Steinberg did in his "View of the World from 9th Avenue."[51]

While this new symbolic genre of world maps continued to thrive in the art market of the late Ming and the Qing, a case can also be made that much synergy was generated in the encounter between Chinese and Jesuit quantitative cartography. This synergy was manifested in the emergence of three more or less concurrent developments. First, the late Ming and early Qing saw the rise of a discourse on mathematics and quantitative cartography as a "lost learning" of ancient China, and, among scholars who hesitated to adopt Jesuit cartography, a conscious movement to revitalize the grid-based mathematical cartography as envisaged by Pei Xiu.[52]

Second, we also see attempts to hybridize Chinese grid-based cartography that extracts its data from measurements of road distances and presumes a flat and square earth with Jesuit cartographical methods that in turn presume the sphericity of the earth and use longitudes and latitudes derived from astronomical observations as coordinates. Ricci himself was a pioneer in this development for he had, in the absence of adequate astronomical data, consulted Chinese grid-based maps and information about road distances contained in various Chinese local gazetteers to arrive at estimates of the coordinates of Chinese cities and landmarks shown on his 1602 world map. In an essay on his map, he also provided a formula for converting horizontal and vertical distances on earth into longitudinal and latitudinal readings of different locations, namely, 1 degree of longitude or latitude equals to 250 *li* in vertical or horizontal distance.[53] That formula served as a basis for some mathematically

51 Saul Steinberg, "View of the World from 9th Avenue," appears on the cover of the March 29, 1976 issue of *The New Yorker*. It presents what is sometimes called a "parochial New Yorker's" view of the world seen from Manhattan, looking west across North America and the Pacific Ocean to the Eurasian continent, which consists of three countries: China, Japan, and Russia. I wish to thank my colleague Win-chiat Lee for providing me with this reference.

52 A representative of this movement is Hu Wei, who called for the need to restore the six principles of mapping laid out by Pei Xiu and a massive, empire-wide survey to obtain accurate cartographical data. See Zhang, *Making the New World Their Own*, 248–250.

53 During the Kangxi reign (1702), this figure was readjusted to 200 *li*.

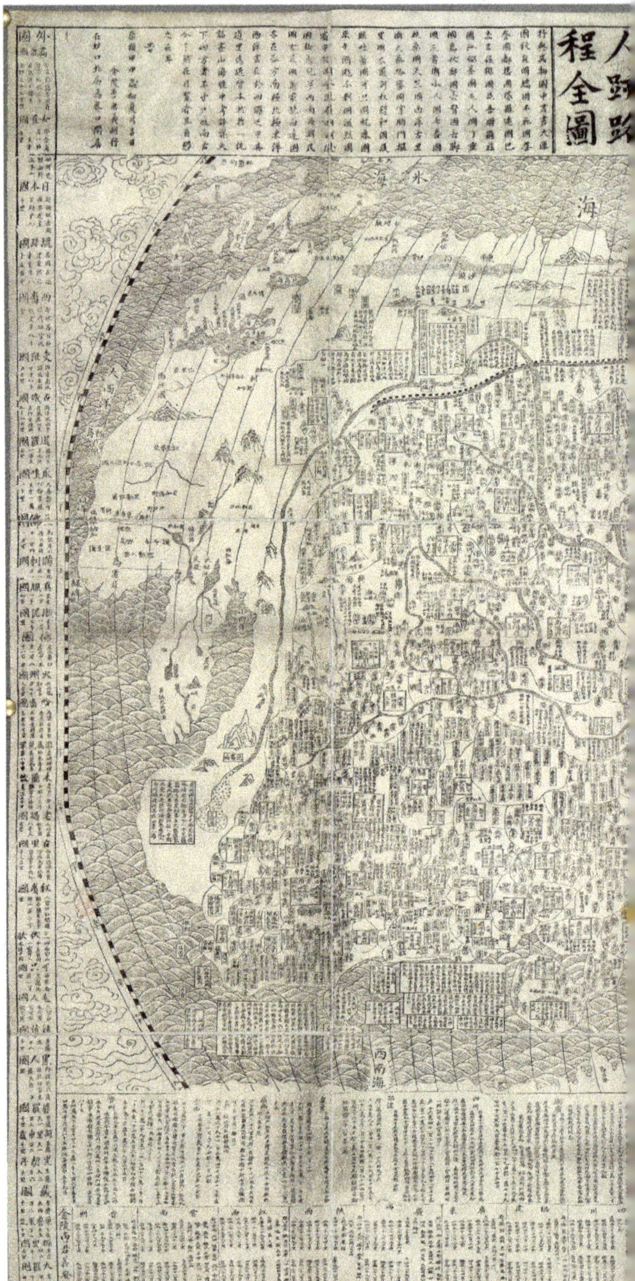

FIGURE 4.8
Hybrid World Map of Cao Junyi 曹君義, *Tianxia jiubian fenye renji lucheng quantu* 天下九邊分野人跡路程全圖 [The Complete Map of the Nine Border Towns and Allotted Fields of All-Under-Heaven and of Human Presence and Travel Routes], 1644. Courtesy of the British Library Board, Shelfmark: Maps *60875.(11).

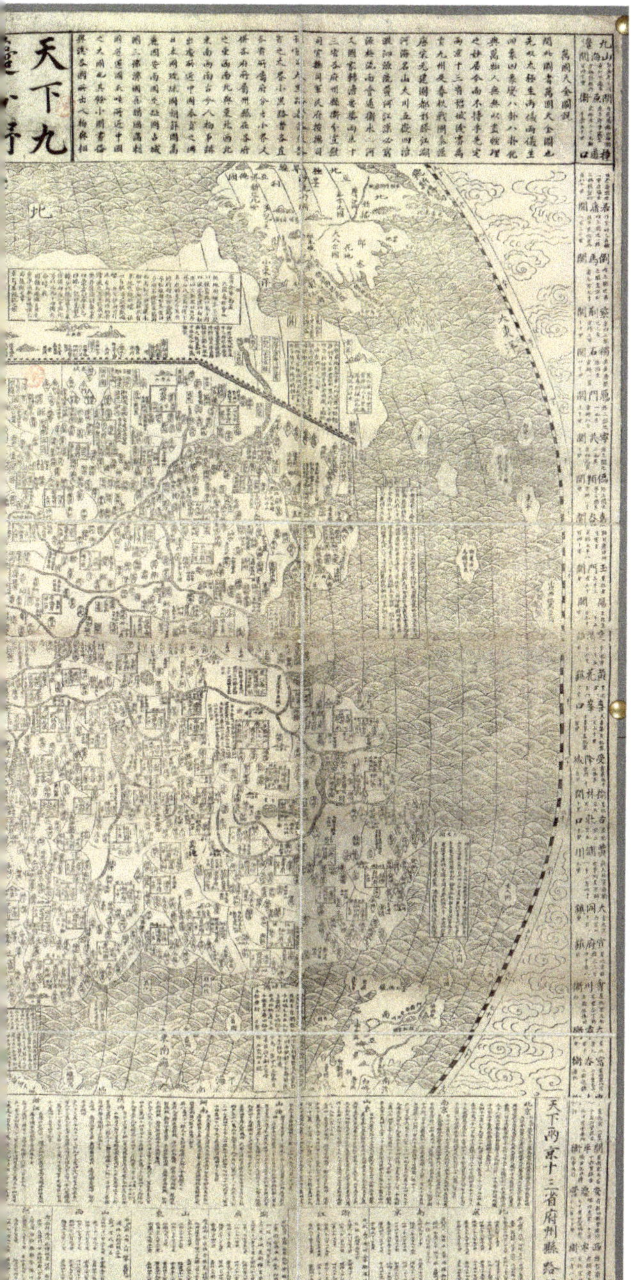
天下九
天下兩京十三省府州縣路程

minded Chinese scholars to comprehend Jesuit cartography and craft their own maps that simulated the ideals of Jesuit mapping.

An example of such hybrid maps is the "Map of the [Tribute of] Yu Shown in its Terrestrial and Corresponding Celestial Dimensions" (Yushu jingtian hedi tu 禹書經天合地圖), possibly drawn by the philosopher Fang Yizhi 方以智 (1611–1671) around 1639, or by his student Jie Xuan before 1666; its only known version carries a caption written by Jie Xuan and appears in a book dated 1675 by Jie's friend You Yi 游藝 (1614–1684), also a student of Fang's (Figure 4.9).[54]

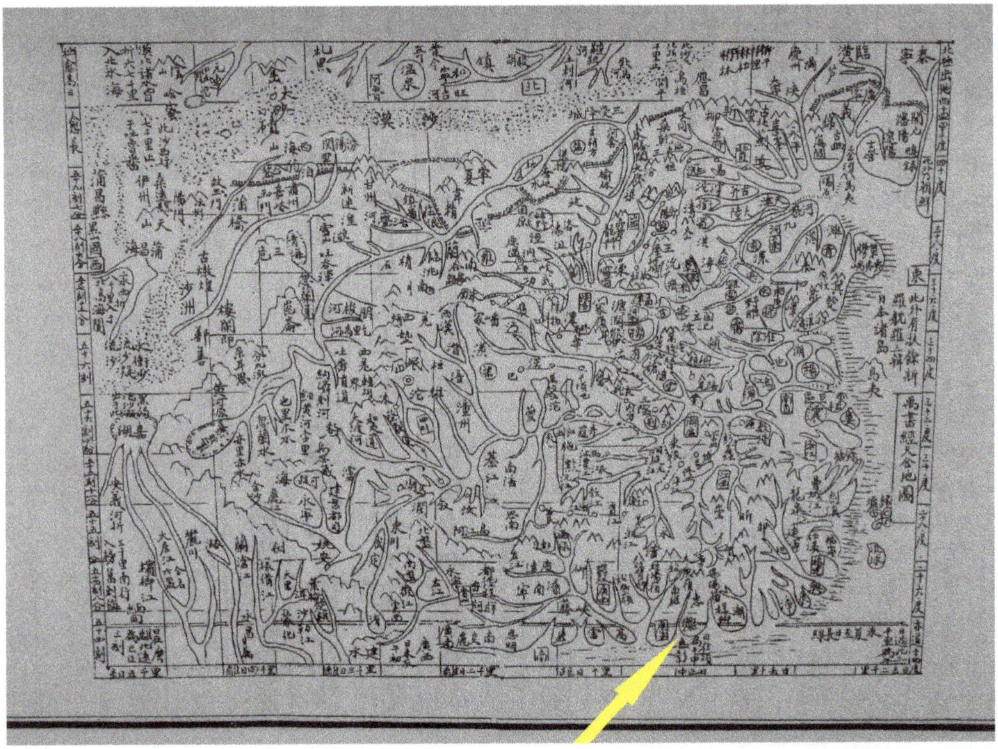

FIGURE 4.9 *Yushu jingtian hedi tu* 禹書經天合地圖 [Map of (the Tribute of) Yu Shown in its Terrestrial and Corresponding Celestial Dimensions], 1675. The yellow arrow points to the image of the sun at the summer solstice. From You Yi 游藝, *Tianjing huowen* 天經或問 [Inquiries into (the Patterns of) Celestial Movements], 1675, *juan* 1, 26a. Anthologized in Zhang Fujiang 張福江, comp. *Siku quanshu tujian* 四庫全書圖鑒 [Selected Illustrations from the Complete Collection of the Four Treasuries] (Beijing: Dongfang chuangshe, 2004), 5:590.

54 For the authorship of this map, see my discussion in Zhang, *Making the New World Their Own*, p. 256, n. 96. I maintain my hypothesis stated there, that Jie Xuan is the author of this map, but with one correction: what You Yi explicitly credited to Jie Xuan is not the map itself, but an essay explaining the map titled "Commentary on *Yushu jingtian hedi tu*" (*Yushu jingtian hedi tu shuo*).

In his caption, Jie explains that the map encompasses both terrestrial and celestial dimensions: terrestrial because it "lays out the Tracks of Yu against the geography of late Ming China, all represented in a grid so that the map captures accurately the configuration of the land," and celestial because it takes account of variations in the lengths of daylight time south to north and in time differences between east and west (or, in his words, the different times at which the sun reached the noon position in different locations on earth east and west). Citing Ricci's formula, Jie explains how these variations are to be calculated. For example, two places that are 7,500 *li* apart east to west would have one-hour difference between them—that is, one hour in the traditional Chinese 12-hour day. The map shows three parameters on its margins: on the right margin, it shows a range of latitudes from 24 to 44 degrees north; on the left margin are marked the lengths of daylight time at summer solstice. On the bottom of the map, in lieu of longitude readings, one finds a set of more cumbersome yet functionally equivalent markings of distances to the position identified as "the mid-day sun position" (*ri zhengzhong* 日正中). This position sits on the Tropic of Cancer and is designated by the drawing of a shining sun. The text below the sun says: "[At this point] the sun is right overhead and casts no shadow." This would be the equivalent of the zero meridian; it is picked as the point of reference probably because it corresponded to the meridian running through the presumed capital of Yu somewhere at the 36-degree latitude. Two longitudinal lines to the east of this are marked as "one thousand *li* past the mid-day sun position" (日去千里) and "two thousand *li* past the mid-day sun position" (日去二千里) respectively. Conversely, the five lines west of this are marked as "one thousand *li* short of the mid-day sun (距日千里) position," "two thousand *li* short of the mid-day sun (距日二千里) position," etc. On the whole, the map demonstrates a relatively accurate grasp of the concepts of longitudes and latitudes, but its rendering of longitudes in terms of the distances between points east and west betrays its ultimate reliance on terrestrial rather than astronomical measurements. The hybridization of Jesuit and pre-existing Chinese cartographical methods was practiced on a large scale during the massive survey of the Qing empire undertaken between 1708 and 1715, commissioned by the Kangxi emperor, and conducted by a large Chinese crew under the leadership of a dozen of Jesuit missionaries, which simultaneously implemented European surveying methods and Chinese methods based on road measurements made with graduated ropes.[55]

The third trend is the relatively slow, but nonetheless significant, shift away from grid-based cartography towards Jesuit cartography among Chinese

55 See Mario Cams, *Companions in Geography: East-West Collaboration in the Mapping of Qing China (c.1685–1735)* (Leiden: Brill, 2017).

cartographers. A major milestone in this development is the creation of the "Map of a Complete View of Imperial Territory" (*Huangyu quanlan tu* 皇輿全覽圖), also known as the Kangxi Atlas, that resulted from the above-mentioned empire-wide survey. While this imperial cartographical project was directed by the Jesuits, by the late eighteenth century an increasing number of Chinese began to practice Jesuit-style cartography on their own, in unofficial capacities. In 1794 Zhuang Tingfu 莊廷尃 created an elaborate world map by modeling and improving on Verbiest's "Complete Geographical Map of the Earth" (*Kunyu quantu* 坤輿全圖, 1674). His example was emulated by Ye Zipei 葉子佩 and a few others in the following decades.[56] Regional geographical maps drawn with longitudes and latitudes also began to appear in local gazetteers around this time. The 1822 version of *Guangdong Provincial Gazetteer* (Guangdong tongzhi 廣東通志) features a general map of Guangdong and other regional maps belonging to this new genre. An editorial comment preceding the section of maps gives a brief review of ancient Chinese cartography as represented by the six principles set out by Pei Xiu:

> During Ming times Westerners arrived in China. They used astronomical measurements to reckon distances on earth and drew maps employing latitudes and longitudes. This they boasted as their ingenious invention. Little did they know that Pei Xiu had already fully expounded on these concepts in his *Geographical Maps of the Tribute of Yu* (*Yugong diyu tu* 禹貢地域圖). Xiu says: "Graduated divisions (*fenlu* 分率), [the first of the six principles], are the means of determining the *guang* 廣 and *lun* 輪 of the map." In his commentary on the *Rites of Zhou*, Ma Rong (馬融, 79–166) says: "*Guang* means east-west, and *lun* means north-south." *North-south would be longitude, and east-west latitude. Thus, we know Pei Xiu's maps had already used longitudes and latitudes.*[57] [Emphasis added.]

---

56 Zhuang Tingfu's 莊廷尃 map is titled *Da Qing tongshu zhigong wanguo jingwei diqiu shi fangyu gujin tu* 大清統屬職貢萬國經緯地球式方輿古今圖 [The Great Qing Dynasty World Map of Tribute-Bearing Countries with Spherical Coordinates, Past and Present]. A corrupted copy of this map is held at the Library of Congress, item id: gm71005053. For a discussion on this map, see Smith, *Mapping China and Managing the World*, 77. Ye Zipei's 葉子佩 map, *Wanguo dadi quantu* 萬國大地全圖 [Complete Map of the Lands of Ten Thousand Countries] (1845), is reprinted and studied in Li Shengwu 李勝伍, *Qingdai guoren huike de shijie ditu—wanguo dadi quantu* 清代國人繪刻的世界地圖—萬國大地全圖 [A World Map Created by an Author of the Qing Dynasty—"Complete Map of the Lands of Ten Thousand Countries"] (Beijing: Zhongguo dabaike quanshu chubanshe, 2002).

57 Ruan Yuan 阮元 (editor-in-chief) and Chen Changqi 陳昌齊, et al. (lead authors), (*Daoguang*) *Guangdong tongzhi* （道光）廣東通志 [Chorography of Guangdong], *juan* 83,

With this "discovery" of the concepts of latitude and longitude in Pei Xiu's grid-based cartography, a signature feature of Renaissance European cartography embodied by the Ricci maps finally found its own place in China.

The above discussion serves to demonstrate how the global integration of conceptions of space occurred within the context of early modern China. The close examination of the Ricci map of 1602 brings out the ways in which larger forces of the contact zone—the social historical circumstances of the Jesuit mission; the Ming empire's conflicts with the Portuguese, Spaniards, and other maritime Europeans; and the parallel trends in the intellectual lives of Renaissance Europe and late Ming China—intersected with Ricci's personal agency to shape the outcomes of his mapmaking. On the other hand, the longitudinal survey of the afterlives of Ricci maps and Jesuit cartography from the late sixteenth to the early nineteenth century, in comparison with their counterparts in Tokugawa Japan, sheds light on how the particular brands of Renaissance European understanding and the representations of the world fashioned by Ricci and later Jesuits in China in turn brought both opportunities and risks to Chinese world geographers and cartographers. More importantly, it shows how Ricci's rhetorical strategy of accommodation sparked off a new discourse on the Chinese origination of Western learning, which was fueled further by the ideological collisions between Chinese and European ethnocentrisms and by the empire-building of the Chinese state. Within this complex web of local and global currents, individual Chinese scholars managed to salvage or manufacture conceptual parallels to Renaissance European notions of the globe, the multiple continents and oceans, and terrestrial longitudes and latitudes, among others, within their own inherited traditions—which indeed turned out to be much richer and more complex than what they, and we, had previously understood. The result was a global integration whereby some seminal ideas of European origin became centrally embedded in the tapestry of a genuine Chinese cultural renewal, rather than a conversion at the scale that the Jesuits were hoping for. The interdependent and interlinking knowledge-making of the Jesuits and of the Chinese scholars in this story, and the surprising twist to its ending, lends further support to Subrahmanyam's

in *Xuxiu siku quanshu* 續修四庫全書 [Sequel to the *Imperial Library of Four Treasuries*] (Shanghai: Shanghai guji chubanshe, 1995–1999), vol. 670: 603. Ruan and Chen were prominent representatives of evidential classical scholarship, and both of them and the cartographer for this gazetteer, Li Mingche 李明徹, engaged Jesuit astronomy and cartography in their own writings. This comment, made in the name of the editorial board of the gazetteer, likely reflected a consensus shared by these three scholars and others in this circle.

characterization of global early modernity as one of "connected histories" and "separate destinies."[58]

Acknowledgements

I am grateful for the generous visiting fellowship from the Institute for Research in Humanities (IRH) at Kyoto University, Japan, and the Stroupe Faculty Fellowship from Wake Forest University that made this project possible. Part of the research for this article was conducted at the IRH in summer 2017. I wish to thank Professor TAKEDA Tokimasa 武田時昌, for his hospitality and helpful advice, and other colleagues at the IRH, especially LIU Qing 劉青, for facilitating my work in many ways.

Bibliography

Arai Hakuseki 新井白石. *Sairan Igen* 采覧異言 [Gleanings of Sayings about Foreign Lands], 1713. Reprinted edition of 1881, book 1. Hathitrust.org. Accessed 4/18/2017. http://hdl.handle.net/2027/keio.10812709705.

Bentley, Jerry. "Early Modern Europe and the Early Modern World." In *Between the Middle Ages and Modernity: Individual and Community in the Early Modern World*, edited by Charles H. Parker and Jerry H. Bentley, 13–31. Plymouth, UK: Rowman and Littlefield, 2007.

Brook, Timothy. "Completing the Map of the World, 1644: Cartographic Exchange between China and Europe." Paper presented at Academia Sinica, Taipei, on October 21, 2019. Accessed on 8/6/2020. http://mingching.sinica.edu.tw/en/Academic_Detail/808.

Brook, Timothy. "Europaeology? On the Difficulty of Assembling a Knowledge of Europe in China." In *Christianity and Culture: Japan and China in Comparison: 1543–1644*, edited by M. Antoni J. Ucerler, S.J., 269–293. Rome: Institutum Historicum Societatis Iesu, 2009.

Cams, Mario. *Companions in Geography: East-West Collaboration in the Mapping of Qing China (c.1685–1735)*. Leiden: Brill, 2017.

Cheng Dachang 程大昌 (1123–1195). *Yugong shanchuan dili tu* 禹貢山川地理圖 (Illustrations and Maps of Mountains and Rivers in the "Tribute of Yu"). Photo-reproduced edition. Beijing: Zhonghua shuju, 1985.

58 Subrahmanyam, "Connected Histories."

D'Elia, S.J., Pasquale M. "Recent Discoveries and New Studies (1938–1960) on the World Map in Chinese of Father Matteo Ricci, S.J." *Monumenta Serica* 20 (1961): 82–164.

Dong Han 董含. *Sangang shilue* 三岡識略 [Cursory Records from Sangang], *juan* 2. Entry on "Island of Yakṣas" [*yecha dao*]. 1844 edition. The State Library of Berlin, p. 76 v. Accessed on 11/20/2017. http://digital.staatsbibliothek-berlin.de/werkansicht?PPN=PPN3346157571&PHYSID=PHYS_0076&DMDID=DMDLOG_0006.

Early Modernities. Special issue of *Daedalus*, vol. 127, no. 3 (Summer 1998).

Elman, Benjamin. "A Jointly Regional-Global Approach to Rethinking Early Modern East Asian History." In *The 'Global' and the 'Local' in Early Modern and Modern East Asia*, edited by Benjamin A. Elman and Chao-Hui Jenny Liu, 66–78. Leiden: Brill, 2017.

Fujitani, James. "The Ming Rejection of the Portuguese Embassy of 1517: A Reassessment." *Journal of World History* 27, no. 1 (2016): 87–102.

Gong Yingyan 龔纓晏. "Xiancun zuizao de Li Madou shijie ditu yanjiu" 現存最早的利瑪竇世界地圖研究 [A Study of the Earliest Extant World Maps Made by Matteo Ricci]. *Lishi dili* 歷史地理 [Historical Geography] (2018: no. 2): 1–12.

Han Qi 韓琦. "Manila dafanchuan maoyi dui ming wangchao de yingxiang" 馬尼拉大帆船貿易對明王朝的影響 [The Impact of the Manila Galleon Trade on the Ming Dynasty]. *Shijie jinday shi yanjiu* 世界近現代史研究 [Studies in Modern World History] 10 (2013): 41–71.

Hart, Roger. *Imagined Civilizations: China, the West, and Their First Encounter*. Baltimore: Johns Hopkins University Press, 2013.

Hong Weilian 洪煨蓮. "Kao Li Madou ditu" 考利瑪竇地圖 [An Inquiry into Matteo Ricci's Maps]. *Yugong banyue kan* 禹貢半月刊 [Yugong Biweekly Journal] 5, no. 3–4 (1936): 1–50.

Hostetler, Laura. *Qing Colonial Enterprise: Ethnography and Cartography in Early Modern China*. Chicago: University of Chicago Press, 2001.

Hsia, Ronnie. *A Jesuit in the Forbidden City*. Oxford: Oxford University Press, 2010.

Inagaki Shisen 稲垣子戩. *Kon'yo zenzu setsu* 坤輿全圖説 [Legends on the Complete Terrestrial Map]. Kyoto: Kyoto University Library. Prefaced 1802 (n.p.).

Jin Guoping 金國平. *Xili dongjian: Zhongpu zaoqi jiechu zhuixi* 西力東漸：中葡早期接觸追昔 [Eastward Expansion of Western Powers: Historical Memories of Early Sino-Portuguese Contacts]. Macao: Macao Foundation, 2000.

Liang Zhou 梁輈. *Qiankun wanguo quantu gujin renwu shiji* 乾坤萬國全圖古今人物事跡 [A Comprehensive Map of Heaven, Earth, the Myriad Countries, and Ancient and Modern Persons and Artifacts].

Liu, Yu. *Harmonious Disagreement: Matteo Ricci and His Closest Chinese Friends*. New York: Peter Lang Publishing, 2015.

Lu Ciyun 陸次雲. *Bahong yishi; Bahong huangshi* 八紘譯史；八紘荒史 [Translated Accounts of Things within the Eight Corners of the World; Accounts of the

Wilderness within the Eight Corners of the World]. Combined modern reprinted edition. Taipei: Guangwen shuju, 1969.

Ma Duanlin 馬端臨. *Wenxian tongkao* 文獻通考 [Comprehensive Examination of Literature], *juan* 347. In *Wenyuange siku quanshu* 文淵閣四庫全書 [Imperial Library of Four Treasures, based on the copy preserved at the Pavilion of Literary Profundity], 1773–1782, edited by Ji Yun 紀昀 (1724–1805) et al. Photolithographically reproduced edition, vol. 616. Taipei: Taiwan shangwu yinshuguan, 1983–1986.

Mish, John L. "Creating an Image of Europe for China: Aleni's *Hsi-Fang Ta-Wen* 西方答問. Introduction, Translation, and Notes." *Monumenta Serica* 23 (1964): 1–87.

Mudimbe, V.Y. *The Invention of Africa: Gnosis, Philosophy, and the Order of Knowledge*. Indianapolis: Indiana University Press, 1988.

Pang Naiming 龐乃明. *Mingdai zhongguoren de ouzhou guan* 明代中國人的歐洲觀 [Ming Dynasty Chinese Views of Europe]. Tianjin: Tianjin renmin chubanshe, 2006.

Parker, Charles H. *Global Interactions in the Early Modern Age, 1400–1800*. Cambridge, UK: Cambridge University Press, 2010.

Pliny the Elder (23–79 CE). *The Natural History of Pliny*, translated by John Bostock and H.T. Riley, vol. 2. London: Henry G. Bohn, York Street, and Covent Garden, 1855–57.

Porter, David. *Comparative Early Modernities, 1100–1800*. New York: Palgrave Macmillan, 2012.

Pratt, Mary Louis. "Arts of the Contact Zone." In *Professing in the Contact Zone: Bringing Theory and Practice Together*, edited by Janice M. Wolff, 1–18. Urbana, IL: National Council of Teachers of English, 2002.

Pratt, Mary Louise. *Imperial Eyes: Studies in Travel Writing and Transculturation*. London and New York: Routledge, 1992.

Ptak, Roderich. "Gouguo, the 'Land of Dogs,' on Ricci's World Map." *Monumenta Serica* 66, no. 1 (2018): 71–89.

Ricci, S.J., Matteo and Nicolas Trigault. *China in the Sixteenth Century: The Journal of Matthew Ricci: 1583–1610*, translated from Trigault's Latin text by Louis J. Gallagher, S.J. New York: Random House, 1942/1953.

Ricci, S.J., Matteo. *Konyo bankoku zenzu* 坤輿萬國全圖. Tohoku University Library, http://www.i-repository.net/contents/tohoku/kano/ezu/kon/kon.html.

Ricci, S.J., Matteo. *Kunyu wanguo guantu* 坤輿萬國全圖 [Complete Geographical Map of Ten Thousand Countries]. The Hong Kong Maritime Museum 香港海事博物館. Google Arts & Culture. https://www.google.com/culturalinstitute/beta/asset/kunyu-wanguo-quantu-complete-map-of-all-nations-on-earth-c-17th-century-left-half/dwEyNpMhRJtBcw.

Ricci, Matteo. *Li Madou zhongwen zhuyi ji* 利瑪竇中文著譯集 [The Chinese Works and Translations of Matteo Ricci]. Edited by Zhu Weizheng 朱維錚. Shanghai: Fudan University Press, 2001.

Ricci, S.J., Matteo. *Liangyi xuanlan tu* 兩儀玄覽圖 [A Mysterious Visual Map of the Two Forms]. 1603. Soongsil Christian Museum in Seoul, South Korea.

Ruan Yuan 阮元, Chen Changqi 陳昌齊, et al. *(Daoguang) Guangdong tongzhi* （道光）廣東通志 [Chorography of Guangdong], *juan* 83. In *Xuxiu siku quanshu* 續修四庫全書 [Sequel to the *Imperial Library of Four Treasuries*]. Photolithographically reproduced edition, vol. 670. Shanghai: Shanghai guji chubanshe, 1995–1999.

Sekisui Nagakubo. *Chikyu bankoku sankai yochi zenzu* 地球萬國山海輿地全圖 [Complete Map of Ten Thousand Countries, Mountains, and Seas on Earth] ca. 1785. Accessed on 11/10/2017 at World Digital Library: https://www.wdl.org/zh/item/14749/.

Smith, Richard J. *Chinese Maps*. New York: Oxford University Press, 1996.

Smith, Richard J. *Mapping China and Managing the World: Culture Cartography and Cosmology in Late Imperial Times*. New York: Routledge, 2013.

Steinberg, Saul. "View of the World from 9th Avenue." *The New Yorker* March 29, 1976.

Subrahmanyam, Sanjay. "Connected Histories: Notes Towards a Reconfiguration of Early Modern Eurasia." *Modern Asian Studies* 31, no. 3 (1997): 735–762.

Taylor, G.R. "A Letter Dated 1577 from Mercator to John Dee." *Imago Mundi* 13 (1956): 56–68.

Unno Kazutaka 海野一隆. "Ri matou <Kon'yo bankoku zenzu> no syohan" 利瑪竇《坤輿萬國全圖》の諸版 [The Various Editions of Matteo Ricci's *Kon'yo bankoku zenzu*], *Touhou gakuhou* 東洋学報 [Journal of Oriental Studies] 87, no. 1 (2005): 101–143.

Wei Jun 魏濬. "Lishuo huangtang huoshi" 利說荒唐惑世 [Ricci's Theory Is Absurd and Deludes the World]. In *Shengchao poxie ji* 聖朝破邪集 [Collected Essays Exposing Heterodoxy] edited by Xu Changzhi 徐昌治, *juan* 3, 1639. Reprinted in *Jidu jiao yu zhongguo wenhua congkan zhi yi* 基督教與中國文化叢刊之一 [Series in Christianity and Chinese Culture], no. 1, edited by Xia Guiqi 夏瑰琦, 183–186. Hong Kong: Jiandao shenxue yuan, 1996.

Wills, John E. *China and Maritime Europe, 1500–1800: Trade, Settlement, Diplomacy, and Missions*. Cambridge: Cambridge University Press, 2011.

Xiao Jiefu 蕭萐父. "The Rough Road to China's Philosophical Enlightenment." In *Chuisha erji* 吹沙二集 [Blowing off Sand [to Sift Gold], Second Collection], 553–599. Chengdu: Bashu shushe, 1999.

Xiong Mingyu. "Daoyi zhuan" 島夷傳 (Records of Foreign Islanders). In *Luxue lou ji* 綠雪樓集 [Collections of Works Written in the Green Snow Villa]. *Siku jinhui shu congkan* 四庫禁毀書叢刊 [Books Marked as "To be Banned or Destroyed" by Compilers of the Imperial Library of Four Treasuries], vol. 185: 176–187. Beijing: Beijing chubanshe, 2000.

Xiong Renlin. *Diwei* 地緯 [Geography of the Earth] 1624. In *Hanyu tong* 函宇通 [Penetrating [the Secrets of] the Universe]. Edited by Xiong Zhixue 熊志學, preface dated 1648. Copy held at the US Library of Congress.

Ye Zipei 葉子佩. "Wanguo dadi quantu" 萬國大地全圖 [Complete Map of the Lands of Ten Thousand Countries], 1845. Reprinted in *Qingdai guoren huike de shijie ditu—wanguo dadi quantu* 清代國人繪刻的世界地圖—萬國大地全圖 [A World

Map Created by an Author of the Qing Dynasty—"Complete Map of the Lands of Ten Thousand Countries"], edited by Li Shengwu 李勝伍. Beijing: Zhongguo dabaike quanshu chubanshe, 2002.

Yee, Cordell. "Taking the World's Measure: Chinese Maps between Observation and Text." In *Cartography in the Traditional East and Southeast Asian Societies*, edited by J.B. Harley and David Woodward, 96–127. Vol. 2, bk. 2 of *The History of Cartography*. Chicago: University of Chicago Press, 1994.

Zhang Fujiang 張福江, comp. *Siku quanshu tujian* 四庫全書圖鑒 [Selected Illustrations from the Complete Collection of the Four Treasuries]. Beijing: Dongfang chuangshe, 2004.

Zhang, Qiong. "A Triangular Cartographic Encounter: The Strange Journeys of the Yakṣas in Ming-Qing China and Tokugawa Japan, via the Ricci Maps." Paper presented at the international workshop, "Cartographic Operations: Art, Science and Politics in South East Asia," co-organized by Christine Anne Habbard of SUTD, and Jean-Marc Besse, CNRS, Paris, and held at the National Library in Singapore, 17 May 2018.

Zhang, Qiong. *Making the New World Their Own: Chinese Encounters with Jesuit Science in the Age of Discovery*. Leiden: Brill, 2015.

Zhao Yongfu 趙永復. "Li Madou 'Kunyu wanguo quantu' suo yinyong de zhongguo ziliao" 利瑪竇坤輿萬國全圖所引用的中國資料 [The Chinese Sources Incorporated into Matteo Ricci's "Complete Geographical Map of Ten Thousand Countries"]. *Lishi dili yanjiu* 歷史地理研究 [Research in Historical Geography] 1 (1986): 200–207.

Zhou Yunzhong 周運中. "Li Madou *Yutu zhi* yiwen kaoshi ji qita" 利瑪竇〈輿圖志〉佚文考釋及其他 [An Examination of Some Fragments from Matteo Ricci's *Yutu zhi* (World Map with Legends) and Other Relevant Issues]. *Ziran kexueshi yanjiu* 自然科學史研究 [Studies in the History of Natural Sciences] (2010: no. 4) 437–445.

Zhuang Tingfu 莊廷旉. "Da Qing tongshu zhigong wanguo jingwei diqiu shi fangyu gujin tu" 大清統屬職貢萬國經緯地球式方輿古今圖 [The Great Qing Dynasty World Map of Tribute-Bearing Countries with Spherical Coordinates, Past and Present], 1794. Copy held at the US Library of Congress, accessible online at: https://www.loc.gov/item/gm71005053/.

CHAPTER 5

The Introduction of Ricci's World Maps into Edo Period Japan: A Detailed Comparative Investigation of Maps

Aoyama Hiro'o

1 World Maps in Early Modern Japan

The production of world maps in Japan changed significantly during the early modern period, namely the end of the sixteenth century to the mid-nineteenth century. World maps until then were Buddhist, created based on the Buddhist world view and showing India in the center and China and Japan in East Asia. However, direct contacts between Japan and Europe started with the coming of the Portuguese in 1543 and the arrival of Francis Xavier in Japan in 1549. In subsequent decades, trade with the so-called '*Nanban* [western] ships'—primarily from Portugal and Spain—flourished and much European culture was introduced. This also included world maps and globes, which enabled Japanese to obtain knowledge of the world from Europe.

Thus, already before the seventeenth century, a number of map styles were available to be consulted in Japan. The infusion of geographic knowledge from Europe would, however, soon be curtailed. In order to thoroughly prohibit Christianity and control trade, the Tokugawa shogunate, established in 1603, prohibited Japanese from traveling to and returning from overseas, recognized trade only with China and Holland, and limited that trade to ports in Nagasaki. This so-called "closed country" structure was completed by 1641 and continued for more than two hundred years until the opening of the country through the signing of the Treaty of Peace and Amity between the United States of America and the Empire of Japan in 1854.

Along with this, importing books written in China by Jesuits and Chinese Christians was also prohibited in 1630. This included the works of Matteo Ricci, who had died in 1610. When the prohibition edict was reinforced in 1685, importing European books on natural science also became difficult.

Yet, world maps brought from Europe prior to the closure continued to circulate. Some were duplicated by Japanese painters and created as folding screens. More than a few included *jinbutsu zu* 人物圖 [illustrations of peoples]

© AOYAMA HIRO'O, 2024 | DOI:10.1163/9789004684782_007
This is an open access chapter distributed under the terms of the CC BY-NC-ND 4.0 license.

depicting ethnic groups in various regions of the world. More than thirty such world maps known as *Nanban*-style world maps are still extant. For example, a world map also showing the four big cities (Lisbon, Seville, Rome and Constantinople) of the world (Kobe City Museum collection) was based on a world map produced by Pieter van der Keere in 1609.[1]

In addition, until the establishment of the closed country structure, Japanese had engaged in trade with Southeast Asia. The Japanese were greatly interested in navigational charts and duplicated Portuguese portolan charts of the region. As practical aids to navigation, these charts depicted realistic shorelines from East Asia to the Indian Ocean and even contain depictions of banks, coral reef islands and other features.

The *Kunyu wanguo quantu* 坤輿萬國全圖 [Complete Terrestrial Map of All Countries] produced by Matteo Ricci, in cooperation with Li Zhizao 李之藻, in Beijing in 1602 (hereafter referred to as the 1602 Ricci World Map) had also been brought to Japan, probably soon after publication, but no definite proof of the details of its transmission exists. The earliest positive evidence indicating importation to Japan is the hand-drawn copy deposited in the Tamba Sasayama City Museum of History and Art (hereinafter referred to as the Sasayama Ricci World Map) thought to be copied by hand about 1675 (private collection) (Figure 5.1).

Since the 1602 Ricci World Map was inscribed with *kanji* (Chinese characters), it was also probably very readily accepted in early modern Japan, which was a part of the *kanji* cultural sphere. As I relate in detail in this study, many hand-drawn copies of this map were created and numerous published versions were produced based on them. More than twenty hand-drawn copies have been confirmed to date, many of them containing important information for research addressing the 1602 Ricci World Map.

In the history of maps in Japan, one leading characteristic of the early modern times is that maps came to be published individually. Maps prior to that were published as illustrations in books; the publication of individual single-sheet maps began during the 1620s. The first publication of an urban map was in 1624–26 and the first publication of a map of Japan was in 1624. The first publication of a world map, the *Bankoku Sōzu* 萬國総圖 [Map of All the Countries] (Figure 5.2) (Shimonoseki City Museum of History collection),

1 TAKAHASHI Tadashi, "Nanban toshizu byōbu kara Kaeriusu sekaizu e" 南蛮都市図屏風からカエリウス (Pieter van der Keere) 世界図へ [From *Nanban* City View Folding Screens to World Maps by Pieter van der Keere], in *Ezu no kosumorojii* 絵図のコスモロジー [Cosmology of Picture Maps], ed. Katsuragawa Ezu Kenkyūkai 葛川絵図研究会 (Kyoto: Chijin Shobō, 1988) 1: 248–264.

FIGURE 5.1 Sasayama Ricci World Map. Private collection (deposited in the Tamba Sasayama City Museum of History and Art). The museum calls this map the "坤輿万国全図屏風" [Kon'yo Bankoku Zenzu folding screen], thereby distinguishing it from the other Sasayama World Map in its collection (shown in figure 5.15).

FIGURE 5.2 *Bankoku Sōzu* 萬國総圖, 1645, Shimonoseki City Museum of History collection.

was in 1645. It pairs up with a *jinbutsu zu* depicting pairs of people from various parts of the world.

When maps began to be published, they were produced in large quantities compared to the age of hand-drawn maps. Printing allowed them to be disseminated widely in society, thus also promoting world geographic knowledge. Maps encountered the age of printing and various types of maps were printed. As a result, the social impact of maps increased.

Beginning in the eighteenth century, the leadership of Japan started to feel constrained by the results of the ban on European books. Specifically, when planning calendar reform, the eighth Shogun, Tokugawa Yoshimune, became aware how research on European scholarly learning was obstructed. He thus eased the ban in 1720. His release of the ban on books other than those addressing Christian doctrine opened the way to full-scale academic research on Europe, otherwise known as *Rangaku*. The *Zhifang waiji* 職方外紀 [Unofficial Accounts of Foreign Countries] (published in 1623) by Giulio Aleni was among the works that then became available.

As *Rangaku* advanced, the most recent geographical knowledge was brought from Europe and *Rangaku*-style world maps were produced mainly as double hemisphere world maps on the basis of the most recent world maps brought from Europe. The earliest to be published was a world map produced by Shiba Kōkan 司馬江漢 (1747–1818) in 1792. Takahashi Kageyasu 高橋景保 (1785–1829), then the government's chief astronomer, later published the *Shintei Bankoku Zenzu* 新訂萬國全圖 [Newly Revised Map of All the Countries] in 1816.

Meanwhile, world maps based on the world maps produced before the relaxation of the ban were also printed. Harame Sadakiyo 原目貞清 (years of birth and death unknown) published the *Yochi Zu* 輿地圖 [World Map] (Figure 5.3) in 1720 and, at the end of the eighteenth century, Nagakubo Sekisui 長久保赤水 (1717–1801) published the *Chikyū Bankoku Sankai Yochi Zenzusetsu* 地球萬國山海輿地全圖說 [Map with an Account of All the Countries, Lands and Seas in the World] (also known as the *Kaisei Chikyū Bankoku Zenzu* 改正地球萬國全圖, hereinafter referred to as the Sekisui World Map) (Figure 5.4). According to the conventional view, both of them were so-called Ricci-style world maps. The full-scale dissemination of this style of world map began after this time.

The Buddhist world maps from earlier ages also continued to be produced as before. In 1710, the *Nansenbushū Bankoku Shōka no Zu* 南瞻部洲萬國掌菓之圖 [Visualized Map of All the Countries in Jambudvīpa] was the first Buddhist world map to be issued as a separate publication, and Buddhist world maps continued to be published until the end of early modern times in the

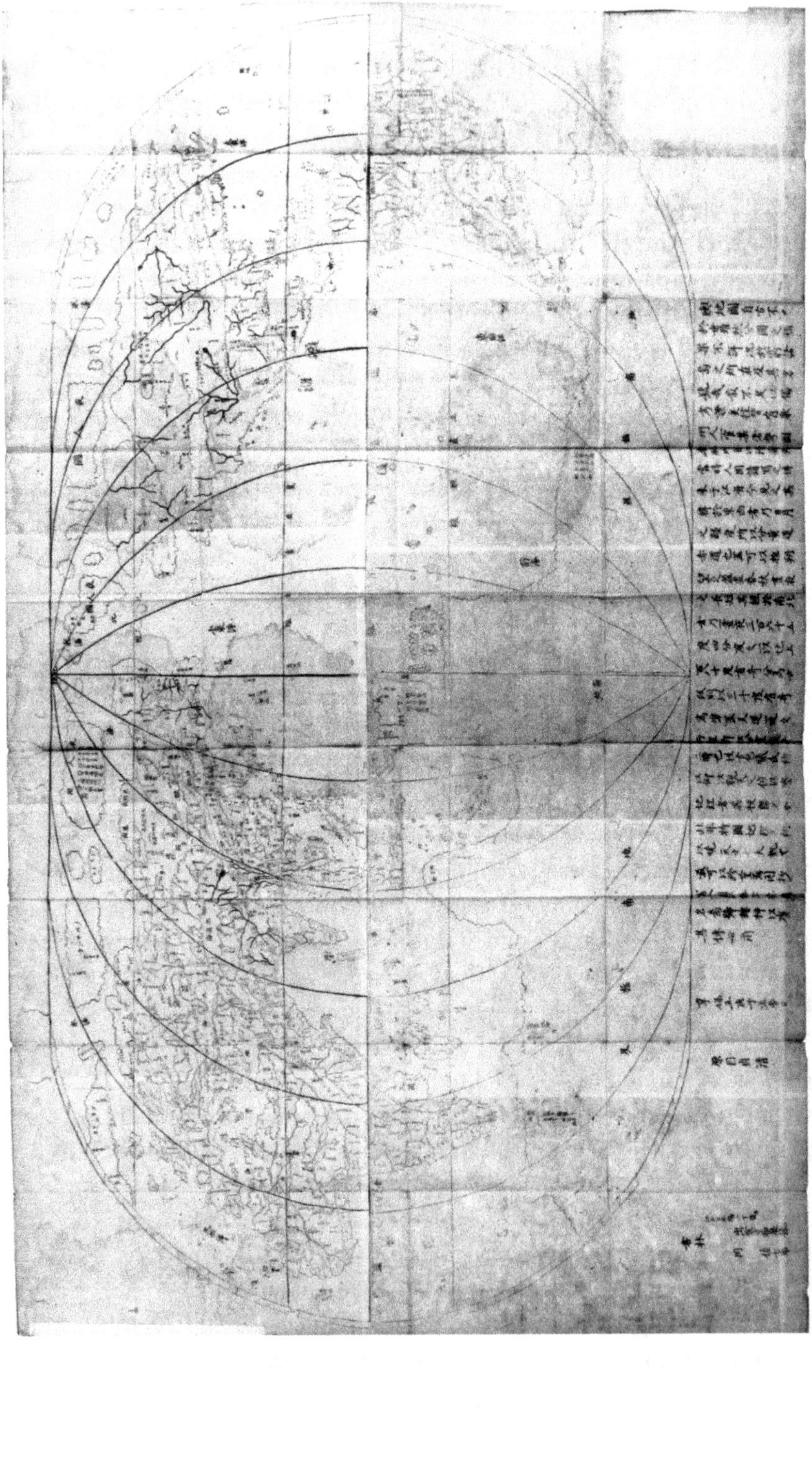

FIGURE 5.3 *Yochi Zu* 輿地圖, 1720, Kobe City Museum collection.

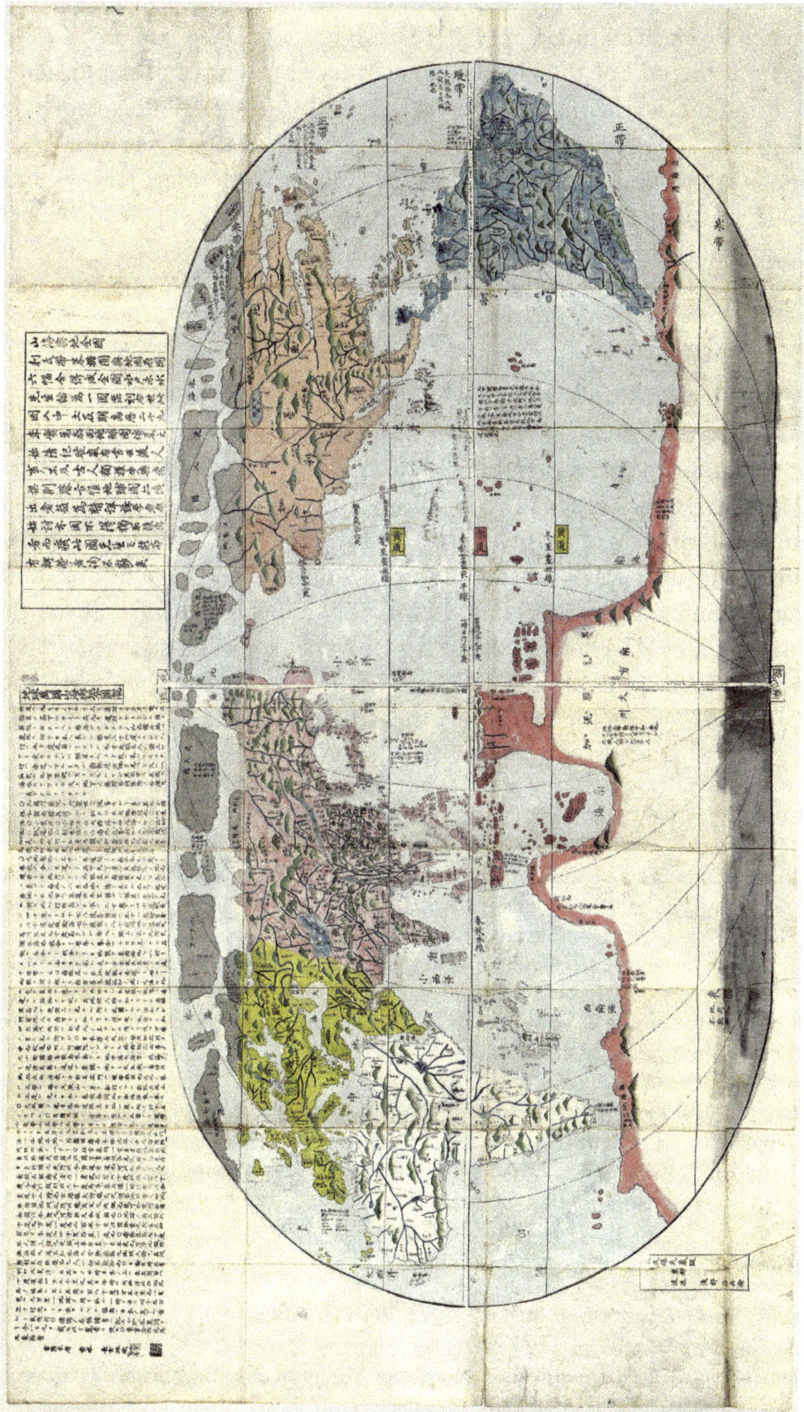

FIGURE 5.4 Sekisui World Map (revised edition) 地球萬國山海輿地全圖説 [Map with an Account of All the Countries, Lands and Seas in the World], late eighteenth century, National Museum of Japanese History collection.

mid-nineteenth century. They had considerable significance in expressing resistance to Western culture.[2]

Thus, from the end of the eighteenth century, three types of world maps existed: Buddhist world maps, Ricci-style world maps, and *Rangaku*-style world maps that entered Japan after 1720. Those referred to here as Ricci-style world maps are world maps based on the 1602 Ricci World Map. However, as explained in detail below, two types of Ricci's world map existed: the 1602 Ricci World Map and the map of the eastern and western hemispheres. We have to distinguish the two in discussing the history of world maps in Japan.

2 The First Edition of the 1602 Ricci World Map and Its Revisions

Currently, a total of six examples thought to be the first edition of the 1602 Ricci World Map exist. One is in the Vatican Library; three are in Japan (in the Miyagi Prefectural Library,[3] Kyoto University Library,[4] and the National Archives of Japan[5]); one was previously in the collection of Philip Robinson;[6] and one recently came into possession of the James Ford Bell Trust and is currently stored at the James Ford Bell Library at the University of Minnesota (hereinafter referred to as the Minnesota Ricci World Map)[7] (Figure 5.5). They are all the same edition (hereinafter referred to as the extant original edition).

2 MUROGA Nobuo and UNNO Kazutaka, "Edojidai kōki ni okeru bukkyō kei sekaizu" 江戸時代後期における仏教系世界図 [Buddhist World Maps in the Late Edo Period], *Chirigakushi Kenkyū* 地理学史研究 2 (February 1962): 189–219.

3 Miyagi Prefectural Library, "Eichi No Mori Web," https://eichi.library.pref.miyagi.jp/da/detail?data_id=041-70138-1-p1 (accessed June 2, 2022).

4 Kyoto University Library, "Rare Materials Digital Archive," https://rmda.kulib.kyoto-u.ac.jp/en/item/rb00013547 (accessed June 2, 2022).

5 National Archives of Japan, "DIGITAL ARCHIVE," https://www.digital.archives.go.jp/img.L/4162882; https://www.digital.archives.go.jp/img.L/4162883; https://www.digital.archives.go.jp/img.L/4162884; https://www.digital.archives.go.jp/img.L/4162885; https://www.digital.archives.go.jp/img.L/4163700; https://www.digital.archives.go.jp/img.L/4163701; https://www.digital.archives.go.jp/img.L/4163702; https://www.digital.archives.go.jp/img.L/4163703; https://www.digital.archives.go.jp/img.L/4163704; https://www.digital.archives.go.jp/img.L/4163705; https://www.digital.archives.go.jp/img.L/4163706; https://www.digital.archives.go.jp/img.L/4163707 (accessed June 2, 2022).

6 Sotheby's, *The Library of Philip Robinson Part II The Chinese Collection* (London: Sotheby's, 22nd November 1988), 79–80.

7 The James Ford Bell Library at the University of Minnesota, https://umedia.lib.umn.edu/item/p16022coll251:8823 (accessed June 2, 2022).

Recent study, however, has pointed out that the extant original edition is not the true first edition.[8] That is to say, it is considered to be the second revised version. Although the true first edition and the first revised version are not extant today, their existence can be traced from many hand-drawn copies of the 1602 Ricci World Map that are still extant in Japan. Based on the above, and adding new findings, we consider in detail the different versions of the 1602 Ricci World Map in this section.

Detailed observation of the above-mentioned six examples, namely the extant original edition, shows evidence of revision in the printing blocks.[9] The first evidence is in Li's preface, which is inscribed on the Pacific Ocean. Specifically, the texts, "唐賈南皮畫寸" [the Tang (dynasty cartographer) Jia Nanpi (i.e., Jia Dan 賈耽 (730–805)'s (method of) drawing inches to divide miles], in the first line and, "異人異書世不易遘" [(As for) different peoples and different writings, (in) the world it is not easy to encounter (them)], in the twenty-second line have obviously been altered through inlay work, which is a revision technique in which the portion of the wood block to be revised is removed and new wood pieces are inlaid and re-engraved (Figure 5.6). Second, signs of the removal of an explanatory note off the shore of Portugal can be seen (Figure 5.7). A few *kanji* in the removed note are just barely legible. Third, although not previously indicated, signs of the removal of five *kanji* located virtually in the center of the Red Sea can be seen (Figure 5.8). The final *kanji* was probably "河," meaning river.

If no map had been printed prior to these alterations, one could say that the extant original edition is the first edition. If these alterations were made after printing some maps, however, that would suggest it was a revised version.

The following facts are revealed when viewing hand-drawn copies[10] of the 1602 Ricci World Map existing in Japan with regard to the above alterations in the printing blocks.[11] First, the kanji text in two segments in Li's preface

---

8 Unno Kazutaka, "Ri Matō Kon'yo Bankoku Zenzu no shohan" 利瑪竇『坤輿万国全図』の諸版 [Different Editions of Matteo Ricci's *Kunyu wankuo quantu*], *Toyo Gakuhō* 東洋学報 87, no. 1 (June 2005): 101–143.

9 Unno Kazutaka, *Chizu no bunkashi: sekai to Nihon* 地図の文化史－世界と日本－ [Cultural History of Maps: The World and Japan] (Tokyo: Yasaka Shobō, 1996), 140–145; idem, "Different Editions of Matteo Ricci's *Kunyu wankuo quantu*."

10 For instance, two examples of the hand-drawn copy are available for viewing on the following websites: Miyagi Prefectural Library, "Eichi No Mori Web," https://eichi.library.pref.miyagi.jp/da/detail?data_id=041-70139-1-p1 (accessed June 2, 2022) and Tohoku University Library, "Image database of the Kano Collection," https://www.i-repository.net/contents/tohoku/kano/ezu/kon/kon.html (accessed June 2, 2022).

11 Aoki Chieko, "Nihon ni genzon suru Kon'yo Bankoku Zenzu shozu ni tsuite" 日本に現存する「坤輿万国全図」諸図について [Extant *Kunyu wankuo quantu* in Japan],

FIGURE 5.5 Minnesota Ricci World Map 坤輿萬國全圖 *Kunyu wanguo quantu*, 1602. James Ford Bell Library, University of Minnesota.

坤輿萬國全圖

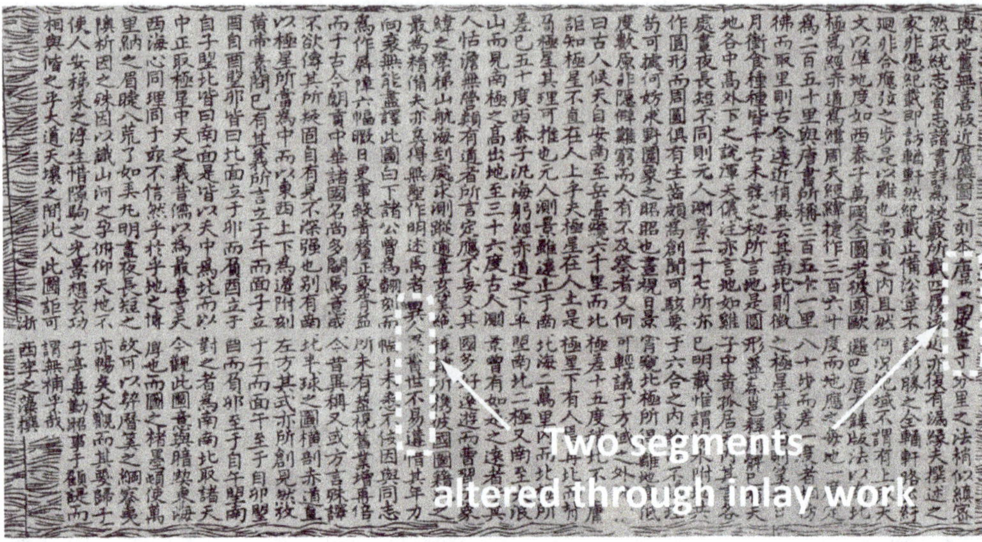

FIGURE 5.6 Li's preface on the extant original edition of the 1602 Ricci World Map. Detail from 坤輿萬國全圖（版本） *Kunyu wanguo quantu* (*banben*/printed edition). Miyagi Prefectural Library collection.

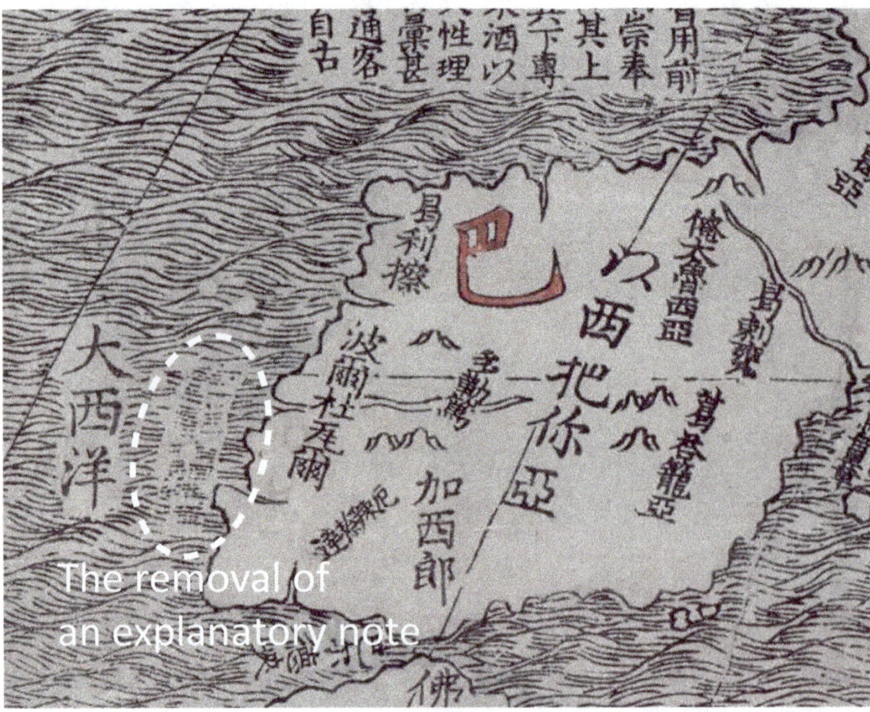

FIGURE 5.7 Area around the Iberian Peninsula in the extant original edition of the 1602 Ricci World Map. Detail from 坤輿萬國全圖（版本） *Kunyu wanguo quantu* (*banben*/printed edition). Miyagi Prefectural Library collection. Signs of the removal of an explanatory note off the shore of Portugal can be seen.

FIGURE 5.8 Area around the Red Sea in the extant original edition of the 1602 Ricci World Map. Detail from 坤輿萬國全圖（版本） *Kunyu wanguo quantu* (*banben*/printed edition). Miyagi Prefectural Library collection. Signs of the removal of five kanji at the center of the Red Sea can be seen.

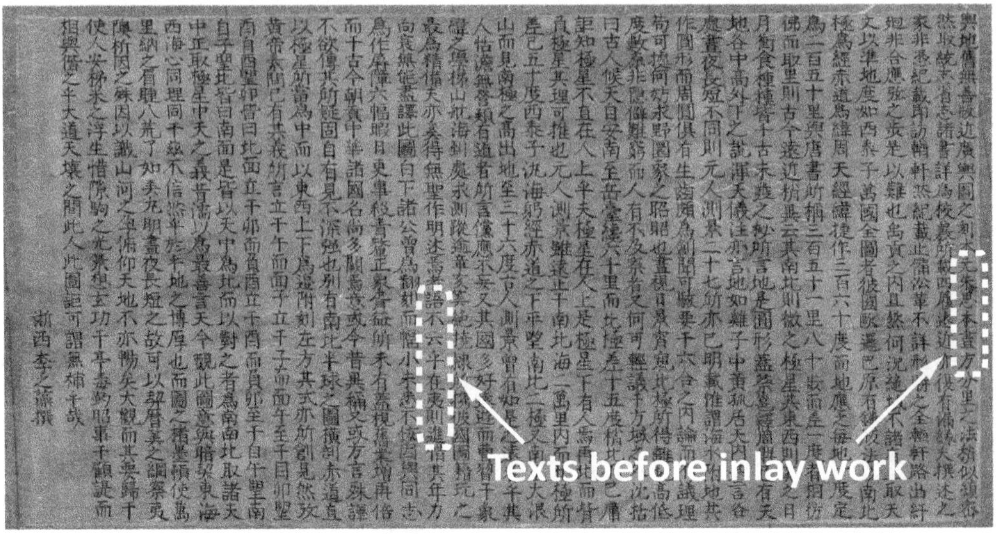

FIGURE 5.9 Li's preface on the hand-drawn copy of the 1602 Ricci World Map. Detail from 坤輿萬國全圖（写本着色） *Kunyu wanguo quantu* (*xieben*/color manuscript). Miyagi Prefectural Library collection.

differs from the above-mentioned examples. Namely, "元朱思本畫方" [the Yuan (dynasty geographer) Zhu Siben drew the grid] and "語不云乎在夷則進" [do the words say nothing? When (he) is in a foreign land, (he) carries (it) out] (Figure 5.9). Second, the explanatory note, "拂郎機乃回回誤稱本名波尓杜曷尓" [Folangji is the mistaken name given (to the Portuguese) by the Muslims. The original name (of the Portuguese) is Bo-er-du-he-er], appears in the same location off the shore of Portugal, coinciding with the kanji that are just barely legible in the extant original edition (Figure 5.10).

Seen in this light, this would seem to indicate that some hand-drawn copies of the 1602 Ricci World Map that exist in Japan represent the content of the Ricci Map prior to the alterations of the printing blocks. That is to say, the original of these hand-drawn copies was made before the alteration of the printing blocks. The *kanji* text on the Red Sea is, however, thought to have been removed even before the first printing, because it does not appear in any of the

Kirishitan Bunka Kenkyūkai Kaihō キリシタン文化研究会会報 102 (October 1993): 1–12; idem, "Nihonjin to Kon'yo Bankoku Zenzu" 日本人と坤輿万国全図 [Japanese People and the *Kunyu wankuo quantu*], in *Sekaizu yūran: Kon'yo Bankoku Zenzu to higashi Ajia* 世界図遊覧―坤輿万国全図と東アジア—[Excursion in World Maps: *Kunyu wankuo quantu* and East Asia], ed. Tsuchiura City Museum (Tsuchiura: Tsuchiura City Museum, 1996), 8–11; Unno Kazutaka, *Cultural History of Maps: The World and Japan*; idem, "Different Editions of Matteo Ricci's *Kunyu wankuo quantu*."

FIGURE 5.10 Area around the Iberian Peninsula in the hand-drawn copy of the 1602 Ricci World Map. Detail from 坤輿萬國全圖（写本着色）*Kunyu wanguo quantu* (*xieben*/color manuscript). Miyagi Prefectural Library collection. Note the explanatory note off the shore of Portugal.

published versions of the 1602 Ricci World Map nor in any of the hand-drawn copies. Its removal was presumably the modification of a somewhat simple mistake in the process of engraving.

Besides these three segments, it is also possible to identify other points in these hand-drawn copies that differ from the extant original edition.[12] To cite the major points, first, the country name for Portugal on the Iberian Peninsula is noted as "拂郎機," Fulangji (Figure 5.10) while it is noted as "波爾杜瓦爾," Bo'erduwa'er (Figure 5.7) in the extant original edition. This change was accompanied by the removal of the explanatory note off the shore of Portugal. Second, the explanation of the eighth layer of heaven in the figure of the nine layers of heaven is given as 7,000 years while it is noted as 49,000 years in the

12 Aoki Chieko, "Extant *Kunyu wankuo quantu* in Japan"; idem, "Japanese People and the *Kunyu wankuo quantu*"; Unno Kazutaka, "Different Editions of Matteo Ricci's *Kunyu wankuo quantu*."

extant original edition. These are also alterations using inlay work and these hand-drawn copies indicate the state prior to the alterations of the printing blocks.

Based on the above, the extant original edition is undoubtedly a revised version with the following four revisions: the two segments in Li's preface; the explanatory note located off the shore of Portugal; the name for the country of Portugal; and the explanation of the eighth layer of heaven in the figure of the nine layers of heaven. From a different perspective, the original map of some hand-drawn copies could be thought to be a true first edition, the existence of which today has still not been confirmed.

However, observing the hand-drawn copy in the Toyo Bunko collection in Tokyo, Li's preface is the same as that in the extant original edition, that is, in the state subsequent to revision, while the explanatory note located off the shore of Portugal and the rendering of "Portugal" are the same as that in other hand-drawn copies, that is, as prior to revision. The figure of the nine layers of heaven cannot be identified because it is lacking in the hand-drawn copy. It is, however, also evidently in a state prior to revision, judging from the hand-drawn copy in the Nanjing Museum collection in China, in which the two above segments are in the same state as this.

Based on this research, Li's preface was revised before the map was printed. In other words, the hand-drawn copy in the Toyo Bunko collection predates the printing of the extant original edition. The notes relating to Portugal and the figure of the nine layers of heaven were later revised and printed as what we recognize as the extant original edition. To recap, the map was revised twice. Based on this content analysis, the first revision is thought to have been done by Li himself, and the latter, or second revision, by Ricci.

A summary of the examination undertaken thus far would proceed as follows: First is the stage with five *kanji* located virtually in the center of the Red Sea—this is referred to as Version 0 because of having not been printed. Next, the map printed with these *kanji* removed is the original map used for virtually all of the hand-drawn copies and is the first edition—this is referred to as Version 1. Continuing, the version with Li's preface revised by inlay work is the original map used for the hand-drawn copy in the Toyo Bunko collection and is the first revised version of the first edition—this is referred to as Version 1a. The map with inlaid revisions by Ricci of the Portugal-related notes and the figure of the nine layers of heaven is the extant original edition, otherwise known as the second revised version of the first edition—or Version 1b. Thus, aside from Version 0, which was not printed, there were three different versions of the 1602 Ricci World Map, i.e., Versions 1, 1a, and 1b. The extant original edition, which has been regarded as the first edition, was actually Version 1b, or the second revised version of the first edition; two other versions had existed prior to that.

TABLE 5.1 Versions of Matteo Ricci's 1602 World Map in Chinese, with location and defining characteristics (Institutions listed are in Japan unless otherwise noted)

Version		Location (and defining characteristics)
V0		not extant
V1	printed	not extant (*Kanji* removed from Red Sea)
V1	hand-drawn	Tamba Sasayama City Museum of History and Art Tsuchiura City Museum
V1	hand-drawn supplemented	Miyagi Prefectural Library Tohoku University Library (and many additional locations)
V1a	printed	no longer extant (Li's preface is revised)
V1a	hand-drawn	Toyo Bunko Nanjing Museum, China
V1b	printed	"Original Extant Edition" (with revisions off the shore of Portugal, and to the layers of heaven): Vatican Library, Vatican Miyagi Prefectural Library Kyoto University Library National Archives of Japan James Ford Bell Library at University of Minnesota, USA etc.
V1b	hand-drawn	Kameyama City Historical Museum Tokushima University Library

3 Hand-Drawn Copies of and Supplements to the 1602 Ricci World Map in Japan

Numerous hand-drawn copies of the 1602 Ricci World Map were produced in Japan. When observing them in light of the revision of the 1602 Ricci World Map examined in the previous section, they are evidently hand-drawn copies of Versions 1, 1a, and 1b. However, faithful hand-drawn copies of these versions are actually very few; most hand-drawn copies have content that differs.

It is natural that the differences include corrected errors in the Japan portion and the addition of *kana* phonetic pronunciations to place names written in *kanji*, or Chinese characters, in the 1602 Ricci World Map, since the hand-drawn copies were produced for the Japanese in Japan. Aside from these are the following three major differences: First, banks, coral reef islands, and so forth have been added in the South China Sea (Figure 5.11), Indian Ocean,

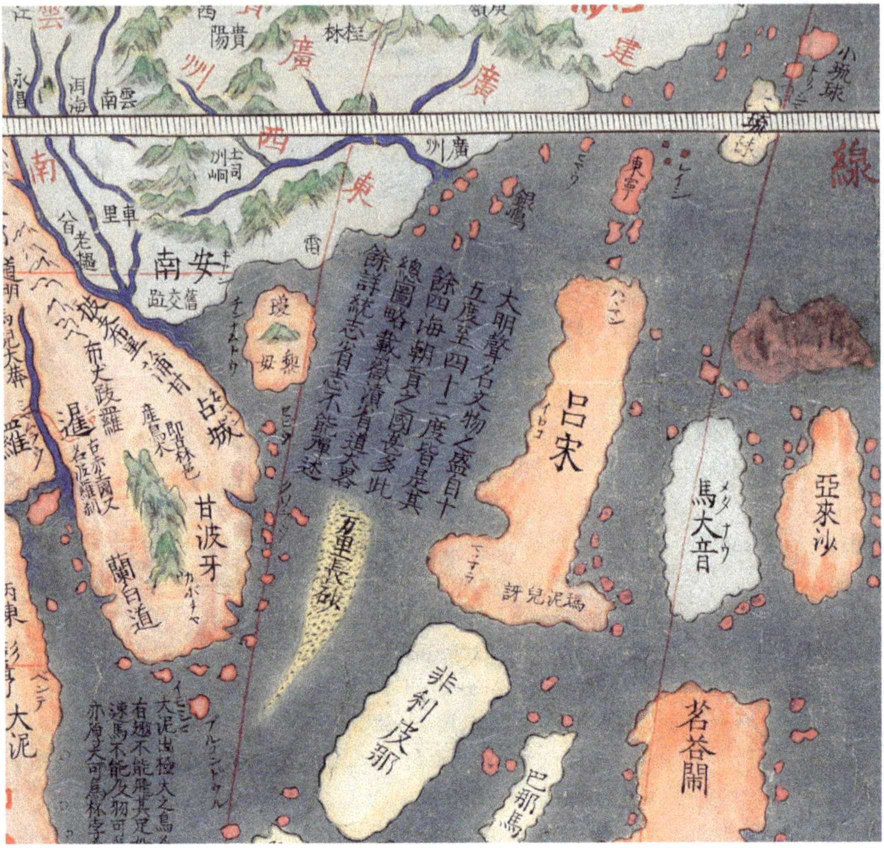

FIGURE 5.11 Area around Southeast Asia in the hand-drawn copy of the 1602 Ricci World Map. Detail from 坤輿萬國全圖（写本着色）*Kunyu wanguo quantu* (*xieben*/color manuscript). Miyagi Prefectural Library collection. The phrase 万里長砂 *wanli changsha*, "thousands of miles of sand," has been added off the eastern coast of the Indochina Peninsula.

Northern Atlantic Ocean off the northeast coast of North America, and off the shore of Peru. Second, a *Kin-jima* (Gold Island) inscription has been added in the sea east of Japan (Figure 5.12) and, third, the name "*Dong-níng* 東寧," Eastern Tranquility, has been appended to an island southwest of Dai-Ryūkyū 大琉球, the Ryukyu Islands (Figure 5.11). If we call the hand-drawn copies with this supplemented information "supplemented hand-drawn copy of Version 1," since they are all based on Version 1 as the original map, then four different types of the hand-drawn copy exist: that of Version 1, Version 1 supplemented, Version 1a, and Version 1b.

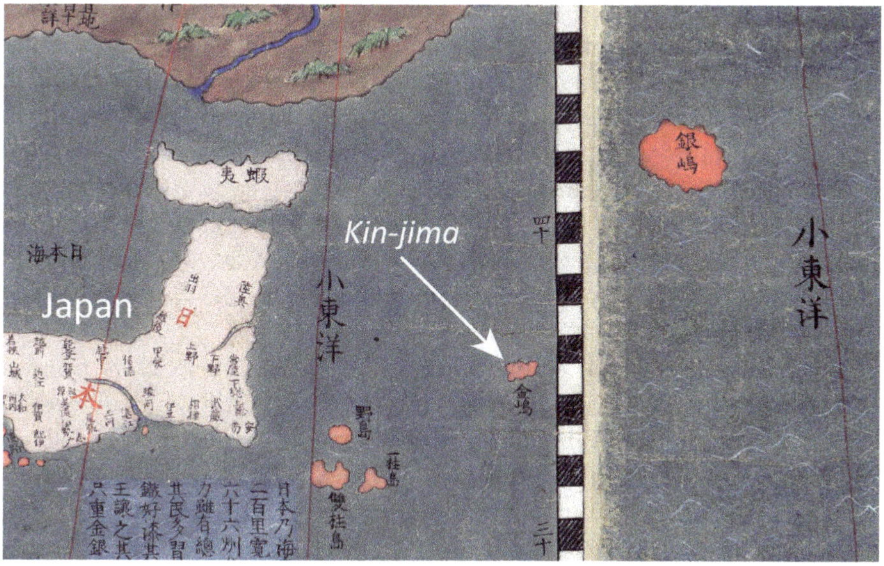

FIGURE 5.12 Area around *Kin-jima*, or "Gold Island," in the hand-drawn copy of the 1602 Ricci World Map. Detail from 坤輿萬國全圖（写本着色）*Kunyu wanguo quantu* (*xieben*/color manuscript). Miyagi Prefectural Library collection.

The hand-drawn copy in the Toyo Bunko collection, a Version 1a hand-drawn copy, was purchased in Beijing and brought to Japan in the early 1930s.[13] This hand-drawn copy did not exist in early modern Japan, as was also the case for the other Version 1a hand-drawn copy (held in the Nanjing Museum collection). Consequently, only three of the four types of hand-drawn copy existed in early modern Japan. Yet, more than twenty hand-drawn copies have been confirmed to date. Most of them are Version 1 *supplemented* hand-drawn copies. In contrast, the only Version 1 hand-drawn copies that are not supplemented are the Sasayama Ricci World Map and the hand-drawn copy previously owned by the Yamamura Family (Tsuchiura City Museum collection), and the only Version 1b hand-drawn copies are the hand-drawn copy previously owned by the Yura Family (Kameyama City Historical Museum collection) and the hand-drawn copy in Tokushima University Library (hereinafter referred to as the Tokushima Ricci World Map), as explained in detail below.

Version 1b of the 1602 Ricci World Map certainly existed in early modern Japan, as seen in the collections of the Miyagi Prefectural Library, Kyoto

13 Akioka Takejirō, *Chizugaku shi* 地図学史 [History of Cartography] (Tokyo: Iwanami Shoten, 1934), 21.

University Library, and the National Archives of Japan. However, there were only a few Version 1b hand-drawn copies. The only example mentioned so far was the hand-drawn copy previously owned by the Yura Family. This research, however, has clarified that the Tokushima Ricci World Map was also a Version 1b hand-drawn copy and that the original map from which it was copied by hand was, very interestingly, the Minnesota Ricci World Map, as detailed below. Only in this case can we clearly identify the direct correspondence between the original and the hand-drawn copy as we have acquired new material for considering the process of reproducing and disseminating the 1602 Ricci World Map in Japan.

Specifically, if we look at the northeast of North America, we can see that abrasions on the original, the Minnesota Ricci World Map, are exactly reproduced on the Tokushima Ricci World Map. The same can be said about the lacunae of the Chinese characters towards the left of these abrasions (Figure 5.13). A careful look at the Chinese character "南," south, in South America of the original shows that it was not printed but was handwritten later. The Tokushima Ricci World Map does not have the Chinese character "南" on it, because it was copied before the Chinese character was written on the original (Figure 5.14). Many such indications found between the two maps suggest that the Tokushima Ricci World Map is definitely a direct hand-drawn copy of the Minnesota Ricci World Map.

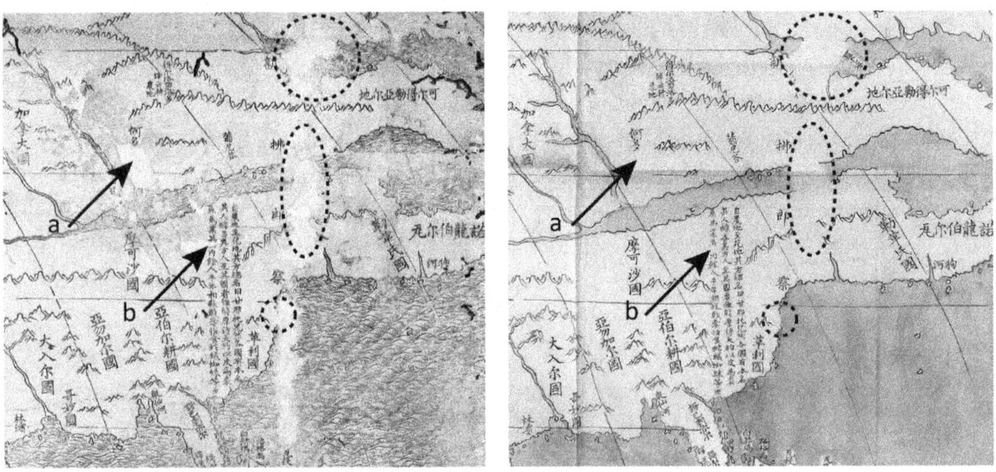

FIGURE 5.13 Comparison of the abrasions and the Chinese characters on the northeast of North America between the Minnesota Ricci World Map (James Ford Bell Library, University of Minnesota) on the left and the Tokushima Ricci World Map (Tokushima University Library collection) on the right. Areas surrounded by dotted lines show the abrasions. Letter a) shows where three Chinese characters are missing. Letter b) indicates where the Chinese character "山" is missing.

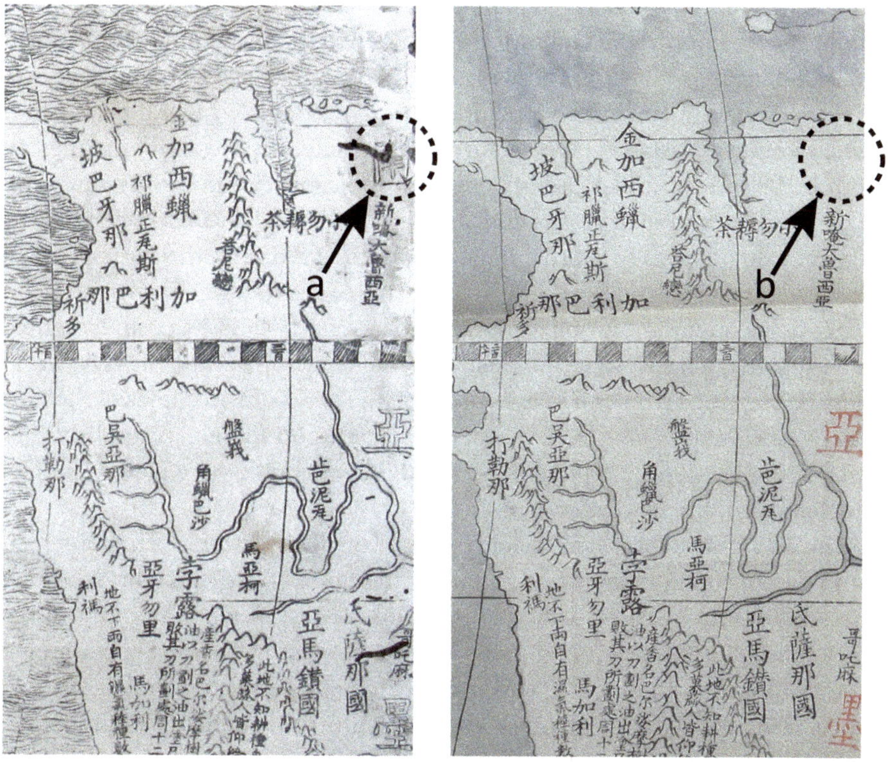

FIGURE 5.14 Comparison of the Chinese character "南," or "south," on South America as depicted in the Minnesota Ricci World Map, left (James Ford Bell Library, University of Minnesota) and the corresponding area in the Tokushima Ricci World Map, right (Tokushima University Library collection). Letter a) indicates where the Chinese character "南" was later handwritten. Letter b) shows where the Chinese character "南" is missing.

The Tokushima Ricci World Map was originally owned by the feudal lord of the Hachisuka Family that ruled Awa and Awaji Provinces, the eastern part of Sikoku and the neighboring Awaji Island, in early modern times. While the exact date of the hand-drawn copy is still unknown, the above findings have revealed that the Minnesota Ricci World Map had definitely existed in early modern Japan, particularly presumably around Awa and Awaji Provinces, and that it had probably been owned by the Hachisuka Family or those who had some kind of relationship with them. As a matter of fact, a certain trader of the map showed me it in Kyoto in 2007 before it was sold, and then told me that it was probably discovered in Sikoku and vicinity which included Awa and Awaji.

The Version 1 supplemented hand-drawn copies, which are more than mere duplications, inform us as follows. First, in regard to the supplemented depictions of banks, coral reef islands, and so forth, it is necessary to be attentive to the maps' relationship to previously produced *Nanban*-style world maps, since banks, coral reef islands, and so forth are depicted in the same manner on both. To be sure, portolan charts were also produced prior to the Version 1 supplemented hand-drawn copies and also depicted such features; however, they generally do not go beyond the range of Southeast Asia and the Indian Ocean. North America and the Northern Atlantic Ocean are not even shown on these maps. Consequently, a strong possibility exists that the Version 1 supplemented hand-drawn copies were supplemented while referring to *Nanban*-style world maps or maps of a similar lineage. Although the hand-drawn copies of the 1602 Ricci World Map have heretofore been looked upon merely as duplications, at least the Version 1 supplemented hand-drawn copies were clearly supplemented by completely different maps.

The addition of *Kin-jima* (Gold Island), on the Version 1 supplemented hand-drawn copies, relates to exploration in search of the island by Sebastián Vizcaíno in 1611. Vizcaíno, who explored the sea east of Japan in quest of the El Dorado of Asia, which had aroused wide interest in Europe, encountered storm winds time after time and returned to Japan without confirming the existence of the island. Then Japanese came to know the legend of El Dorado and think that the Gold Island might exist in the sea east of Japan. Consequently, depiction of Gold Island indicates that the Version 1 supplemented hand-drawn copies were produced subsequent to that. However, the time of their production can be narrowed further by the place name of *Dong-níng*, the third discrepancy.[14] *Dong-níng* 東寧 was a name given to Formosa in 1664, meaning that the Version 1 supplemented hand-drawn copies were produced after that.

How these supplementary depictions come into being must then be considered. The pair of world map folding screens in Tamba Sasayama provides material for considering this issue. One of them is the Sasayama Ricci World Map which is the earliest hand-drawn copy of the 1602 Ricci World Map. As this is a Version 1 hand-drawn copy none of the supplementary depictions given above are present.

The other folding screen of the pair is a world map folding screen (hereinafter referred to as the Sasayama World Map) (Figure 5.15). Although topographical

14 Aoki Chieko, "Extant *Kunyu wankuo quantu* in Japan"; idem, "Mateo Ritchi no sekai chizu: Nihon e no denrai to sono henyō" マテオ・リッチの世界地図—日本への伝来とその変容— [World Maps of Matteo Ricci: Their Introduction into Japan and Transformation], *Sinica* しにか 6, no. 2 (February 1995): 31–39.

descriptions and place names are mostly omitted, the overall configuration of the land masses in this world map is very similar to that of the 1602 Ricci World Map. Differences become conspicuous, however, when observed in detail. For example, as stated above, besides supplementary depictions of banks, coral reef islands, Gold Island, and so forth, mythical Lake Parime is clearly depicted immediately below the Equator in South America, although it was never depicted on the 1602 Ricci World Map (Figure 5.16). The Sasayama World Map was thus very likely influenced by *Nanban*-style world maps.

In addition, islands in Southeast Asia are depicted with realistic shapes, differing greatly from the 1602 Ricci World Map (Figure 5.17). Such realistic geographical knowledge of Southeast Asia may indicate the influence of portolan charts.

The pair of folding screens is to be used as a set, providing the condition for a new single world map which would combine the content of the two. The two screens are presumed to have been produced in about 1675, which is consistent with the time of production of the Version 1 supplemented hand-drawn copies of 1664 or later. Thus, these two world maps may have had some sort of relationship to the production of the Version 1 supplemented hand-drawn copies.

Another currently extant world map very closely resembles the latter of this pair of world map screens: the Sasayama World Map. This world map is thought to have been produced by Shibukawa Shunkai 渋川春海 (1639–1715) (private collection deposited in the Osaka Museum of History) who developed the first major, specifically Japanese calendar, the *Jōkyō* 貞享 (Figure 5.18). Shibukawa was named the government's first chief astronomer in 1684; he also studied *Shintō* and other disciplines. This world map as well may have also been involved in the production of the Version 1 supplemented hand-drawn copies.

Seven, or about one-third, of the more than twenty hand-drawn copies of the 1602 Ricci World Map are located in the Sendai area, most of which are thought to have been produced by the astronomy school of the Sendai-*han* (domain) from the end of the eighteenth century to the mid-nineteenth century. Astronomy in the Sendai-*han* was established by Endō Moritoshi 遠藤盛俊 (1669 or 1672–1734), a pupil of Shibukawa. It was strongly influenced by *Shintō* based on the *On-myō-dō* 陰陽道—the Japanese interpretation of *yin-yang* principles, or the dualism of two basic opposing elements—and, in its latter stage, also had a close relationship to the principles of *Kokugaku* 国学, which sought to clarify the unique spiritual culture of Japan. The 1602 Ricci World Map was considered to show that Japan was land suitable for rice cultivation and human habitation, because it said that Japan was located in the mid-latitudinal belt with a temperate climate. It was, therefore, considered to

FIGURE 5.15 Sasayama World Map, 1675. Private collection (deposited in the Tamba Sasayama City Museum of History and Art). Although topographical descriptions and place names are mostly omitted, the overall configuration of the land masses in this world map is very similar to that of the 1602 Ricci World Map. The museum calls this map the "世界図屏風" [The Sekaizu folding screen], thereby distinguishing it from the other Sasayama World Map in its collection (shown in fig. 5.1).

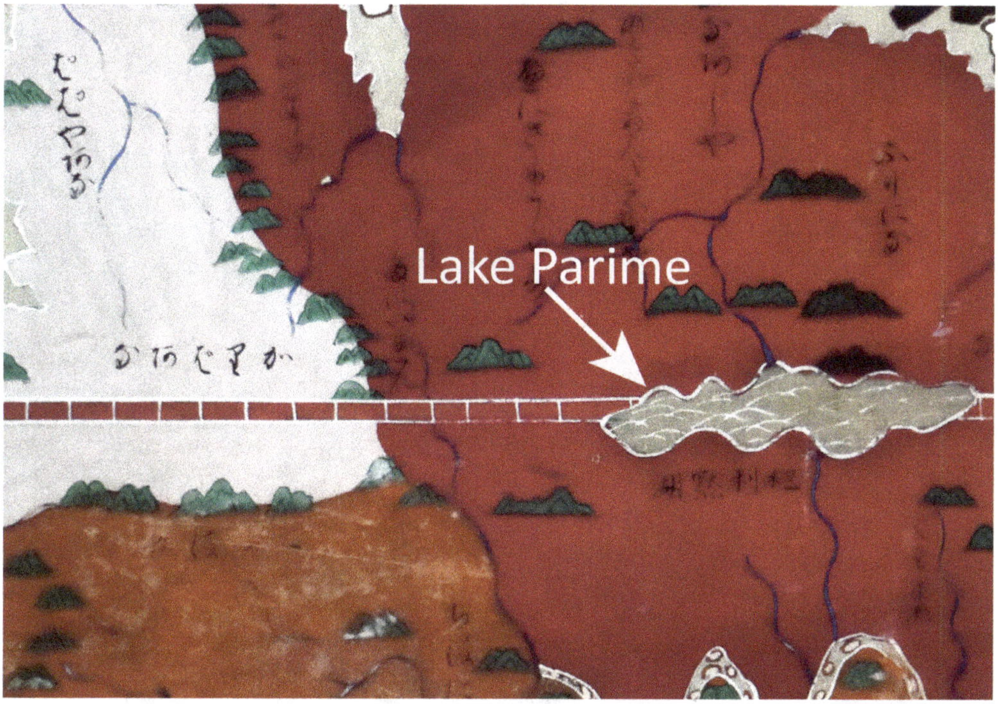

FIGURE 5.16 Lake Parime in South America in the Sasayama World Map, private collection (deposited in the Tamba Sasayama City Museum of History and Art). The mythical Lake Parime, which was absent on the 1602 Ricci World Map, is depicted immediately below the Equator in South America.

correspond with the thought of *Kokugaku* that was based on a view of Japan as the "*Toyo Ashi-hara no Mizuho no Kuni* 豊葦原の瑞穂の国," meaning a land of bountiful rice harvests, and regarded the view as a basis of the unique spiritual culture of Japan. In addition, it was also highly esteemed in terms of rivalry with the *Rangaku* school because of its being written in *kanji*. Accordingly, many hand-drawn copies of it were made by the astronomy school of the Sendai-*han*.[15]

To summarize, these new findings reveal the hand-copied versions of the Ricci Map as having value as historical materials in and of themselves, particularly in revealing what the Tokushima Ricci World Map was copied from, where the Minnesota Ricci World Map was located before now, and when it

15 AOKI Chieko, "Extant *Kunyu wankuo quantu* in Japan"; idem, "World Maps of Matteo Ricci: Their Introduction into Japan and Transformation"; idem, "Japanese People and the *Kunyu wankuo quantu*."

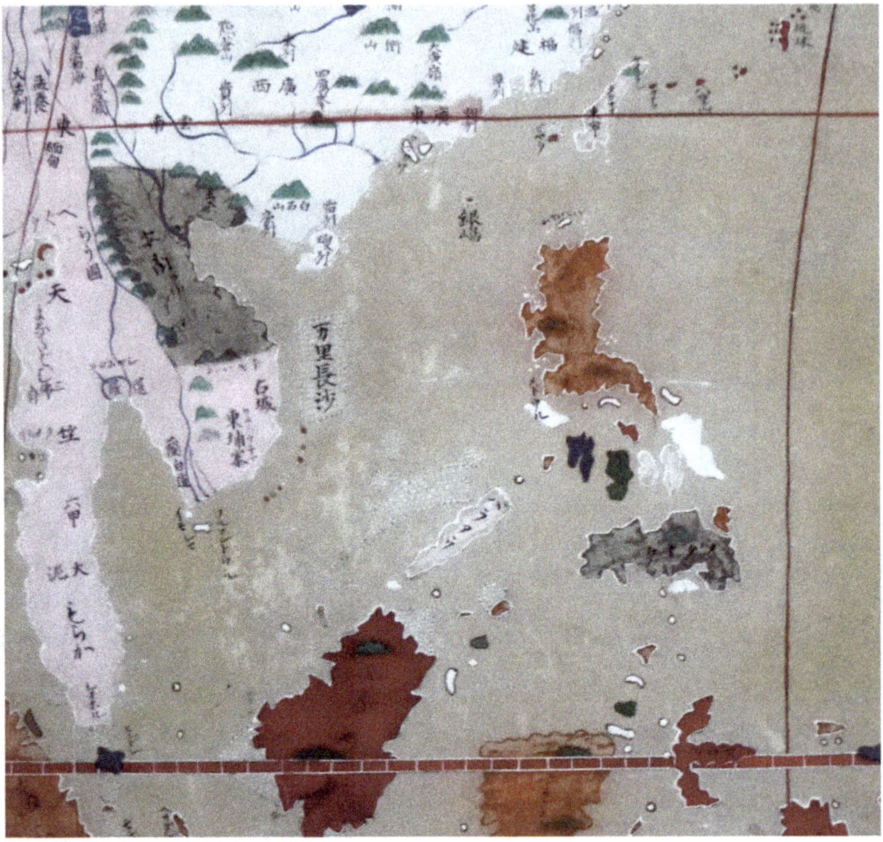

FIGURE 5.17 Area around Southeast Asia in the Sasayama World Map, private collection (deposited in the Tamba Sasayama City Museum of History and Art). Islands in Southeast Asia such as Luzon and Kalimantan are depicted with realistic shapes, unlike in the 1602 Ricci World Map.

existed there. Secondly, we could understand the social appreciation for the 1602 Ricci World Map, namely, how people viewed it in those days, since the Tokushima Ricci World Map was copied carefully, conscientiously and faithfully. Thirdly, inquiring into the background of the two maps would suggest a network of people concerned and the linkage of knowledge in society through the transmission of knowledge of the 1602 Ricci World Map.

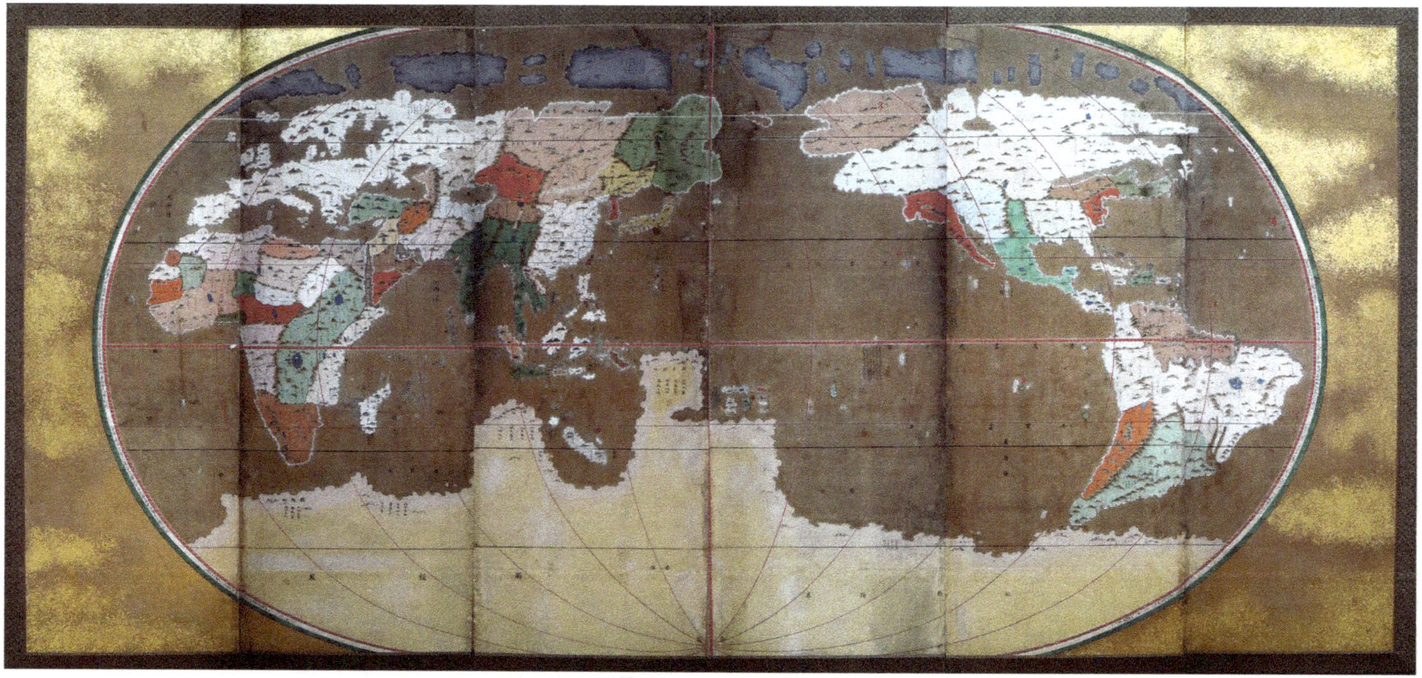

FIGURE 5.18 World map produced by Shibukawa Shunkai, ca. 1698, private collection (deposited in the Osaka Museum of History).

4 Ricci's Map of the Eastern and Western Hemispheres and the Century of the Bankoku Sōzu

The 1602 Ricci World Map is thought to have appeared in Japan soon after publication in Beijing due to the *Bankoku Sōzu* having been produced in 1645, and the assumption that it was based on the 1602 Ricci World Map; this view, however, needs to be reexamined.[16]

One becomes aware when observing the land masses depicted on the *Bankoku Sōzu* that high latitude land masses, especially the Arctic Ocean islands painted black, converge at two points to the east and west of the northern extremity of the map. In addition, an unusually large space also exists between Eurasia and North America. These land masses would not have been depicted in this manner if the map had been produced based on the 1602 Ricci World Map.

The depictions of the *Bankoku Sōzu*, which can be divided into two segments, and the convergence of the Arctic Ocean islands at the center of the northern extremity of each of those two segments, strongly suggest that the model for the *Bankoku Sōzu* was a map of the eastern and western hemispheres depicted as two circles. The two circles seem to have been connected together, creating an oval-configured world map with virtually no regard for projection conversion. That is to say, the two segments are the eastern and western hemisphere respectively, and the large space stems from a space between the two circles.

Seen in this light and considering the time period, the map that served as the original model of the *Bankoku Sōzu* would most likely have been Ricci's map of the eastern and western hemispheres. Although this map no longer exists, it was reproduced in the *Fangyu shenglüe* 方輿勝略 [Compendium of Geography] (Figure 5.19), which was published in China in 1612 and brought to Japan by the mid-seventeenth century.

Observing Luzon and other Southeast Asian islands on the *Bankoku Sōzu*, we see that they are depicted with more accurate configurations than in the 1602 Ricci World Map. These configurations could not have been depicted if they had been based on the 1602 Ricci World Map. They were more than likely produced while referring to practical nautical charts based on realistic geographical knowledge: the portolan charts. In addition, depicting the Maldives,

16 UNNO Kazutaka, "Shōhō kan Bankoku Sōzu no seiritsu to rufu" 正保刊「万国総図」の成立と流布 [Establishment and Distribution of the *Bankoku Sōzu* Published in the *Shōhō* Period], in *Nihon yōgaku shi no kenkyū* 日本洋学史の研究 [Research in the History of Western Studies in Japan], ed. ARISAKA Takamichi (Tokyo: Sōgensha, 1991), 10: 9–75; AOKI Chieko, "World Maps of Matteo Ricci: Their Introduction into Japan and Transformation."

a group of atolls in the Indian Ocean, and Gold Island, in the sea east of Japan, would also have been impossible based on the 1602 Ricci World Map. In particular, the former was more than likely produced while referring to *Nanban*-style world maps. Some place names on the *Bankoku Sōzu* also have *kana* phonetic pronunciations drawn directly from the original European languages instead of the *kanji* place names from the 1602 Ricci World Map. Multiple indications exist that the *Bankoku Sōzu* was produced based on sources of information differing from those of the 1602 Ricci World Map.

It follows from this that the *Bankoku Sōzu* was produced mainly on the basis of Ricci's map of the eastern and western hemispheres, being supplemented by maps, information, and so forth already available in Japan, including portolan charts and *Nanban*-style world maps. The *Bankoku Sōzu* can, therefore, be classified as a world map within the lineage of Ricci's map of the eastern and western hemispheres rather than one of the lineages of the 1602 Ricci World Map.

Although the *Bankoku Sōzu* was a published map, some of the place names are handwritten. Because the map was produced for learning in the private school of Higuchi Kentei 樋口謙貞 (1601–84) the intent was to have the pupils studying world geography add in some of the place names later at the time of their completing the school course.[17] It was, therefore, used only in the school but not widely disseminated in broader society. In contrast, maps not expecting additional entries, that is, wood blocks fully printed from the beginning—full sheet printed versions—were subsequently published in succession for use by the public.

As a result, the *Bankoku Sōzu* and other world maps of the same lineage were distributed widely in society from the mid-seventeenth to the mid-eighteenth century, through their publication in at least several versions, in major cities such as Edo (present Tokyo), Kyoto, and Osaka, and as illustrations in dictionaries for commoners, called *Setsuyō-shū* 節用集, which had a large circulation. This era of wide distribution came to be known as the Century of the *Bankoku Sōzu*. Ricci's map of the eastern and western hemispheres had a strong, but not exclusive, impact on Japan through these world maps in the early Tokugawa.

The *Bankoku Sōkai Zu* 萬國総界圖 "General World Map" first published in 1688 with a second edition in 1708, is a world map produced by Ishikawa Ryūsen 石川流宣 (years of birth and death unknown), an *ukiyo-e* artist, long been considered to be an inferior imitation of the *Bankoku Sōzu*. Close observation reveals numerous points differing from the *Bankoku Sōzu*, which are particularly evident in the depiction of East Asia. In the *Bankoku Sōkai Zu*, East Asia

17 Unno Kazutaka, "Establishment and Distribution of the *Bankoku Sōzu* Published in the *Shōhō* Period."

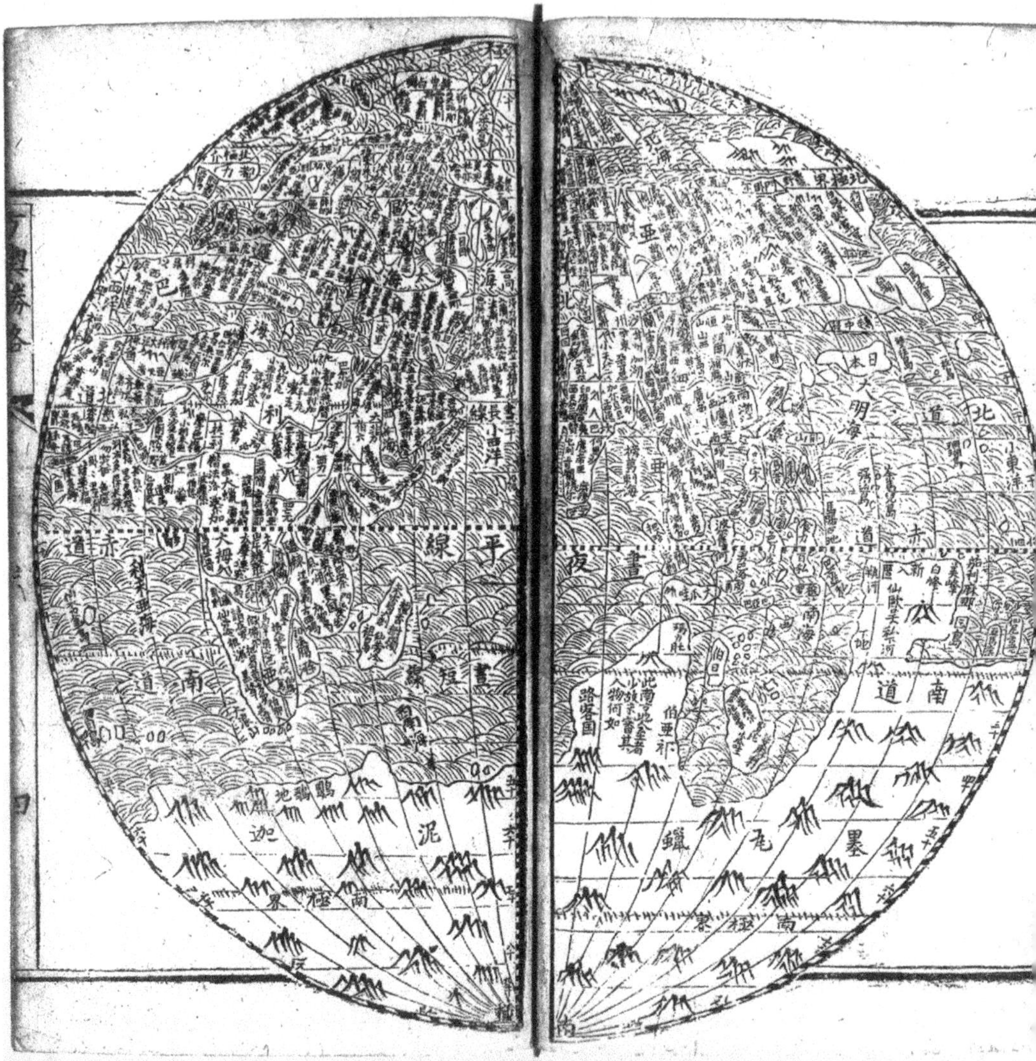

FIGURE 5.19 Map of the eastern and western hemispheres published in the 方輿勝略 *Fangyu shenglüe*, [Compendium of Geography], 1612, National Diet Library collection. Thought to have been brought to Japan by the mid-seventeenth century, the *Fangyu shenglüe* became part of the basis for the *Bankoku Sōzu*. *Fangyu shenglüe* shows the hemisphere with the New World on the left and the hemisphere with the Old World on the right. However, in this chapter, the left and right sides are swapped to make it easier to compare with the *Bankoku Sōzu*.

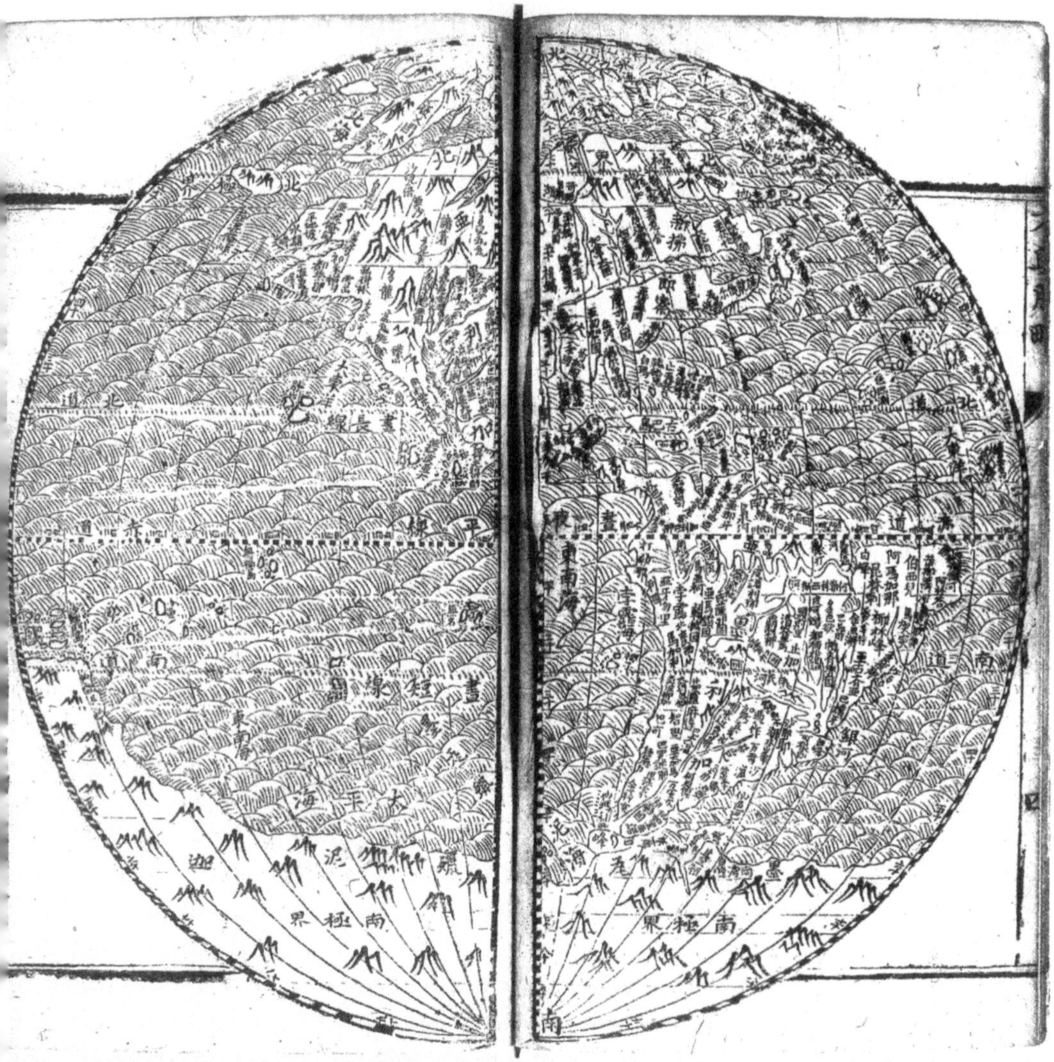

is enlarged exaggeratedly; in contrast, Magellanica is shrunk. Moreover, an island named Etchūgo 越中後 is located off the northern coast of the Sea of Japan, although this cannot be found in the *Bankoku Sōzu*. Taking the above evidence into consideration, the *Bankoku Sōkai Zu* was not exclusively based on the *Bankoku Sōzu*.

The island named Etchūgo did not actually exist and came about since Ricci mistakenly gave the name to the northern island of Japan, Yezo (Hokkaido), on his map. Subsequently, only the name was cut off from the real island. However, precisely because the island is fictional, it helps clarify the genealogical relationships of the maps. Thus, focusing on this island, we become aware that, besides the *Bankoku Sōkai Zu*, a map of the eastern and western hemispheres published in the *Huiji yutu beikao quanshu* 彙輯輿圖備攷全書 "Reference Work on the Maps of China" (1633) is the only other map that shows the island at this location. This map is also based on the map in the previously mentioned *Fangyu shenglue* and, due to changes in the configuration of land masses in the transcription process, some portions of the map contain unique depictions; one is the island of Etchūgo, and another is the exaggerated size of East Asia. The map is, therefore, definitely in the lineage of Ricci's map of the eastern and western hemispheres. It was reprinted in Japan in 1661 and a strong possibility exists that Ishikawa Ryūsen saw it when he was producing the *Bankoku Sōkai Zu*. While referring to the *Bankoku Sōzu*, he also had access to a map of the hemispheres from the Ricci lineage and from it added partial revisions.[18]

5 The 1602 Ricci World Map as a Classic and Its Dissemination

The question arises of when and how the influence of the 1602 Ricci World Map was expressed. According to the established view, the earliest example, aside from the *Bankoku Sōzu* and maps of its lineage, is probably the *Sekai Bankoku Chikyū Zu* 世界萬國地球圖 "Map of the World," which was produced by Inagaki Kōrō 稲垣光朗 (years of birth and death unknown), a pupil of Higuchi Kentei, in 1708. This map is thought to have been based on the maps of the northern and southern hemispheres positioned above and below on the sixth scroll of the 1602 Ricci World Map.

18 Aoyama Hiro'o, *Zen kindai chizu no kūkan to chi* 前近代地図の空間と知 [Space and Knowledge of Pre-modern Maps] (Tokyo: Azekura Shobō, 2007), 141–149.

However, differences between the *Sekai Bankoku Chikyū Zu* and the double sphere map on the 1602 Ricci World Map extend throughout the maps, including their depictions of the Arctic Ocean islands, Scandinavian Peninsula, Italian Peninsula, east coast of Siberia, Persian Gulf, northeast coast of North America, mythical Lake Parime in northern South America, Madagascar Island, and islands east and west of the large peninsula of Magellanica. Thus, the *Sekai Bankoku Chikyū Zu* can certainly not be considered to be a copy of the map of the northern and southern hemispheres of the 1602 Ricci World Map.

Continuing, the *Yochi Zu* (Figure 5.3) produced by Harame in 1720 is thought to be based on the main map located at the center of the 1602 Ricci World Map. More precisely, it is probably based on the Version 1 supplemented hand-drawn copies with a partial revision of the islands of Southeast Asia, the depiction of atolls in the Indian Ocean, and place names written in *katakana*.

The previously mentioned Sasayama World Map is one that, while closely related to the Version 1 supplemented hand-drawn copies, depicts land masses differently. When the Sasayama World Map and the Version 1 supplemented hand-drawn copies are compared to the *Yochi Zu*, one becomes aware that the depictions in the *Yochi Zu* accord with those in the former and differ from those in the latter. The differences between the *Yochi Zu* and the Version 1 supplemented hand-drawn copies include, for example, the existence or nonexistence of mythical Lake Parime and the depiction of the islands of Southeast Asia, Madagascar, the northwest coast of Africa, the Italian Peninsula, the Scandinavian Peninsula, the Kola Peninsula, and Newfoundland.

The *Yochi Zu* is more similar in content to the Sasayama World Map than to the Version 1 supplemented hand-drawn copies. As indicated above, the Sasayama World Map, while related to the 1602 Ricci World Map, features revised land masses based on *Nanban*-style world maps or portolan charts. It is also presumed to have been involved in establishing the Version 1 supplemented hand-drawn copies. Therefore, the argument that a partially revised map from among the Version 1 supplemented hand-drawn copies was the original model of the *Yochi Zu* is not appropriate.

The Sekisui World Map (Figure 5.4) produced by Nagakubo Sekisui was published at the end of the eighteenth century and has generally been thought to be produced based on the *Yochi Zu*. Sekisui's preface imprinted on it consists of text virtually identical to the *Yochi Zu*. However, this map also includes content that could not have been depicted only on the basis of the *Yochi Zu*. For example, the names of Asia, Europe, Libya, North America, South America, and Magellanica, and the explanatory note regarding Lilliput, the land of dwarfs, in northern Europe and other places as well appear on his map, although they do not appear on the *Yochi Zu*. In addition, unlike the 1602 Ricci World Map, the name *Da dong yang* 大東洋 [Great Eastern Ocean], is written on the Atlantic Ocean in the *Yochi Zu*. In Sekisui's map, however, it reverts to the eastern Pacific Ocean, as in the 1602 Ricci World Map.

Thus, when producing his World Map, Sekisui made numerous changes while taking the *Yochi Zu* as a source. This is evident from the revision notes that Sekisui wrote in red ink on the original copy of the *Yochi Zu* that he himself owned (Meiji University Library collection). These notes display the titles of geographical works such as *Zhifang waiji* and *Sairan Igen* 采覧異言 [Collected Views and Strange Words] as well as the names of geographers such as Nishikawa Joken 西川如見 (1648–1724).

Sekisui also added his own unique revisions. If we observe the Sekisui World Map in detail, we notice that Sakhalin (Karafuto 樺太) Island and the Kuril (Chisima 千島) Islands are depicted based on new geographical knowledge. This reflects the achievements of the northern area surveys conducted by the Tokugawa shogunate in 1785 and 1786.

While the Sekisui World Map itself was reprinted, world maps of its lineage were also issued by many publishers in other forms, such as miniaturized, abbreviated, and pirated versions. They were available in rural areas, to commoners, and even to children. For example, Kan'no Hachirō 菅野八郎 (1810–88) wrote a note to people in the area of his rural homeland saying "Go and get the *Bankoku Yochi Zenzu* 万国輿地全図—a world map of the Sekisui lineage." This shows that such a map was available in rural areas. There is also a scrapbook which had such a map collected by a certain commoner (Figure 5.20), and a popular dictionary entitled *Setsuyō-shū* that had such a map. These show that such a map was available to commoners. Moreover, there is an edited version of such a map for children (Figure 5.21). In this manner, the Sekisui World Map was disseminated in society in Japan from the end of the eighteenth century to the mid-nineteenth century.

This dissemination of world maps in general was because of mounting interest in world geography, which owed to tensions in international relations such as the southward advance of Russia. Sekisui maps were especially popular, however, largely due to his reputation. In short, besides the name that he had already gained for himself because of his maps of Japan and China, Sekisui was looked upon as a person who further advanced orthodox knowledge in Japan as a student of Confucianism and a scholar serving the Mito-*han*, which was one of the three Tokugawa family domains and whose school influenced thought nationwide.

Thus, the Sekisui World Map became a world map representing Late Tokugawa or early modern Japan and contributed greatly to enhancing

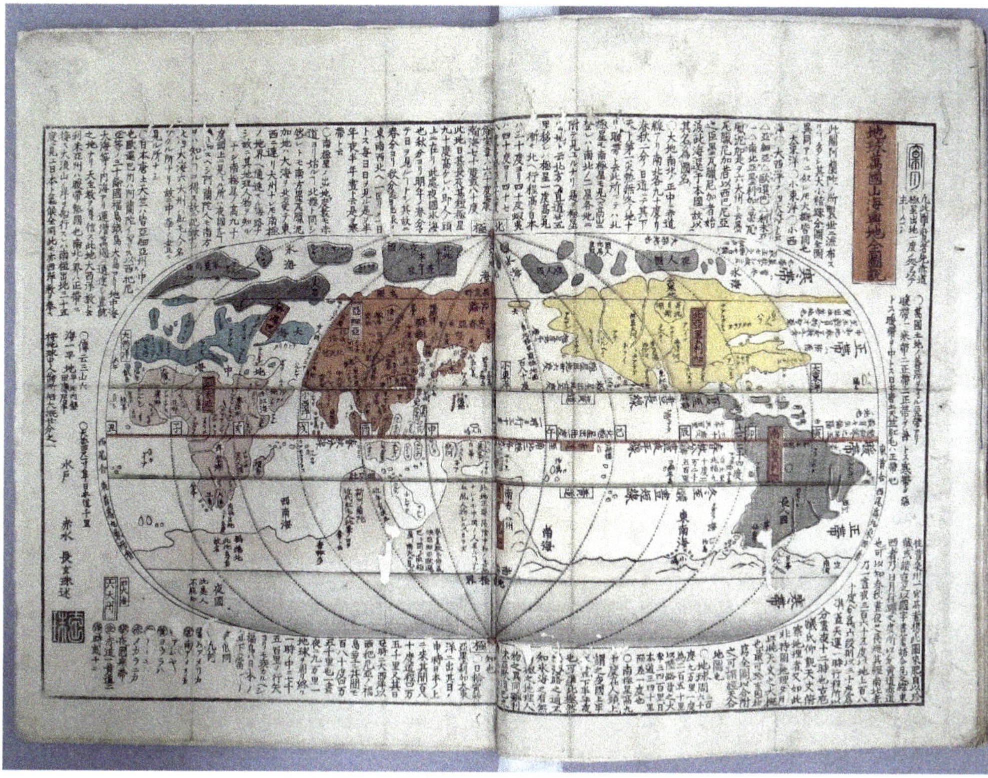

FIGURE 5.20 World map of the Sekisui lineage from a scrapbook made by a commoner: *Futokoroni-Tamaru-Morokuzu* 懐溜諸屑, late nineteenth century, National Museum of Japanese History collection.

knowledge of the world among the public. That also meant that the 1602 Ricci World Map had penetrated into Japanese society.

The original source of the Sekisui World Map, in spite of the partial revisions that had been made, was a world map from nearly two hundred years before and criticism of its being out of date was inevitable. Moreover, *Rangaku* learning had also advanced by then and the publication of *Rangaku*-style world maps based on the latest world maps produced in Europe had already started.

How intellectuals at the time likely viewed this out-of-date map, the Sekisui World Map, is revealed in the epigraph of a *Rangaku* scholar newly appended to the revised version. The epigraph, by Katsuragawa Hoshū 桂川甫周 (his autonym is Kuniakira 國瑞, 1751–1809), expounded on the significance of the Sekisui World Map as a reprint of a classic after first indicating that it was based on an old map: the 1602 Ricci World Map. That is to say, the map had significance precisely because it was out of date. The underlying cause of this

FIGURE 5.21 World map of the Sekisui lineage edited for children: *Bankoku Yochi Sankai Zusetsu* 万國輿地山海圖説, mid-nineteenth century, National Museum of Japanese History collection.

estimation is the current thoughts of those days, particularly the trend of esteeming classics, which had gained ground since the beginning of the eighteenth century.

Another instance clarifies this further. The *Kon'yo Zenzu* 坤輿全圖 "Map of the Earth" (Figure 5.22) produced by Inagaki Shisen 稲垣子戩 (1764–1836) was published in 1802, exactly two hundred years after the publication of the 1602 Ricci World Map. This map was intended to reproduce faithfully the 1602 Ricci World Map with some place names corrected through a comparative investigation of several hand-drawn copies; most of the place names are compiled in a separate volume. In brief, faithfully reproducing the classic map without adding any new geographical knowledge was emphasized. In addition, the faithful reproductions came to be published and circulate. This suggests that there was social demand for such a map and clearly indicates that the 1602 Ricci World Map had been steadily gaining status as a classic work and was ultimately looked upon as such.

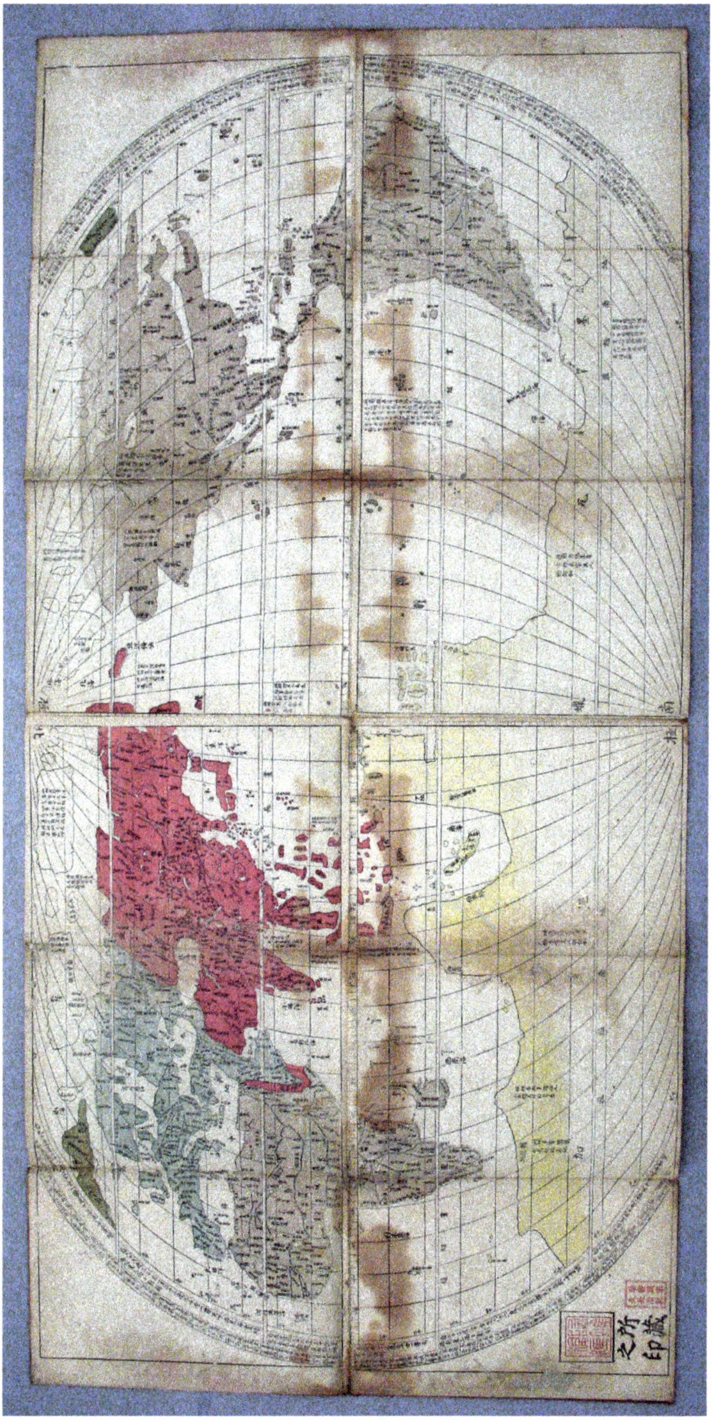

FIGURE 5.22 *Kon'yo Zenzu* 坤輿全圖, 1802, National Museum of Japanese History collection.

The *Kon'yo Zenzu* also has features designed to facilitate understanding, such as transcriptions of Chinese characters in the Japanese *katakana* syllabary, Japanese translations of the original Chinese, and added illustrations. The map is also smaller, with place names and notes in a separate set of booklets so as to make it more practical. Improving the simplification and the operability of the map must have made it more acceptable and helped promote the popularization of this classic map.

6 Conclusion

It seems then that at least three different published versions of the 1602 Ricci World Map exist: a first version (Version 1), a first revised version (Version 1a), and a second revised version (Version 1b). We now know that the extant original edition that has thus far been regarded by scholars as the first version is actually Version 1b. The actual first version has not yet been confirmed as extant. It is, however, presumed to have been in existence because hand-drawn copies are extant today. Such hand-drawn copies having been made in Japan suggests the possibility that the first version was brought to Japan already during the short time between when Version 1 and Version 1a were produced, probably within one year from the publication of the former in 1602.

The hand-drawn copies, however, progressed beyond being mere duplicates of the 1602 Ricci World Map. While taking it as a source, the hand-drawn copies were partially supplemented by geographical data from *Nanban*-style and other world maps dating from the end of the sixteenth century to the early seventeenth century. Hand-drawn copying and partially supplementing characterize the manner of accepting the 1602 Ricci World Map in Japan. We must note that the school of Sino-Japanese astronomy in Japan continuously made hand-drawn copies, and that the impact of the hand-drawn copies on society was limited because they were handwritten and produced within that school.

In contrast, some publications did have an impact on society. The first was the *Bankoku Sōzu*, published in 1645. In particular, the so-called full sheet printed versions of it played a major role in stimulating this impact on society because they were available to the public. While partially using information obtained elsewhere, the *Bankoku Sōzu* and other maps of the same lineage basically used Ricci's map of the eastern and western hemispheres as their model. Consequently, among Ricci's world maps, his map of the hemispheres impacted Japanese society first.

The influence of the 1602 Ricci World Map persisted until the mid-nineteenth century through the Sekisui World Map published at the end of the eighteenth century and other maps. The Sekisui World Map, in particular, was published in large numbers through at least several tens of versions, including its revised versions and those available to a wide range of people, such as people in rural areas, commoners, children and so on. As a result, these derivations of Ricci's map developed a broad reach throughout society. The above-mentioned fact means that the 1602 Ricci World Map was naturalized in Japanese society over the passage of two centuries as the nucleus of geographical knowledge brought to Japan from Europe from the latter half of the sixteenth century through the seventeenth century. Consequently, the map gained status as a classic work from the end of the eighteenth century to the early nineteenth century.

As indicated above, Ricci's map of the eastern and western hemispheres and the 1602 Ricci World Map had impact on Japanese society from the mid-seventeenth century to the mid-eighteenth century and from the end of the eighteenth century to the mid-nineteenth century, respectively. The underlying ideologies of the latter included philological positivism focusing on text as methodology, and classicism emphasizing classics as the source of knowledge, both of which had predominated since the beginning of the eighteenth century. Thus, immediately prior to establishing modern educational system in 1872, the 1602 Ricci World Map played an important role in providing orthodox geographical knowledge and establishing national standards for education among the Japanese people regarding world geography.

Acknowledgement

This work was supported by JSPS KAKENHI Grant Number JP16K03210.

Bibliography

AKIOKA Takejirō. *Chizugaku shi* 地図学史 [History of Cartography]. Tokyo: Iwanami Shoten, 1934.

AOKI Chieko. "Mateo Ritchi no sekai chizu: Nihon e no denrai to sono henyō" マテオ・リッチの世界地図ー日本への伝来とその変容ー [World Maps of Matteo Ricci: Their Introduction into Japan and Transformation]. *Sinica* しにか 6, no. 2 (February 1995): 31–39.

AOKI Chieko. "Nihon ni genzon suru Kon'yo Bankoku Zenzu shozu ni tsuite" 日本に現存する「坤輿万国全図」諸図について [Extant *Kunyu wankuo quantu* in Japan]. *Kirishitan Bunka Kenkyūkai Kaihō* キリシタン文化研究会会報 102 (October 1993): 1–12.

Aoki Chieko. "Nihonjin to Kon'yo Bankoku Zenzu" 日本人と坤輿万国全図 [Japanese People and the *Kunyu wankuo quantu*]. In *Sekaizu yūran: Kon'yo Bankoku Zenzu to higashi Ajia* 世界図遊覧ー坤輿万国全図と東アジアー [Excursion in World Maps: *Kunyu wankuo quantu* and East Asia]. Edited by Tsuchiura City Museum, 8–11. Tsuchiura: Tsuchiura City Museum, 1996.

Aoki Chieko. "Waga kuni ni genzon suru Kon'yo Bankoku Zenzu no kanpon ni kansuru ichi kōsatsu" 我が国に現存する「坤輿万国全図」の刊本に関する一考察 [Consideration of Extant *Kunyu wankuo quantu* in Japan]. *Kyūko* 汲古 23 (July 1993): 100 + 107–113.

Aoyama Hiro'o. *Zen kindai chizu no kūkan to chi* 前近代地図の空間と知 [Space and Knowledge of Pre-modern Maps]. Tokyo: Azekura Shobō, 2007.

Ayusawa Shintarō. "Mateo Ritchi no sekaizu ni kansuru shiteki kenkyū: Kinsei Nihon ni okeru sekai chiri chishiki no shuryū" マテオ・リッチの世界図に関する史的研究ー近世日本における世界地理知識の主流ー [A Historical Research on Matteo Ricci's World Map: On the Main Current of the Knowledge of World Geography in the Age of Tokugawa]. *The Journal of Yokohama Municipal University* 横浜市立大学紀要 18(A-4) (August 1953): 1–239.

Funakoshi Akio. "Kon'yo Bankoku Zenzu to sakoku Nihon: Sekaiteki shiken no seiritsu" 『坤輿万国全図』と鎖国日本ー世界的視圏の成立ー [*Kunyu wankuo quantu* and Japan in the Period of Isolation: Establishment of a Global Perspective]. *Toho Gakuhō* 東方学報 (*Journal of Oriental Studies, Kyoto*) 41 (March 1970): 595–710.

Funakoshi Akio. "Zaika Iezusukaishi sakusei chizu to sakoku jidai no chizu: Kon'yo Bankoku Zenzu, Kōki Zu no hyōka, jūrai no kenkyū o megutte" 在華イエズス会士作成地図と鎖国時代の地図ー「坤輿万国全図」「康熙図」の評価・従来の研究をめぐってー [Maps Produced by Jesuits in China and Maps of the Period of Isolation: Assessment and Conventional Research of the *Kunyu wankuo quantu* and *Kangxi tu*]. *Jimbun-Chiri* 人文地理 [Human geography] 24, no. 2 (April 1972): 59–79.

Hirakawa Sukehiro. *Mateo Ritchi den* マテオ・リッチ伝 [Life of Matteo Ricci]. Vol. 2. Tokyo: Heibonsha, 1997.

Muroga Nobuo and Unno Kazutaka. "The Buddhist World Map in Japan and Its Contact with European Maps." *Imago Mundi* 16 (1962): 49–69.

Muroga Nobuo and Unno Kazutaka. "Edojidai kōki ni okeru bukkyō kei sekaizu" 江戸時代後期における仏教系世界図 [Buddhist World Maps in the Late Edo Period]. *Chirigakushi Kenkyū* 地理学史研究 2 (February 1962): 135–229.

Muroga Nobuo and Unno Kazutaka. "Nihon ni okonawareta bukkyō kei sekaizu ni tsuite" 日本に行われた仏教系世界図について [On Buddhist World Maps in Japan]. *Chirigakushi Kenkyū* 地理学史研究 1 (June 1957): 67–141.

Nanba Matsutarō, Muroga Nobuo, and Unno Kazutaka, eds. *Nihon no kochizu* 日本の古地図 [Old Maps in Japan]. Tokyo: Sōgensha, 1969.

ODA Takeo, MUROGA Nobuo, and UNNO Kazutaka, eds. *Nihon kochizu taisei: Sekaizu hen* 日本古地図大成ー世界図篇ー [Compilation of Old Maps in Japan]. Tokyo: Kōdansha, 1975.

Sotheby's. *The Library of Philip Robinson Part II The Chinese Collection.* London: Sotheby's, 22nd November 1988.

TAKAHASHI Tadashi. "Nanban toshizu byōbu kara Kaeriusu sekaizu e" 南蛮都市図屏風からカエリウス(Pieter van der Keere)世界図へ [From *Nanban* City View Folding Screens to World Maps by Pieter van der Keere]. In *Ezu no kosumorojii* 絵図のコスモロジー [Cosmology of Picture Maps]. Edited by Katsuragawa Ezu Kenkyūkai 葛川絵図研究会 1: 248–264. Kyoto: Chijin Shobō, 1988.

Tsuchiura City Museum, ed. *Sekaizu yūran: Kon'yo Bankoku Zenzu to higashi Ajia* 世界図遊覧ー坤輿万国全図と東アジアー [Excursion in World Maps: *Kunyu wankuo quantu* and East Asia]. Tsuchiura: Tsuchiura City Museum, 1996.

UNNO Kazutaka. "Bankoku Sekai Igyō Zu ni tsuite" 「万国世界異形図」について [The *Bankoku Sekai Igyō Zu*, Map of All the Countries and Pictures of Strange Peoples in the World]. *Biblia* ビブリア 99 (October 1992): 20–33.

UNNO Kazutaka. *Chizu no bunkashi: sekai to Nihon* 地図の文化史ー世界と日本— [Cultural History of Maps: The World and Japan]. Tokyo: Yasaka Shobō, 1996.

UNNO Kazutaka. "Shōhō kan Bankoku Sōzu no seiritsu to rufu" 正保刊「万国総図」の成立と流布 [Establishment and Distribution of the *Bankoku Sōzu* Published in the *Shōhō* Period]. In *Nihon yōgaku shi no kenkyū* 日本洋学史の研究 [Research in the History of Western Studies in Japan]. Edited by ARISAKA Takamichi, 10: 9–75. Tokyo: Sōgensha, 1991.

UNNO Kazutaka. "Ri Matō Kon'yo Bankoku Zenzu no shohan" 利瑪竇『坤輿万国全図』の諸版 [Different Editions of Matteo Ricci's *Kunyu wankuo quantu*]. *Toyo Gakuhō* 東洋学報 (*The Journal of the Research Department of the Toyo Bunko*) 87, no. 1 (June 2005): 101–143.

UNNO Kazutaka. "World Maps Derived from Matteo Ricci." In *Cartography in the Traditional East and Southeast Asian Societies*, edited by J.B. Harley and D. Woodward, 404–410. Vol. 2, Book 2, of *The History of Cartography*. Chicago: The University of Chicago Press, 1994.

CHAPTER 6

Ad Maiorem Dei Gloriam: Jesuit Mapping in China by Giulio Aleni, Francesco Sambiasi, Niccolò Longobardi, Manuel Diaz, and Others

Paola Demattè

Between 1584 and 1610, Matteo Ricci, at the request of Chinese scholars, officials, and even the Ming Wanli emperor, made several different versions of a world map. The first was probably in Latin, but the subsequent ones were in Chinese. The best known—the *Kunyu wanguo quantu* 坤輿萬國全圖 of 1602—also included diagrams and extensive text in Chinese that expounded the Aristotelian-Ptolemaic geocentric cosmography and hinted at Christian beliefs.[1] The death of Ricci in 1610 did not stop this mapping activity. Several other maps, atlases, geographic treatises, and even a globe were created during the late Ming by the Jesuit missionary-scientists Giulio Aleni, Adam Schall von Bell, Francesco Sambiasi, Niccolò Longobardi, and Manuel Diaz the younger. These Europeans collaborated with Chinese scholar-converts like Xu Guangqi 徐光啟, Li Zhizao 李之藻, and Yang Tingyun 楊廷筠, integrating into their maps material compiled by other Jesuits as well as content from various Chinese sources. These maps presented a view of the world that contradicted locally held cosmological beliefs and decentralized China on the spherical earth. Yet, despite vigorous opposition from some quarters, during the late Ming dynasty several Chinese scholars were favorably disposed to the introduction of these foreign ideas and actively supported the Jesuits. The fall of the Ming at the hands of Manchu invaders and the establishment of the Qing dynasty in 1644 changed the situation in dramatic ways.

The Jesuits, who had allied themselves with the new Manchu rulers, obtained control of the astronomy bureau, but lost the support of the Chinese scholar-official elites. Still, even these events did not halt Jesuit mapping in China. Ninety years after Ricci's first map, the Flemish Jesuit Ferdinand Verbiest (1623–88) created the *Kunyu quantu* 坤輿全圖 [Complete Map of the World], 1674, another large-scale planisphere in Chinese for the Kangxi emperor. Though he followed Ricci in many aspects, Verbiest abandoned the

1 Pasquale M. D'Elia, *Il mappamondo cinese del P. Matteo Ricci, S.I.* (Vatican City: Biblioteca Apostolica Vaticana, 1938). Translations are my own unless otherwise noted.

© PAOLA DEMATTÈ, 2024 | DOI:10.1163/9789004684782_008
This is an open access chapter distributed under the terms of the CC BY-NC-ND 4.0 license.

Aristotelian-Ptolemaic conception that the China Jesuits had until then supported, adopting in its stead Tycho Brahe's mixed geocentric system, which afforded better results in terms of astronomical calculations and predictions. He also opted for a bi-hemispheric rather than the original oval projection.

This uninterrupted geographic production bears witness to the continuing role played by cartography in the seventeenth century Sino-European cultural encounter, when missionaries employed aspects of western science to introduce European cosmology and Christianity to China. This paper traces the links between the Ricci and Verbiest maps showing that though created as separate maps, atlases, or globes, these works were parts of a publishing continuum that was collectively carried on in service to both the Christian aims of the mission and the reformist ideals of Chinese converts and sympathizers.

1 Giulio Aleni's Geographic Works: The *Wanguo quantu* and *Zhifang waiji*

Ten years after Ricci's death, Giulio Aleni 艾儒略 (1582–1649), who had arrived in China in 1613, followed in his footsteps making a world map titled *Wanguo quantu* 萬國全圖 [Complete Map of Ten Thousand Countries], 1620. Though much reduced in size, Aleni's map echoes Ricci's *Kunyu wanguo quantu* on many aspects. The layout of the continents with China positioned towards the center is the same, as is the choice of oval projection. Many of the names of geographic localities are also identical, but there are some variations. The character *xian* 仙 [immortal] that Ricci had used to translate 'Saint' for place names was changed by Aleni into *sheng* 聖 [wise, sage]. Aleni changed also the name for New Guinea from Ricci's *Xin Runi* 新入匿 to *Xin Weiniya* 新為匿亞.

At least four examples of Aleni's planisphere survive in two slightly different formats. Two copies are in Milan: one at the Biblioteca Ambrosiana, the other at the Biblioteca Braidense.[2] Two more are at the Biblioteca Apostolica

2 The Ambrosiana copy (Ambrosiana Cimeli, 49.4 × 24.3 cm), in color with the frame and title printed in red, lacks preface and polar projections, and it is unsigned. The Braidense copy (AB 34 XV, 49.4 × 24.3 cm) is made up of three sections: preface (57.8 × 27.4 cm), oval projection (57.8 × 25.8 cm), polar projections (57.8 × 25.8 cm). Pier Francesco Fumagalli "Sinica ambrosiana il fondo antico: libri, cimeli e documenti," *Aevum* 75, no. 3 (2001), 715–736; Hartmut Walravens "Father Verbiest's Chinese World Map (1674)." *Imago Mundi* 43 (1991), 31; Zhou Jiaxiang, "Ancient Chinese books in the Biblioteca Nazionale Braidense of Milan," *Aevum* 77, no. 3 (September–December 2003): 637–671.

Vaticana in Rome.[3] These four maps are nearly identical, except for an important detail: the name of China. The Ambrosiana features *Da Ming yi tong* 大明一統 [Great Ming Domain], whereas the Vaticana and Braidense copies use *Da Qing yi tong* 大清一統 [Great Qing Domain].[4] The Ambrosiana example is then the original version and the earliest of the four. According to D'Elia, it was made between 1613 (the date of Aleni's arrival in China) and 1644 (the date of the fall of the Ming dynasty), probably around 1620. The other three were made after 1644 and before 1649 the year of Aleni's death (unless they are even later reprints). These later versions are framed at the top by an introduction titled *Wanguo quantu xiaoyin* 萬國全圖小引 [Brief Preface to the Complete Map of Ten Thousand Countries] and signed Xihai Ai Rulüe jingti 西海艾儒略敬題 [Respectfully offered by Giulio Aleni of the Western Sea] followed by the Jesuit emblem. The preface is a combination of astronomical and geographic concepts interspersed with references to the Creator, the soul, and Christianity. As elsewhere in Jesuit maps, the emphasis is in transmitting western cosmology: from a reiteration of the European theory of the four elements (earth, water, air, fire) to the idea of the spherical earth and its correspondence to the spherical heaven within which it is nested.

> The creator made the twelve heavens. Fire, air, water, and earth are the four elements. They go from light to heavy mutually interacting. The earth is in the middle of the sky. It is a sphere, but its virtue is square, it is eternally unmoving. North south east west, up down in out, are relative to the position of the person. In reality, there are no directions or center. Earth and heaven are in the same sphere, with corresponding degrees. Therefore, daylight on earth is ruled by heaven. Heaven has an equator and an ecliptic, north and south poles, winter and summer, longitude and latitude, each of 360 degrees. Likewise, the earth is round. Therefore, since the earth's shape is round, to show it as a ball is the most realistic way. Once it is represented on a flat surface there is no way to avoid giving it an elongated shape. It is like cutting an orange peel and spreading it out.

There also is a brief mention of the five continents:

---

3 Barb. Or. 151, fasc. 1a–b, partial copy. See: Henri Cordier, *Bibliotheca Sinica dictionnaire bibliographique des ouvrages relatifs à l'Empire Chinois. 2. éd., revue, corrigée et considerablement augmentée.* (Taipei: Ch'eng-wen, 1966), 3679. A digital copy of the Vatican's version may be also viewed at: https://digi.vatlib.it/view/MSS_Barb.or.151.pt.1, 1r. Accessed 4 April 2023.

4 D'Elia, *Il mappamondo cinese*, 47–48.

> All under heaven is divided into five continents. They are called: Asia, Europa, Libya, America, and Magellanica. In each of these continents there are innumerable countries. There is no space to discuss small countries, so I will give only a cursory description of the largest and most populous ones.

Finally, the conclusion indicates how small humans are and compares man's God-given soul to the universe of creation.

Three years after the publication of the *Wanguo quantu*, Aleni assembled the *Zhifang waiji* 職方外紀 [Unofficial Records on Foreign Countries] (Hangzhou, 1623) a geographical treatise in five *juan* (fascicles) on the lands beyond China. In addition to lengthy texts on the history and sights of the continents, Aleni provided a world map in oval projection nearly identical to the one he had created in 1620; map illustrations for four of the five continents (Asia, Europe, Africa-Libya, Americas); and two polar projections—the southern featuring the representation of Magellanica, the so-called fifth continent.

Significantly, it was with the *Zhifang waiji* that maps made by the China Jesuits started to carry decorative elements, like ships and real or imaginary sea creatures. These images were derived from highly ornamented European maps such as Olaus Magnus' *Carta Marina* (Rome, 1539) or, more likely, Abraham Ortelius' *Theatrum orbis terrarum* (1570–87). Both the *Carta Marina* and Ortelius' map of *Islandia* (Antwerp, 1587) have the prototype of the water erupting scaly sea monster (described in Ortelius as the largest whale known) that appears just off the continent's eastern coast on the *Zhifang waiji* map of Africa.[5] On this same map, above the sea creature, is a lone galleon similar to one that appears in Ortelius' world projection in the 1587 edition of the *Typus Orbis Terrarum*.

Though authored by Aleni, the *Zhifang waiji* had already been planned and conceived by Matteo Ricci as a narrative explanation of the lands shown in his world map. Aleni recounts in his introduction that he based his volume on material that had been assembled by two other Jesuits of the Beijing mission, the Spaniard Diego de Pantoja 龐迪我 (1571–1618) and the Italian Sabatino de Ursis 熊三拔 (1575–1620), who had been working closely with Ricci on

5 This sea creature is marked by the letter H in Ortelius' Islandia map. The H caption states in the English version of 1590: "H. The greatest kind of Whales, which seldome sheweth it selfe; it is more like a little iland, than a fish. It cannot follow or chase the smaller fishes, by reason of the huge greatnesse and waigth of his body, yet he praieth vpon many, which he catcheth by a naturall wile and subtilty which he vseth for to get his food." From http://islandskort.is/en/map/show/2 (accessed April 2, 2023).

astronomical and geographic projects.[6] Specifically, Aleni says that Pantoja had been asked by the Wanli emperor (r. 1572–1620) to write this treatise:

> Later (the emperor) asked another of my confreres, Pantoja, to translate some western maps. From what I understood, he translated them to offer them to the emperor. In the capital, there was much talk about this work, which, however, had never been published and circulated. Eventually, the book reached the emperor and again acquired a degree of notoriety, because thanks to it the sovereign was able to admire the magnificence and infinite variety of all the countries of the world.
>
> I am not very intelligent, however, having enjoyed and admired the work of my confreres, I could not tolerate the idea that it could be lost with the passing of time. Luckily, I have found this old manuscript buried in the dust among moth-eaten books, and with some additions and corrections to the work of my predecessors, I have written this Geography.[7]

Beyond Aleni's Jesuit predecessors, the book was a collaborative project that involved quite deeply three notable Chinese converts: Yang Tingyun, Li Zhizao, and Qu Shigu 瞿式穀. In addition to authoring the prefaces for this first edition of the *Zhifang waiji*, these scholars played an active part in the organization of the manuscript, the selection of topics to discuss, and, no doubt, the polishing of Aleni's writing.

The *Zhifang waiji* was later reissued several times and in different formats. Aleni himself produced a second edition in six *juan*, that was published in Fuzhou in 1626. This was identical to the first text, except for some organizational changes, i.e., an additional preface by the Grand Secretary of Fujian Ye Xianggao 葉向高 and a note on Magellanica based on the experience of Fujianese navigators to the South Seas.[8] Aleni explained that in the new edition an additional *juan* was created simply by dividing the fourth *juan*, which in the original edition discussed both the Americas and Magellanica: thus *juan* four was dedicated to the Americas and *juan* five discussed the hypothetical continent of Magellanica. This choice, he argues, was due to the fact that the Jesuits had always said that there were five continents, but had not provided separate discussion for each of them. To avoid creating confusion he thought

---

6 Eugenio Menegon, *Un solo cielo: Giulio Aleni S.J. (1582–1649): geografia, arte, scienza, religione dall'Europa alla Cina*. (Brescia: Grafo, 1994), 141.

7 Translated from the Italian translation of Paolo De Troia. See, Giulio Aleni, *Geografia dei paesi stranieri. Zhifang waiji*, trans. and ed. Paolo De Troia (Brescia: Fondazione Civiltà Bresciana, Centro Giulio Aleni, 2009), 39–40 and 177.

8 Menegon, *Un solo Cielo*, 144.

each continent should be discussed separately.[9] This edition was included in later compendia, such as the *Tianxue Chuhan* 天學初函 [First Collection of Heavenly studies, 1629] edited by Li Zhizao. But these are not the only iterations of the *Zhifang waiji*.

Around 1672, while working on his world map, the *Kunyu Quantu*, Ferdinand Verbiest revised Aleni's *Zhifang Waiji* and retitled it *Kunyu tushuo* 坤輿圖說 [Illustrated Explanations of the World].[10] Though this new version had a different title, the first edition still carried Aleni's preface.[11] However, thereafter Verbiest re-issued the *Kunyu tushuo* introducing new material. This new version in two volumes was included in the *Siku quanshu* 四庫全書 (Wenyuange edition, 1773–1782). To the first volume he added the texts of the explanatory cartouches of his new world map and to the end of the second he appended twenty-one woodcut illustrations that enlivened and visualized the information on continents, people, animals, and customs originally described in Aleni's *Zhifang waiji*.[12] These images showed exotic land and marine animals such as the American turkey or the African giraffe, mythic creatures like the unicorn and the mermaids, as well as a galleon, the Seven Wonders of the Ancient World, and lastly the Colosseum. Verbiest derived most of his land animals from the illustrations in Konrad Gesner's *Historiæ animalivm* (Zürich, 1551–1587), the standard sixteenth-century source on natural history, and the sea animals from Ortelius.[13] His sources for the seven wonders may have been the decorations on the lower margins of the *Nova totius terrarum orbis geographica ac hydrographica tabula* from Johannes Blaeu's 1630 World Atlas.

---

9 The second edition of the *Zhifang waiji*, which became the standard one, is generally organized as follows: world map in oval projection; map of the northern hemisphere; map of the southern hemisphere with Magellanica; prefaces by Li Zhizao, Ye Xianggao, Yang Tingyun, and Aleni; short notes by Qu Shigu and Xu Xuchen (許胥臣); introductory note by Aleni of the Western Sea and Yang Tingyun of the Eastern Sea (Explanation of the degree system on the map of the five continents); Map of Asia in two pages; *juan* 1 (Asia); Map of Europe in two pages; *juan* 2 (Europe); Map of Africa in two pages; *juan* 3 (Libya = Africa); Map of the Americas in two pages; *juan* 4 (Americas); *juan* 5 Magellanica; *juan* 6 (Explanation of the Four Seas). Based on the copy at the Bibliothèque Nationale (Chinois 1519–1520).

10 Bernard Hung-kay Luk, "A study of Giulio Aleni's Chih-fang wai-chi," *Bulletin of the School of Oriental and African Studies* 1 (1977): 58–84, esp. 76.

11 A copy of the first edition of Aleni's *Kunyu tushuo* is in the Vatican Library (Borg. Cin. 350, fasc. 30).

12 The order of the texts is different, see: Ferdinand Verbiest, *Kunyu tushuo*, in *Siku quanshu*, ed. Ji Yun and Yong Rong, Wenyuange edition, 1773–1782 (Taipei: Commercial Press, 1983) 594: 729–741.

13 Walravens, "Chinese World Map", 34.

Several years after the completion of the *Zhifang waiji*, Aleni published the *Xifang dawen* 西方答問 [Questions and Answers on the West], 1673, a treatise focused on Western countries that was specifically meant to glorify European culture and Christianity.[14] Though not a geography, this book provided additional, if often biased information, on lands that the Chinese knew little about. Interest for this kind of material continued in later times and in 1669 Ferdinand Verbiest in collaboration with Gabriel de Magalhaens (1609–77) and Ludovico Buglio (1606–82) revised and re-published the *Xifang dawen* under the new title of *Yulan Xifang yaoji* 御覽西方要記 [Notes on the West for the Emperor's Reading].

2 The Lacquered Globe of Manuel Diaz and Niccolò Longobardi

A lacquered terrestrial globe inscribed with legends and place names and decorated with sea monsters and sailing ships now at the British Library adds a third dimension to the geographic and missionary activities of the Jesuits in China (Figure 6.1).[15] It was created by the Jesuits Emmanuel Diaz the younger 楊瑪諾 (1574–1659) and Niccolò Longobardi 龍華民 (1559–1654) in 1623, the same year that Aleni completed the first version of the *Zhifang waiji*.[16] According to records at the British Library, this piece (supposedly the earliest surviving Chinese terrestrial globe) was acquired in 1938 by Sir Percival David from a Parisian dealer, who had in turn discovered it in Beijing in the late 1930s. The globe was said to have been located in the imperial palace where it had probably been made during the late Ming period (the erasure of the character Ming 明 on the map of China indicates that, though made during the Ming, the globe continued to be used also during the subsequent dynasty). The original owner must therefore have been someone with close associations with the Ming court: a member of the imperial family, a court official, or even the emperor himself. However, the motor behind the project was

14 Luk, "A study," 76; John L. Mish, Ai Ju-lio, and Chiang Te-ching, "Creating an image of Europe for China: Aleni's Hsi-Fang Ta-Wen 西方答問," *Monumenta Serica* 23 (1964): 1–87.

15 British Library: Maps G.35.

16 Yolande Jones, *Chinese and Japanese Maps: An Exhibition Organised by the British Library at the British Museum, 1 February–31 December 1974*. Catalogue by Yolande Jones, Howard Nelson, Helen Wallis (London: British Museum 1974), C7; Richard J. Smith, *Chinese Maps: Images of "All under Heaven."* (New York: Oxford University Press, 1996), 50; British Library website http://www.bl.uk/collection-items/chinese-terrestrial-globe. Currently, the globe is undergoing studies in view of restoration. Its surface has deteriorated over the years and the paint has peeled off in several spots, such as the south American continent and the cartouche signed by Longobardi and Diaz.

probably someone else: in the explanatory cartouche placed in the southern hemisphere the name of the Creator—*zaowuzhu* 造物主—is preceded by an honorific empty space. This suggests that the person who ordered or coordinated the making of the globe was a high-ranking official, who was also a convert to Christianity.[17] Given his involvement at court and his proximity to the Jesuits, most likely this person was Paulus Xu Guangqi 徐光啓, the celebrated minister, scholar, and convert to Christianity who had closely collaborated with Matteo Ricci on matters of geography and astronomy. In fact, sometime after the ascent to the Ming throne of the Tianqi emperor 天啓 (r. 1620–27), both Longobardi and Diaz were invited to work at court by Xu Guangqi and other converted mandarins with the excuse that their services were needed to reform the army, particularly the artillery. From 1621, Diaz was even a guest at Xu Guangqi's country villa in the outskirts of Beijing. Those purported military activities were never implemented because the missionaries were more interested in other matters.[18] Nonetheless it was probably during that time at the Ming court that Longobardi and Diaz began working on the globe.

Niccolò Longobardi, a Sicilian nobleman, arrived in China in 1597. In 1609, he was called to Beijing by Matteo Ricci himself, whom he had the opportunity to get to know well (and respectfully disagree on theological issues relating to Confucian rites). At the latter's death, Longobardi succeeded him as the Superior General of the China mission. Emmanuel Diaz the younger, a Portuguese, arrived in China in 1610 and initially lived in the southern provinces. He authored numerous religious treatises in Chinese, but is known especially for the *Tianwen lüe* 天問略 [Questions about the Heavens, 1615], an illustrated booklet that described Ptolemaic astronomy and introduced Galileo's telescope discoveries.

The globe, a hollow structure except for the poles that are made of a solid block of wood, has a diameter of 59 cm (23 inches) and it is drawn on a scale of 1:21 million. Today, it is set on a modern tilted mount to represent the actual inclination of the earth on the ecliptic plane.[19] However, originally it was probably mounted in accordance with the principles of Ptolemaic astronomy.

---

17 Jones (*Chinese and Japanese Maps*, C7) argues that since *zaowuzhu* (and 造化者 *zaohuazhe*, author of change) is a very ancient Chinese expression going back to the 4th century BCE it does not necessarily refer to the Christian God. However, given that the authors of the globe were Jesuits it is hard to believe that it might indicate something else.

18 Louis Pfister, *Notices biographiques et bibliographiques sur les Jésuites de l'ancienne mission de Chine, 1552–1773* (Shanghai: Imprimerie de la mission catholique, 1932–34), 106–111.

19 The globe is said to be mounted at 22.5 degrees approximating the ecliptic inclination which is 23.5 degrees. Helen M. Wallis and E.D. Grinstead, "A Chinese Terrestrial Globe, A.D. 1623," *British Museum Quarterly* 25, no. 3/4 (June 1962): 91, footnote 11.

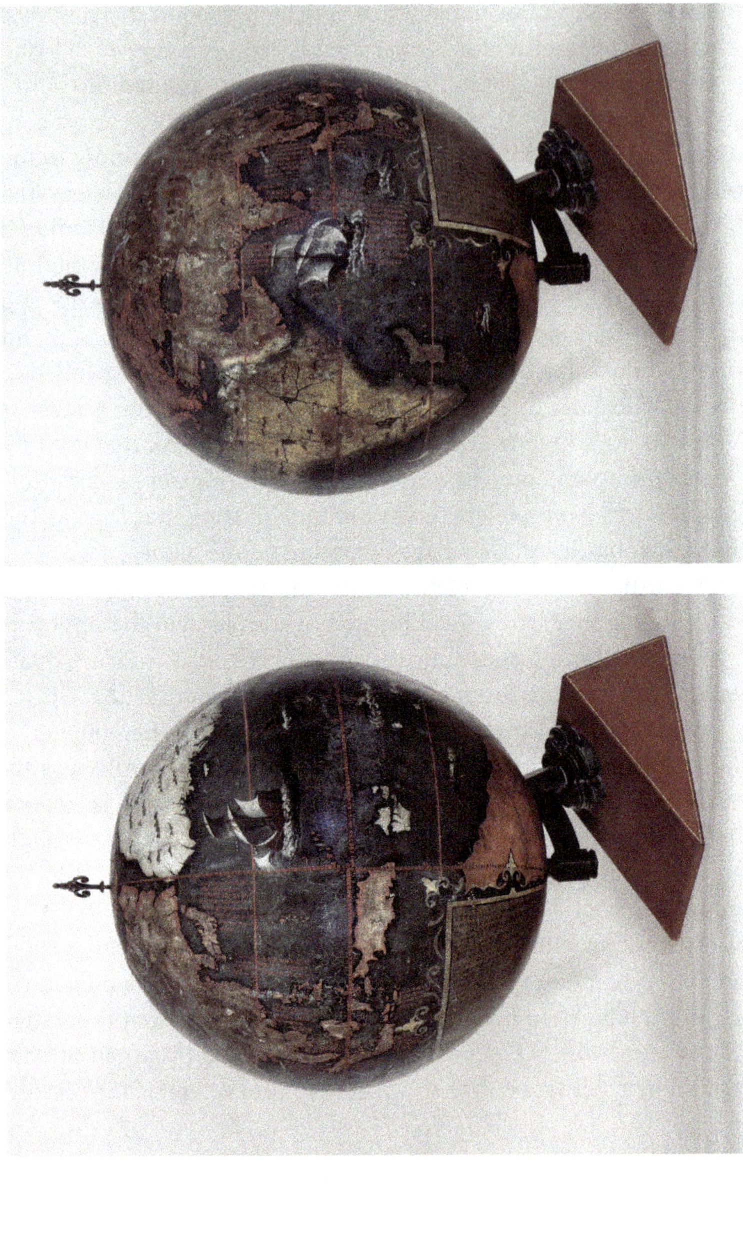

FIGURE 6.1 Lacquered terrestrial globe, Manuel Diaz the Younger and Niccolò Longobardi. Left: view centering on the Pacific Ocean; right: view centering on the Indian Ocean. The British Library: Maps G. 35. © Courtesy of the British Library Board.

Though there is no information about it, the original structure may have been similar to those used around the same time in Europe, such as the one that supports the early seventeenth century globe by Willem Jansz Blaeu in the collection of the British Library.[20] These globes are set into wooden stands with a horizon ring supported by pillars emerging from a base in the middle of which is inserted the (meridian) ring that holds the sphere and allows it to be rotated. A picture of a terrestrial globe published in the eighteenth century in the Qing ritual encyclopedia *Huangchao liqi tushi* 皇朝禮器圖式, 1759, may illustrate such a mounting.[21]

Alternatively, it is possible that the mounting was similar to that of another Chinese globe, the Rosthorn of the MAK Museum in Vienna, which is held by a meridian ring set into a vertical wooden frame. In the preface, the authors explain the structure of the original support (part of which is apparently still inside the globe):[22]

> We placed the poles within the ring and let the ring rest on the meridian. [We set the globe on the mount so that] part of it emerges above the top surface of the [wooden] frame, which emulates the horizon. Turning the ring displays north south changes, and turning the globe shows east west differences.
>
> 立極于環。環方子午。露球櫝面，面擬地平。環轉而南北之變彰。球轉而東西之辨著。[23]

The surface of the globe is covered by geographic and decorative features painted with lacquer. Two red lines mark the tropics of Cancer and Capricorn and a wider one graduated and numbered identifies the equator. The prime meridian is likewise marked in red and set off the coast of Africa, between the

20 This globe, *Nova et accurata Terrae Marisq. Sphaera, denuo recognita et correcta à Gulielmo Blaeu* (Maps G.6.b), was published in 1606 by Blaeu and revised in the 1620s by Willem Schouten and Jacob La Maire to include new discoveries in south America (https://www.bl.uk/collection-items/blaeus-terrestrial-globe).

21 Helen M. Wallis and E.D. Grinstead, "A Chinese Terrestrial Globe, A.D. 1623," *British Museum Quarterly* 25, no. 3/4 (June 1962): 86. Helen Wallis, "The Influence of Father Ricci on Far Eastern Cartography," *Imago Mundi* 19 (1965): 42.

22 Tom Harper, Curator of Antiquarian Mapping at the British Library, informed me that an X-ray revealed that part of the supporting structure (internal pole) of the globe appears to be still inside. I am grateful for a 2001 transcription of the text and rough translation by Katherine Ou Sham of the Hong Kong Museum of Art that was given to me by the British Library.

23 I would like to thank Prof. Qiong Zhang for her help in translating this passage, parts of which appeared to me rather obscure.

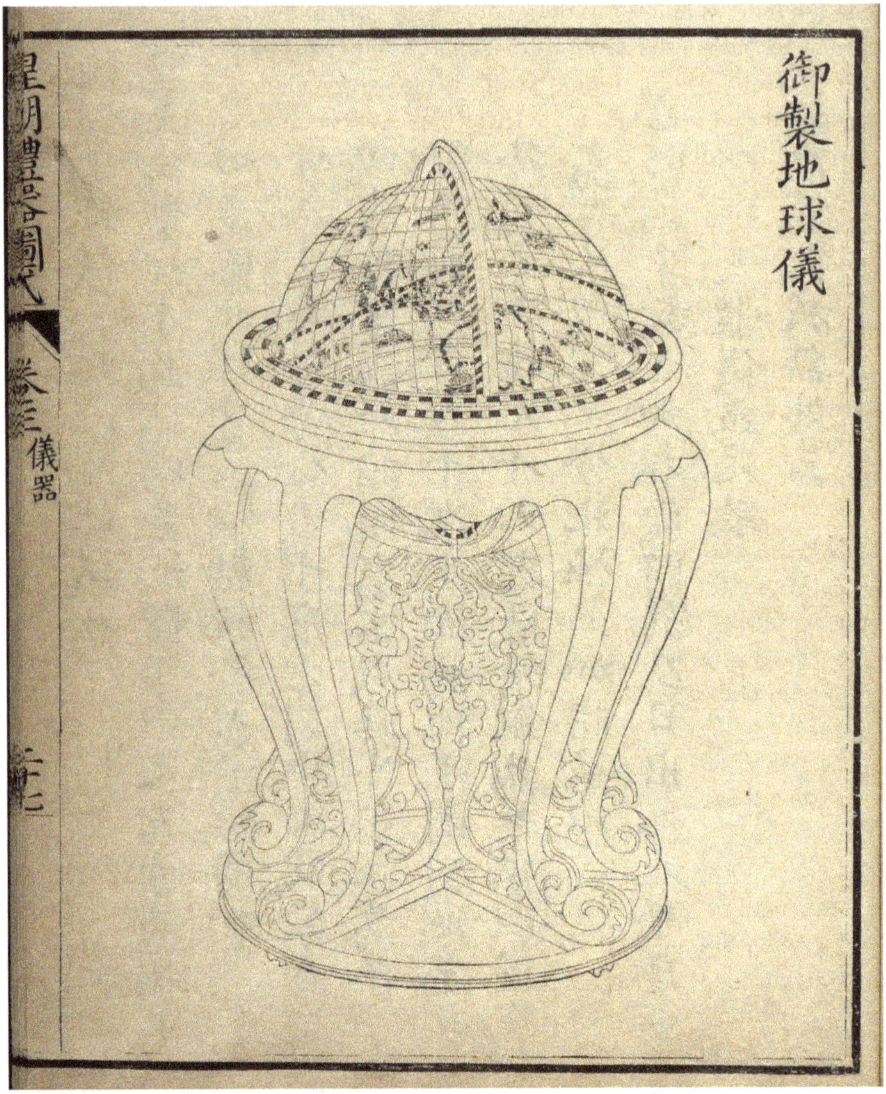

FIGURE 6.2 Illustration of a terrestrial globe published in the Qing ritual encyclopedia, *Huangchao liqi tusi* 皇朝禮器圖式, *juan* 3.

Azores and the Canary Islands, probably at the so-called Fortunate Isles. On the opposite side, the meridian runs over the eastern tip of New Guinea and the western edge of Alaska. The longitude and latitude grid is only partially visible: parallels and meridians, which were originally drawn with a very fine red line, either faded or were covered by later restorations. The five continents were painted and outlined in different colors, though today is it difficult to determine the original hues. Asia is yellow with red coastal outlines. Africa is of a light cream color with greenish edges. The Americas are white with a slightly darker shade at the coastlines. Europe appears to be red. The contours of the continents are up to date with contemporary knowledge and the latest discoveries and are improved from Aleni's maps. Particularly significant is the representation of New Guinea. Unlike Ricci, Aleni had already shown it to be an island, an advance that had been introduced in world cartography following the explorations of Luis Vaez de Torres (fl. 1605–7) in 1606. Longobardi and Diaz followed Aleni in presenting New Guinea as an island, but gave a more realistic representation of its contours.

The globe was made during a time of continued effort by the Jesuits to propagate the idea of a spherical earth and its connections with the works of Ricci and Aleni are evident.[24] Globes had been introduced to China before by Arab and Persian astronomers, but the Jesuits appear to have been unaware of them. Ricci made three globes (now lost) during his lifetime specifically to visually reinforce this concept.[25] The theme of earth sphericity was reiterated also by Aleni who explained it in both the *Zhifang waiji* and the *Wanguo quantu*. The globe's cartouche addresses the same issue and related concepts:

> Thereupon, we took the poles as a reference for longitude and the equator for latitude. We examined north and south and tested east and west. Based on these results, we measured the distance in *li* and drew boundary lines. Having completed our project, we made a map to show it. Still, fearing that it was not enough to demonstrate the theory of the sphericity of the earth, we made a globe as an example. That being set, from the poles we marked the five (temperature) belts and used different colors to distinguish the five continents.

---

24 On the introduction of the theory of the sphericity of the earth, see: Chu Ping-yi 祝平一, "Kua wenhua zhishi chuangbo de ge'an yanjiu – Ming mo Qing chu guanyu diyuanshuo de zhengyi, 1600–1800" 跨文化知識傳播的個案研究—明末清初關於地圓說的爭議 [The Formation of Factual Knowledge in Trans-cultural Scientific Transactions: The Debate over the Sphericity of the Earth in China, 1600–1800," *Zhongyang Yanjiuyuan lishi yuyan yanjiusuo jikan* 中央研究院歷史語言研究所集刊 [Bulletin of the Institute of History and Philosophy, Academia Sinica] 69, no. 3 (1998): 589–670.

25 Ricci was originally convinced that his terrestrial globes were the first ever made in China, but on visiting the Nanjing College of Mathematics in 1600 he realized that the Chinese had made both celestial and terrestrial globes as early as the 13th century under the influence of Arab science (the earliest Chinese globe is supposed to have been made in 1267). Wallis and Grinstead, "A Chinese Terrestrial Globe, A.D. 1623," *British Museum Quarterly* 25, no. 3/4 (June 1962): 83–91.

於是準兩極以為經，循赤道以為緯。審定南北測驗東西。按復計里稽線畫疆。全志既成平圖則現。而猶恐以乎證園之球說也，爰製圓球嚴然儀體。方因極定五帶攸分。以色殊五州有別。

The text also elaborates on the experience of western travelers who having sailed west, arrived in the east and ending in the same point they had left demonstrated that the earth is a sphere:

> Asia is in the east. Europe is in the west. North America is to the east of Asia and to the west of Europe.
>
> 亞細亞在東。歐羅巴在西。北亞墨利加在亞細亞之東。又在歐羅巴之西。

Interesting is the reference to earth's magnetism which is mentioned to explain (albeit incorrectly) why everything is pulled to the center of the earth and why there is no up and down on the earthy globe.

> For this reason, the center of the earth is the lowest point. Everything heavy collects there. An approximate explanation is that it is like a needle being attracted by a magnet.
>
> 故地心為最下。有重諸類，無不就之。是乃本性針向磁石下面猶X僅可略喻耳。

Beyond the introductory cartouche, other inscriptions are written in black on the landmasses and in red on the sea surface. Due to the deterioration of the paint these are at times difficult to interpret. One of the more readable is the notice on Japan copied verbatim from Ricci's map:

> Japan is one big island in the middle of the sea 3200 *li* long and no more than 600 *li* wide. Nowadays, there are sixteen prefectures, each with a governor. They like to show off their strength. Even though they have a king the power always goes to the ministers. The population does a lot of fighting and little studying. The land has silver and iron, as well as good lacquer. If their king has a son, the power is given to him when he reaches thirty years of age. In general, in this country, they don't value precious stones, but only gold, silver and antique pottery.
>
> 日本乃海內一大島長三千二百里寬不過六百里。今有六十六州，各有國主。俗尚強力，雖有總王權常在強臣。其民多習武，少習文。土產

銀鐵，好漆。其王生子年三十以王讓之。其國大抵不重寶石，只重金銀及古窯器。

Unlike its map antecedents, the seas on the globe designed by Longobardi and Diaz are liberally covered with images of ships and fantastic sea creatures that are not directly relevant to geographic knowledge but have significant aesthetic and symbolic value. Five galleons with full wind in their sail are positioned in the North and South Pacific, Indian Ocean, North and South Atlantic. In their vicinity are at least ten sea creatures (four in the Pacific, two in the Indian Ocean, two in the south Atlantic, and two in the north Atlantic) adapted from similar images in Ortelius' atlas or in other European maps. Interestingly, there are no animals on land.

The use of ornamental or perhaps symbolic elements such as these appears to have emerged on Chinese Jesuit maps around the 1620s with Aleni's *Zhifang waiji*.[26] This trend may have been set in motion by the seizing by Chinese authorities of decorated European maps, but it was adopted by the missionaries to drive some larger points.[27] It seems likely that sea monsters and ships functioned on two levels. The first were supposed to attract attention towards the fabulous and terrifying world of the vast oceans. The second indicated the possibility of circumnavigating the spherical earth and symbolized the travels and the presence of European explorers, navigators and the missionaries themselves who had spread all over the world. Together they seem to suggest that the missionaries came from a faraway land braving all sorts of dangers to inform the Chinese about the true doctrine that was governing the universe.

3 Adam Schall von Bell's Terrestrial Gores

Globes, being more complex and expensive to produce than printed maps, appear to have been made only rarely by the China Jesuits. For these reasons, few have survived. However, a map made by the German Jesuit Adam Schall von Bell (1591–1666) has clear connections with the European tradition of globe-making. Schall made this book size version of a world map for his treatise on the armillary sphere, the *Huntian yi* 渾天儀, which was published in the *Chongzhen Lishu* 崇禎曆書 [Chongzhen reign treatises on calendrical

26 Ricci's maps including the well-known 1602 version featured no animals or ships, though these elements appeared on later reprints or copies of Ricci's map, such as the one in the collection of the Nanjing Museum or the Whaling Museum in New Bedford, MA.

27 Eugenio Menegon, "New Knowledge of Strange Things: Exotic Animals from the West," *Gujin lunheng. Disquisitions Past and Present* 15 (October 2006): 39–48.

astronomy], 1635, a collection of European-influenced works on astronomy edited by Xu Guangqi.[28] Unlike the other Jesuit maps, which are mainly some version of the oval projection, Schall's map is in the unusual format of the extended interrupted version with twelve symmetric gores, very similar to Martin Waldeseemüller's *Terrestrial globe gores* of 1507. Effectively, it is a globe map in paper format, a template that someone interested in making a globe could trace.[29] This world map appears alongside a map of the celestial sphere in the same interrupted format.

4 Francesco Sambiasi World Map

Francesco Sambiasi 畢方濟 (1582–1649) reached Macao in 1610, the year of Ricci's death, and entered China somewhat surreptitiously in 1613. He settled shortly in Nanjing before making his way to Beijing where he stayed until 1616, when the expulsion of missionaries from major cities forced him to go first to Shanghai (where he stayed at Xu Guangqi's home) and then to Hangzhou where he was hosted by Li Zhizao. When the Manchus invaded north China, Sambiasi remained loyal to the Ming moving south with the remainder of the court. In fact, he was even appointed Ming ambassador and sent to Macao to ask for Portuguese help in defeating the Manchus. During the last fight of the Ming in south China, Sambiasi was wounded. It was during these tumultuous times in Canton that he produced an oval planisphere modeled on Aleni's first world map, with some additions and innovations that are probably to be attributed to his having had access to the latest Portuguese maps during his trips to Macao. The map titled *Kunyu quantu* 坤輿全圖 (like the 1674 world map by Verbiest) was finished in 1647 or 1648, shortly before Sambiasi's death in 1649.

There are two known versions of this map, which survives in six copies. The three maps of the first group (A: University of Ghent library, the Vatican Library, and the Bibliothèque Nationale de France) are characterized by the presence of the title, a long introductory text, and a compass rose at the center of the map. Those of the second group (B: Archivio di Stato di Torino (Turin, Italy),

28 Unno Kazutaka 海野一隆, *Tōzai chizu bunka kōshōshi kenkyū* 東西地図文化交渉史研究 [Monographs on the history of cartographical exchange between the East and the West] (Osaka: Seibundo 2003): 101–114; Xu Guangqi [Ming], *Chongzhen lishu*, ed. Pan Nai 潘鼐 (Shanghai: Guji 2009).

29 Elly Dekker, "Globes in renaissance Europe," in *The History of Cartography*, edited by J.B. Harley and David Woodward (Chicago: University of Chicago Press, 1994), 2.2: 141–143.

Vienna, and Tokyo) lack the title, the introductory text, and the compass rose.[30] Other than that, all maps are identical, though only those from Ghent, Torino, and Tokyo are hand-painted. Sambiasi's map includes decorative elements like dragons around the title and preface, ships and sea-creatures in the seas, as well as six cosmological and astronomical diagrams on the structure of the cosmos, the spherical earth, the phases of the moon, and the eclipses.[31] The text set above the oval projection in three of the extant maps is a detailed discussion of all the evidence that points to the sphericity of the earth. Much of it is derived from Sabatino de Ursis (1575–1620) *Biaodu shuo* 表度說 [On the Gnomon, Beijing] 1614. Unlike the introduction on Aleni's map, which places much emphasis on religion, this one is focused on giving a series of scientific demonstrations relating to the time of sunrise and sunset in different parts of the earth, the earth's circular shadow on the eclipsed moon, and the changing angles of celestial poles at different latitudes to show that the earth is indeed a sphere. Three diagrams, referred to in the text, complete the narrative: the first shows a spherical earth, the second a square earth as envisioned by traditional Chinese views, and the last features a composite one showing both square and spherical options together.[32] Four diagrams are arranged at the corners of the map showing: the position of the earth within the crystalline spheres (upper right corner); the phases of the moon (lower right corner); the solar and lunar eclipses (lower left corner); and an armillary sphere (upper left corner). An additional two diagrams are placed on the land of Magellanica on both sides of the South Pole. One illustrates the complex relationships of the Four Elements with climate (warm, cold, dry, humid), the other explains how anywhere on earth heaven is always above and nobody can fall off the earth. Finally, a cartouche with an elaborate frame carries Sambiasi's signature and his affiliation: 耶穌會畢方濟撰 [Compiled by Francesco Sambiasi of the Society of Jesus]. The cartouche takes different forms in the two versions of the map.

---

30 Reproductions of the Ghent and Turin maps may be appended to Ann Hierman, Paolo De Troia, and Jan Parmentier, "Francesco Sambiasi, a Missing Link in European Map Making in China?" *Imago Mundi* 61, no. 1 (January 2009): 29–46. https://doi.org/10.1080/03085690802024158. The Bibliothèque national exemplar may be viewed on Gallica at https://gallica.bnf.fr/ark:/12148/btv1b72001882.

31 Ann Heirman, "An introduction to the world map of Francesco Sambiasi (1582–1649)," *Annali dell'Istituto Universitario Orientale* 60–61 (2000–2001): 365–373; Musée Royaux d'Art et d'Histoire ed. *Chine ciel et terre: 5000 ans d'inventions et de découvertes*. Exhibition Cat. (Brussels: Musées Royaux d'Art et d'Histoire, 1988), 406–407, n. 285; Walravens, "Father Chinese World map," 31; Jones, *Chinese and Japanese maps*, C8.

32 These images and also the text reappear in one of the large cartouches that frame Verbiest's map of 1674.

In terms of geographic contours, the Sambiasi planisphere is more advanced than Ricci's and Aleni's maps. This is visible in particular in the contours of New Guinea. Note that the names given to New Guinea differ on all these maps (in Ricci's maps the name is 新入匿 *Xin Runi*, in Sambiasi's 諾瓦記匿亞 *Nuowa Jiniya*, and in Aleni's 新為匿亞 *Xin Weiniya*). Sambiasi's map has more decorative elements than previous maps, but the sources of his images appear to have remained the same as those of Aleni's.

5 European and Asian Cartographic and Geographic Sources

From the late sixteenth to the seventeenth century, the Jesuits working in China had access to a variety of geographic resources from Asia and Europe. At the same time, they were the producers and distributors of geographic knowledge on both continents. The earliest and most influential European sources used by the Jesuits for their maps were the various edition of Abraham Ortelius' (1527–1598) atlas *Theatrum Orbis Terrarum* (1570, 1598, and posthumously 1601, 1612), and the maps and atlases by Gerardus (1512–1594) and Rumold Mercator (1545–1599), in particular the world map (*Nova et Aucta Orbis Terrae Descriptio ad Usum Navigantium Emendate Accommodata*, 1569) and the world atlas (*Atlas Sive Cosmographicae Meditationes de Fabrica Mundi et Fabricati Figura*, 1595 and 1602). Later, other sources included the works of Jodocus Hondius (*Atlas minor*, 1607), Petrus Plancius and eventually the Blaeus (Willem Janszoon, Joan, and Cornilis) in particular the *Theatrum Orbis Terrarum, sive Atlas Novus* (1635, 1655) and the *Geographia Blavania* (1667).[33] For the *Zhifang waiji*, Aleni may also have made use of the *Moderne Tavole di Geografia* (Venice, 1597), by the Italian mathematician Giovanni Antonio Magini (1555–1617). Other, more detailed navigational maps were probably accessible from Portuguese mariners that the Jesuits could contact in Macao or on the coast of Fujian. We know little about these latter sources, but Aleni and others mention that Chinese authorities would sometime seize maps found aboard foreign ships and give them to the Jesuits to translate.

The Jesuits and their collaborators made use of several Asian sources, not only for the Chinese landmass and coast, but also for better information on the outline of the Korean peninsula, the Japanese archipelago, and parts of Southeast Asia. Some early Chinese maps, such as the *Yuji tu* 禹跡圖 [Map of the Tracks of Yu], which was carved into stone in 1137 and installed in the Xi'an

33 Leo Bagrow, *History of Cartography*, 2d ed., rev. and enl. by R.A. Skelton (Chicago: Precedent, 1985): 179–181; Luk, "A Study," 78.

Stele Forest were very detailed and influential, but rather old by the time the Jesuits came around. An obvious source for China is Luo Hongxian's 羅洪先 (1504–1564) *Guangyu Tu* 廣輿圖 (1579), an atlas with more than forty maps of Chinese regions that contained also information on lands beyond China gained from sources related to the Ming admiral Zheng He's 鄭和 (1371–1433) naval expeditions to south east Asia, India, Persia, Arabia and east Africa.[34] Luo's atlas was used by Ricci for his maps and later also by Fathers Michal Boym and Martino Martini for their atlases of China in European languages. Martini was also a committed collector of geographic information in the field, which he no doubt shared with his confreres.[35] Between the 1640s and the 1650s, he traveled high and wide to collect first-hand data on the provinces of China that he used to compile his *Novus Atlas Sinensis* (Amsterdam, 1655), which was included as part six of Blaeu's *Theatrum Orbis Terrarum sive Novus Atlas*.

Another likely Chinese source is the *Da Ming hunyi tu* 大明混一圖 [Map of the Great Ming domain], *c.*1389 or 1402, a world map that shows an oversize China at the center surrounded clockwise by Mongolia, Korea, Japan, Southeast Asia, India, Arabia, Africa and even Europe.[36] The *Da Ming hunyi tu* shows great detail in the representation of southern China and places the Ming capital at Nanjing indicating that it was made in the early Ming period, before the seat of power was moved to Beijing. The map carries over a thousand place names in Chinese as well as Manchu, the latter having been added in the Qing period, an indication that the map was still in full use.

Other sources on Asian geography available to the Jesuits were Korean and Japanese. Though not a direct source for the maps discussed above, the *Honil kangni yoktae kukto chito* 混一疆理歷代國都之圖 [Map of Integrated Lands and Regions of Historical Countries and Capitals], which was originally drafted in 1402 and is colloquially known as the *Kangnido*, gives an idea of the geographic information on Korea that would have been available at the time

34 On the impact of Zheng He's travels on Chinese records, see: Hui Chun Hing 許振興 "The '*Huangming zuxun*' and Zheng He's Voyages to the Western Oceans," 《皇明祖訓》與鄭和下西洋 *Journal of Chinese Studies* 51 (2010): 67–85.

35 The atlas of China of Martini, contained seventeen maps (a general one of China and Japan, one for each of the fifteen Chinese provinces and one of Japan). Martini used mostly Chinese sources. David E. Mungello, *Curious Land: Jesuit Accommodation and the Origins of Sinology* (Honolulu: University of Hawaii Press, 1989): 116–124.

36 The map currently in the collection of First Historical Archives of China (中國第一歷史檔案館), Beijing, is painted on silk and measures cm 386 by 456. Liu Yingsheng 劉迎勝 and Yang Xiaochun 楊曉春, "Da Ming hun yi tu" *yu* "Hun yi jiang li tu" *yanjiu: zhong gu shidai houqi dongya de huanyutu yu shijie dili zhishi* [Research on the *Da Ming hun yi tu* and *Hun yi jiang li tu*: World Maps and World Geographical Knowledge in Asia during the Late Middle Ages] (Nanjing: Fenghuang chubanshe, 2011).

the Jesuits were operating in East Asia.[37] In fact, though the *Kangnido* shares many elements of world geography with the *Da Ming hunyi tu*, from which it is supposed to have mostly derived, it gives greater space and a wealth of detail about the Korean peninsula that are not present on Chinese maps.[38]

In addition, a number of Chinese sources may have provided useful information on south east Asia. The earliest is Zhao Rukuo's 趙汝适 (1170–1231) *Zhufan zhi* 諸蕃志 [Description of Barbarian Nations] a late Song (13th century) geography containing material borrowed from Arab maps and treatises that describe foreign nations as far as the Indian Ocean, Arabia, Africa, and Europe. Another is Ma Huan's *Yingyai shenglan* 瀛涯勝覽 [Survey of the ocean shores], *c.*1436, which illustrates the countries touched during the Zheng He's expeditions (this text survives only in later editions and remaking).[39] Finally, there were probably other less conventional sources, such as maps privately drafted by independent coastal traders or even pirates. The most compelling example of this type of source is the so-called Selden map, which was recently rediscovered in the collection of the Bodleian Library, Oxford University, where it had been since 1659. This map, which was likely created for Chinese merchants from the Fujian area, marks various sea routes and ports in China, Japan and Southeast Asia outlining a network of merchant shipping that linked all the major ports of east Asia.[40]

6 Conclusion: Authors, Authorship, Identity, and Transculturation

The seventeenth century Jesuit process of mapping the earth in China was a collective effort: the contributions of different individuals merged into a larger project *ad maiorem Dei gloriam*, to the greater glory of God. Though different traditions of map-making informed the production of these works, ultimately,

37 As there are several systems of romanization for Korean, the *Kangnido* is also sometimes referred to as the *Kangni to*, as in Chapter 8 of this volume, or *Gangnido* as in Chapter 9.

38 The Kangnido today survives in four copies all of them are in Japan. The oldest is the 15th century the Ryukoku Kangnido. The others, like the Hingkoji Kangnido, are 16th century copies. Angelo Cattaneo, "World Cartography in the Jesuit Mission in China: Cosmography, Theology, Pedagogy," in *Education for New Times: Revisiting Pedagogical Models in the Jesuit Tradition. Macau Ricci Institute Studies*, ed. Artur K. Wardega s.j., 6 (Macau: Macau Ricci Institute, 2014): 76–77; Kenneth R. Robinson, "Chosŏn Korea in the Ryūkoku 'Kangnido': Dating the Oldest Extant Korean Map of the World (15th Century)," *Imago Mundi* 59 no. 2 (2007): 177–192.

39 Bertuccioli, "Europe seen from China," 20–21.

40 Robert Batchelor, "The Selden Map Rediscovered: A Chinese Map of East Asian Shipping Routes, c.1619," *Imago Mundi* 65, no. 1 (2013): 37–63; Timothy Brook, *Mr. Selden's Map of China: decoding the secrets of a vanished cartographer* (New York: Bloomsbury Press, 2013).

we can say that these traditions were re-elaborated in the interest of a common Jesuit cartographic and ideological project. From this perspective, it can be argued that there was only one map that changed through time acquiring new information and varying in style but maintaining a unity of intent.[41] Still, it is also clear that there were different personalities at work who asserted their individuality: it is significant that all maps, atlases, and also the globe are signed and that authorship is clearly demarcated both in the main texts and in the prefaces. Authorship was intended differently in the past, but in these works it is explicit who was taking credit for what: some European Jesuits generated content, other collaborated or allowed the use of their previously published material. Chinese scholars co-translated and helped the Jesuits write in elegant classical language, but more importantly they wrote prefaces and facilitated the publication and distribution of these works among the literati class and the court. They also helped the Jesuits access specialized Chinese material, such as local maps, atlases, and texts that improved their mapping of China and East Asia.

Much ink has been spilled in trying to assess what was the real impact of European scientific culture on China. In the past, it was uncritically assumed that China was passively on the receiving end of the equation. Recently, more emphasis has been placed on the Chinese contributions in the exchange and on the collaborative nature of the enterprise. The late Ming was a period rife with intellectual ferment that aimed to reform the by then ossified Neo-Confucian educational and knowledge system. The *Shixue* 實學 or Practical Scholarship movement emerged from this reformist context and some of its members were attracted by the novelty of European science expounded by the Jesuits, which they thought could provide a useful support to their efforts.[42] The objectives of European priests and Chinese scholars or converts were not identical and not always stable. For the priests, the goal was mainly the conversion of the Chinese people, but the acquisition of knowledge to be dispersed in Europe was an important second. For Chinese scholars, the aim was principally the reform of the Ming system of knowledge, and particularly the promotion of

41 Angelo Cattaneo, "World Cartography in the Jesuit Mission in China. Cosmography, Theology, Pedagogy," in *Education for New Times: Revisiting Pedagogical Models in the Jesuit Tradition*, ed. Artur K. Wardega S.J. (Macau: Macau Ricci Institute, 2014), 71–86, argues that even the much-celebrated Ricci maps were not just Ricci's but the fruit of a larger collaborative effort.

42 Benjamin A. Elman, *Classicism, Politics, and Kinship: The Ch'ang-chou School of New Text Confucianism in Late Imperial China* (Berkeley: University of California Press, 1990); Lionel M. Jensen, *Manufacturing Confucianism: Chinese Traditions & Universal Civilization* (Durham: Duke University Press 1997).

Chinese scientific knowledge, which they shared with the Jesuits. At the same time, for the Chinese scholar-converts Christianity was a path to reform.

Though different, the objectives and personalities of the various parts were not always stable. In the course of these interactions, authors, makers, and facilitators were all changed by their experience of working together and sharing diverse cultural experiences. So even though the Jesuits are often unproblematically presented as foreigners, the situation was more complex. The Jesuits included not only European priests but also many local converts and collaborators. Furthermore, as pointed out by Qiong Zhang in her discussion of the concept of the "contact zone", the European Jesuits were transformed by their experience of life in China.[43] If not fully Chinese, many had spent most of their lives in China and were aware that they were not going to go back to their home countries. Once they obtained the permits to reside and operate in China, they were effectively Chinese subjects and they were recognized to be so by many. Guo Qingluo 郭青螺 (1542–1618), the viceroy of Guizhou, who had published Ricci's map in one of his geographic studies (the *Junxian shiming* 郡縣釋名 or the *Gujin junguo minglei* 古今郡國名類) declared that Matteo Ricci, having passed more than twenty years in China, could no longer be considered a foreigner.[44] Aleni confirms this in his preface to the *Zhifang waiji*, where he says:

> For his majesty the emperor, in the perfection and harmony of his wisdom, there were no foreigners or guests, thus my confrere Ricci was able to donate him the annotated map of the ten-thousand nations.[45]

The Chinese collaborators of the missionaries were equally transformed by the experience. As such, we can say that the experience of seventeenth-century Sino-European cultural exchange is effectively a case of transculturation, the beginning of a mutual transfer of culture that changed both Europe and China.

---

43 Qiong Zhang, *Making the new world their own: Chinese encounters with Jesuit science in the age of discovery* (Leiden: Brill, 2015).

44 D'Elia, *Il mappamondo cinese*, 76.

45 Based on De Troia Italian translation, 39.

Acknowledgements

Research for this paper was carried out with support from the Humanities Fund of the Rhode Island School of Design, Providence, RI. My greatest thanks go to all those who helped me gain access to study the original materials: Monsignor Pier Francesco Fumagalli (Biblioteca Ambrosiana, Milan), Dr. Aldo Coletto (Biblioteca Braidense, Milan), Dr. Monica Grossi (Archivio di Stato, Turin), Dr. Tom Harper (British Library, London), Dr. Federica Orlando (Biblioteca Apostolica Vaticana) and all the staff at these institutions and elsewhere who facilitated my research. Thanks are due also to Laura Hostetler who edited this book and the staff of the Ricci Institute.

Bibliography

Aleni, Giulio. *Geografia dei paesi atranieri alla Cina. Zhifang Waiji.* Edited and translated by Paolo De Troia. Brescia: Fondazione Civiltà Bresciana, 2009.

Aleni, Giulio. *Xifang dawen* 西方答問 [Questions and Answers on the West]. Jinjiang: Jinjiang Jingjiaotang, 1637.

Aleni, Giulio. *Zhifang waiji* 職方外紀 [Unofficial Records on the Foreign Countries]. Annotated by Xie Fan. Beijing: Zhonghua, 1996.

Bagrow, Leo, and R.A. Skelton. *History of Cartography.* 2nd revised edition. New Brunswick, NJ: Transaction Publishers, 2010.

Batchelor, Robert. "The Selden Map Rediscovered: A Chinese Map of East Asian Shipping Routes, c.1619." *Imago Mundi* 65, no. 1 (2013): 37–63.

Bertuccioli, Giuliano. "Europe Seen from China." In *Scholar from the West. Giulio Aleni s.j. (1582–1649) and the Dialogue between Christianity and Chin*, edited by Tiziana Lippiello and Roman Malek, 19–28. Fondazione Civiltà Bresciana, 9. Brescia: Fondazione civiltà bresciana, 1997.

Brook, Timothy. *Mr. Selden's Map of China: Decoding the Secrets of a Vanished Cartographer*. New York: Bloomsbury Press, 2013.

Cattaneo, Angelo. "World Cartography in the Jesuit Mission in China: Cosmography, Theology, Pedagogy." In *Education for New Times: Revisiting Pedagogical Models in the Jesuit Tradition*, edited by Artur K. Wardega, s.j., 71–86. *Macau Ricci Instititute Studies* 6. Macau: Macau Ricci Institute, 2014.

Chu Ping-yi 祝平一. *Kua wenhua zhishi chuangbo de ge'an yanjiu – Ming mo Qing chu guanyu diyuanshuo de zhengyi* 跨文化知識傳播的個案—明末清初關於地圓說的爭議 [The Formation of Factual Knowledge in Trans-cultural Scientific Transactions: The Debate over the Sphericity of the Earth in China, 1600–1800]. *Lishi yuyan yanjiusuo jikan* 中央研究院歷史語言研究所集刊 [Bulletin of the Institute of History and Philosophy, Academia Sinica] 69, no. 3 (1998): 589–670.

Cordier, Henri. *Bibliotheca Sinica: dictionnaire bibliographique des ouvrages relatifs à l'Empire Chinois. 2. éd.,* revisé, *corrigée et considerablement augmentée*. Taipei: Ch'eng-wen, 1966.

D'Elia, Pasquale M. *Il mappamondo cinese del P. Matteo Ricci, s.i.* Vatican City: Biblioteca Apostolica Vaticana, 1938.

Dekker, Elly. "Globes in renaissance Europe." In *Cartography in the Traditional East and Southeast Asian Societies*, edited by J.B. Harley and David Woodward, 135–173. Vol. 1, bk. 2 of *The History of Cartography*. Chicago: University of Chicago Press, 1994.

De Ursis, Sabatino. *Biaodu shuo* 表度說 [On the Gnomon]. In *Siku quanshu*, edited by Ji Yun and Yong Rong. Vol. 543. Wenyuange edition, 1773–1782. Taipei: Commercial Press, 1982.

Elman, Benjamin A. *Classicism, Politics, and Kinship: The Ch'ang-chou School of New Text Confucianism in Late Imperial China*. Berkeley: University of California Press, 1990.

Fumagalli, Pier Francesco. "Sinica ambrosiana il fondo antico: libri, cimeli e documenti." *Aevum* 75, no. 3 (2001): 715–736.

Gessner, Konrad. *Historiæ animalivm*. Zurich: Apvd Christ, 1551–1587.

Heirman, Ann. "An introduction to the world map of Francesco Sambiasi (1582–1649)." *Annali dell'Istituto Universitario Orientale* 60–61 (2000–2001): 365–373.

Hui Chun Hing 許振興. "The '*Huangming zuxun*' and Zheng He's voyages to the Western Oceans." 《皇明祖訓》與鄭和下西洋. *Journal of Chinese Studies* 51 (2010): 67–85.

Jensen, Lionel M. *Manufacturing Confucianism: Chinese Traditions & Universal Civilization*. Durham, NC: Duke University Press, 1997.

Jones, Yolande. *Chinese and Japanese Maps: An Exhibition Organised by the British Library at the British Museum, 1 February–31 December 1974*. Catalogue by Yolande Jones, Howard Nelson, Helen Wallis. London: British Museum, 1974.

Li Zhizao. *Tianxue chuhan*. 天學初函 [Ming, 1626]. Taipei: Taiwan Xuesheng Shuju, 1965.

Liu Yingsheng 劉迎勝 and Yang Xiaochun 楊曉春. "Da Ming hun yi tu" *yu* "Hun yi jiang li tu" *yanjiu: zhonggu shidai houqi dongya de huanyutu yu shijie dili zhishi* 《大明混一图》与《混一疆理图》研究：中古时代后期东亚的寰宇图与世界地理知识 [Research on the *Da Ming hun yi tu* and *Hun yi jiang li tu*: World Maps and World Geographical Knowledge in Asia during the Late Middle Ages]. Nanjing: Fenghuang, 2011.

Luk, Bernard Hung-kay. "A Study of Giulio Aleni's Chih-fang Wai-chi." *Bulletin of the School of Oriental and African Studies* 40, no. 1 (1977): 58–84.

Menegon, Eugenio. *Un solo cielo: Giulio Aleni S.J. (1582–1649): geografia, arte, scienza, religione dall'Europa alla Cina*. Brescia: Grafo, 1994.

Menegon, Eugenio. "New Knowledge of Strange Things: Exotic Animals from the West." *Gujin lunheng. Disquisitions Past and Present* 15 (October 2006): 40–48.

Mish, John L., Ai Ju-lio, and Chiang Te-ching. "Creating An Image of Europe for China: Aleni's Hsi-Fang Ta-Wen (西方答問)." *Monumenta Serica* 23 (1964): 1–87.

Mungello, David E. *Curious Land: Jesuit Accommodation and the Origins of Sinology*. Honolulu: University of Hawaii Press, 1989.

Pfister, Louis. *Notices biographiques et bibliographiques sur les Jésuites de l'ancienne mission de Chine. 1552–1773*. Shanghai: Imprimerie de la mission catholique, 1932–1934.

Robinson, Kenneth R. "Chosŏn Korea in the Ryūkoku 'Kangnido': Dating the Oldest Extant Korean Map of the World (15th Century)." *Imago Mundi* 59, no. 2 (2007): 177–192.

Unno Kazutaka 海野一隆. *Tōzai chizu bunka kōshōshi kenkyū* 東西地図文化交渉史研究 [Monographs on the History of Cartographical Exchange between the East and the West]. Osaka: Seibundo, 2003.

Verbiest, Ferdinand. *Kunyu tushuo* 坤輿圖說 [Illustrations of the World Map]. In *Siku quanshu*, edited by Ji Yun and Yong Rong, 594: 729–792. Wenyuange edition, 1773–1782. Taipei: Commercial Press, 1983.

Wallis, Helen. "The Influence of Father Ricci on Far Eastern Cartography." *Imago Mundi* 19 (1965): 38–45.

Wallis, Helen M., and E.D. Grinstead. "A Chinese Terrestrial Globe, A.D. 1623." *British Museum Quarterly* 25, no. 3/4 (June 1962): 83–91.

Walravens, Hartmut. "Father Verbiest's Chinese World Map (1674)." *Imago Mundi* 43 (1991): 31–47.

Xu Guangqi [Ming]. *Chongzhen lishu* 崇禎历書, edited by Pan Nai 潘鼐. Shanghai: Guji 2009.

Yee, Cordell D.K. "Traditional Chinese cartography and the myth of Westernization." In *Cartography in the Traditional East and Southeast Asian Societies*, edited by J.B. Harley and David Woodward, 170–202. Vol. 1, bk. 2 of *The History of Cartography*. Chicago: University of Chicago Press, 1994.

Zhang, Qiong 張琼. *Making the new world their own: Chinese encounters with Jesuit science in the age of discovery*. Leiden: Brill, 2015.

Zhou, Jiaxiang. "Ancient Chinese books in the Biblioteca Nazionale Braidense of Milan." *Aevum* 77, no. 3 (2003): 637–671.

CHAPTER 7

The World Revealed: Science, Mythology, and the Natural World in Ferdinand Verbiest's *Kunyu Quantu* 坤輿全圖 (1674)

Mark Stephen Mir

> I joy, that in these straits, I see my west;
> For, though their currents yield return to none,
> What shall my west hurt me? As west and east
> In all flat maps (and I am one) are one,
> So death doth touch the resurrection
>
> JOHN DONNE (1572–1631)[1]

The world maps created in China by the Jesuit missionaries Matteo Ricci, Giulio Aleni, Francesco Sambiasi, and Ferdinand Verbiest during the course of the seventeenth century are important milestones in the history of cartography, geography, and cosmology, as well as landmarks of Sino-Western cultural exchange. Nonetheless, the methods used by the Jesuits did not disrupt nor supersede traditional Chinese cartography of the period, and scholars still argue about the depth and permanence of their influence. Blockprint maps using traditional Chinese techniques continued to be produced for the next two centuries, but after the Qing dynasty established itself in 1644, maps produced for the private use of the imperial Court would indeed rely on newer technologies like precision timepieces and theodolites for surveying and copperplate engraving for printing which allowed large maps of fine detail to be produced. Verbiest was an accomplished scientist and his inventiveness and

1 From "Hymn to God, my God in my Sickness." Variously dated 1623, 1630, 1631. This stanza is excerpted from John Carey, ed., *John Donne, Selected Poetry. Oxford World Classics* (Oxford: Oxford University Press, 1998), 210. Donne was a contemporary of Giulio Aleni and the concept of cartographic representation of a human being as a world within itself was common. See Niayesh, Ladan, "All flat maps, and I am one: Cartographic References in the Poems of John Donne," *Études Épistémè*, nº 10 (automne 2006).

© MARK STEPHEN MIR, 2024 | DOI:10.1163/9789004684782_009
This is an open access chapter distributed under the terms of the CC BY-NC-ND 4.0 license.

accuracy are remarkable for the period, perhaps even the equal of Tycho Brahe, whose instruments, observations, and theories he put into practice.[2] Yet an analysis of Verbiest's famous world map also reveals a disparity between accurate and precise knowledge of observable and measurable natural processes such as lunar or solar events, eclipses, and planetary motion and geological or atmospheric phenomena not easily measured or observed. In these cases, Verbiest viewed and presented a "scientific" explanation through a classical Renaissance or Aristotelian lens for earthquakes, tides, mountains, the elements, or even the reality of unicorns and sea monsters. These theories that we would now term "unscientific" were often in harmony with contemporary Chinese science. The 1674 world map produced by Ferdinand Verbiest reflects an amalgam of Eastern and Western knowledge and includes elements of the world maps produced by the "China Jesuits" that invite the viewer to explore and discover, and be awed and astonished at the world around them.

World maps of the early modern period fit into a special category of knowledge. As opposed to utilitarian, local, strategic, or administrative maps, world maps were not confined to geographical information alone but were a mix of the real and the speculative, an Imago Mundi that included many non-cartographic details. Often profusely illustrated with descriptions of distant places, peoples, animals, products, religions, and customs, world maps contained information not only assembled by cartographers or navigators but also obtained from sailors, merchants, missionaries, and travelers, and might further include local myths and legends with descriptions of fabulous creatures and rare treasures. In both East and West these maps (or parts of them) were collected in atlases, gazetteers, geographies, pharmacopeias, and bestiaries. For Europeans during the 16th through the 18th centuries, the missionaries of the Society of Jesus served as intellectual intermediaries between distant civilizations and cultures. The Jesuits were the primary European source of information about China and East Asia, and their published books, letters, atlases, and reports were widely collected and avidly read. During the early modern period the Jesuits served in the reverse role as well, introducing European science, philosophy, history, and literature as a foundation for the introduction of Christianity to a Chinese audience. The ability of the Jesuits to communicate in classical Chinese meant that the literati in Korea and Japan could also read and comprehend this information. Theology combined with science in Jesuit efforts at evangelization and maps were important tools in this process.

2 Including the equatorial armillary, an astronomical instrument Brahe designed but never actually built. See Allen Chapman, "Tycho Brahe: Instrument Designer, Observer and Mechanician," *Journal of the British Astronomical Association* 99, no. 2 (1989): 70–77.

During the late Ming and early Qing period in China (ca. 1550–1800), science in Europe was undergoing radical changes, and the tension between reason and faith, as exemplified by the cases of Giordano Bruno and Galileo Galilei, played out against the backdrop of the old China mission. After the official formation of the Society of Jesus by Ignatius of Loyola in 1540, the Jesuits established a network of colleges, the most prestigious being the Collegio Romano,[3] where Jesuits preparing for their missions were trained in mathematics and the sciences by important scholars such as Christof Clau (1537–1612), better known by his Latin honorific, Clavius.[4] China-bound Jesuits followed the same basic curriculum prescribed in the Ratio Studiorum, which included Euclidian geometry, arithmetic, the spherae (astrolabe), cosmography, astronomy, planetary theory, use of astronomical tables, optics, geography, cartography, and timekeeping (i.e., clocks, clockworks, and mechanical escapements). This training provided the technical foundation for the Jesuit China mission. Development of the tools of scientific inquiry coincided with the so-called "Age of Discovery" and advances in navigation and ship construction provided the Jesuits and their sponsors in the Portuguese Padroado with seaborne access to the distant shores of Asia. This opened a cultural window that permitted Europeans their first accurate accounts of the Chinese empire.

This cultural window was enhanced by the widespread availability of books. Sixteenth-century China was by far the largest and most dynamic book market in the world. The early Jesuits soon realized that any missionary enterprise in China would require a literary (specifically, a printed) approach. Alessandro Valignano 範禮安 (1539–1606), then the Jesuit Visitor, included in his mission principles the requirement that missioners learned to read, write, and speak Chinese.[5] It was via the "apostolate through books" that the Jesuits hoped to reach a wide, literate, and influential audience. The most illustrious of the early China Jesuits was Matteo Ricci 利瑪竇 (1552–1610) who arrived in Macao in 1583, only one year after Pope Gregory had approved the adoption of the Gregorian calendar still in use today.[6] Ricci recognized the role of the scholar-official in Chinese society (which had no exact equivalent in European society) and knew that acceptance and respect of the literati class would be a

3 Now known as the Roman College or Gregorian University.

4 Clavius was one of the most important mathematicians and astronomers of the period and a major architect of the Gregorian calendar reform of 1582, coincidentally the same year his student Matteo Ricci arrived in Macao.

5 Josef Franz Schütte, *Valignano's Mission Principles for Japan* (St. Louis: Institute of Jesuit Sources, 1980–1985).

6 First developed by Aloysius Lilius (ca. 1510–1576) and after his death, modified and promoted by Clavius.

requirement for the future success of the mission. With the help of his Chinese colleagues Ricci produced *Jiaoyou lun* 交友論 [Treatise on Friendship]; translated the first six books of Clavius' version of Euclid's *Geometry* into Chinese [*Jihe yuanben* 幾何原本] with Xu Guangqi 徐光啟 (1562–1633); and wrote influential texts on Western mnemonics and memory techniques, ethics, stoic philosophy, mathematics, catechisms, and discourses using classical Chinese examples—all while adapting Christian and Western (Aristotelian) concepts to the Confucian ethical system. Three of the most important of Ricci's Chinese colleagues, Xu Guangqi 徐光啟 (1562–1633), Li Zhizao 李之藻 (1565–1630), and Yang Tingyun 楊廷筠 (1562–1627), also contributed greatly to the cartographic endeavors of the Society of Jesus during the late Ming dynasty. These individuals are collectively known to Chinese Catholics as the "Three Pillars of the Chinese Church."

By the time Ricci was permitted residence in Beijing in 1601, he had become a much sought-after figure. In his personal records he noted his visitors' great curiosity regarding the unusual items displayed at the Jesuit residence: a planisphere, clocks, prisms, a harpsichord, and Western books and paintings, but most of all he noted their fascination with the latest maps by Mercator and Ortelius. As early as 1584, Ricci had made copies of these and added toponyms translated into Chinese. He later added detailed information on East Asia from Chinese sources. This became the first in a series of maps that would eventually evolve into the *Kunyu wanguo quantu* 坤輿萬國全圖 [Complete Map of the Myriad Countries] with editions dating from 1602–1608. It is from this point that later Jesuit missionaries became known for their cartographic and geographic skills, with successive Jesuits such as Giulio Aleni 艾儒略 (1582–1649) and Francesco Sambiasi 畢方濟 (1582–1649), publishing their own world maps prepared for Chinese audiences during the waning years of the Ming dynasty.

Ricci's world map contained the first depiction of the Americas, the western coasts of Africa and Europe, and the vague coastline of a large southern continent (Australia) on a Chinese map. Of particular interest to Ming scholars was the cosmological data included on the map's outer panels, with explanations and diagrams of the motion of the sun and phases of the moon, the change of seasons, solar and lunar eclipses, and other phenomena.[7] Maps made by (or based on) Ricci and Verbiest world maps also included proofs for the sphericity

7 With his associate Li Zhizao, Ricci translated and published Clavius' commentary of Joannes de Sacrabosco's *Tractatus de sphaera* (ca. 1230) as the *Qiankun tiyi* 乾坤體義 [On the Structure of Heaven and Earth, 1608], and Clavius' *Astrolabe*, as *Hungai tongxian tushuo* 渾蓋通憲圖說 [Illustrated Explanation of Cosmological Patterns, 1607]. In these works, he discussed stereographic projections for the astrolabe, fixing the latitude of Beijing at 40° N and the angle of obliquity of the ecliptic as 23½.

of the earth. Placing the Pacific Ocean in the center of the map and inserting smaller detailed sections around the periphery of the map allowed large blocks of text to be included.

After Ricci died in Beijing in 1610, additional Jesuits trained in science, mathematics, and the arts slowly arrived after the long journey from Portugal in response to Niccolò Longobardo's 龍華民 (1565–1655) request for men trained as scientists. In 1618 three Jesuits set sail from Portugal to Macao on the same ship: Giacomo Rho 羅雅谷 (1593?–1638), Johann Terrenz Schreck 鄧玉函 (1576–1630), and Johann Adam Schall von Bell 湯若望 (1592–1666). Schall would serve as the first Jesuit director of the Calendrical Bureau (Liju 曆局), established in 1629 during the late Ming Dynasty, with the intention of adopting Western methods including related mathematical and mechanical technologies.

Since antiquity, China possessed sophisticated knowledge of mathematics and astronomy, and instruments such as armillaries are believed to have been invented there around the first century BCE. Astrological-astronomical predictions were often important affairs of state, and for China in particular, the accuracy of the calendar was an important indicator of dynastic legitimacy. Solar and stellar observations were carefully recorded and analyzed, and many Chinese astronomical texts have survived. Belief in predictive astrology required accurate almanacs announcing stellar events, eclipses, comets, and other phenomena in order to prescribe the auspicious inauguration of many tasks, both personal and public. Military campaigns, marriages, temple dedications, voyages, feast days, burials, and so on required consulting the stars.[8] Omens like comets and earthquakes were of major political significance and the role of the emperor as much as his empire was assumed to be directly affected by the stars. Thus, Schall's position was quite an important one. When the great Flemish astronomer and scientist Ferdinand Verbiest 南懷仁 (1623–1688) arrived in Macao on June 7, 1658, Schall recruited him as an assistant.

Yet, the early decades of the Qing (est. 1644), during which Verbiest arrived in China, were an unstable and politically dangerous time for the Jesuits, who had previously served under and supported the Ming. Their position was not stabilized until 1669 through a series of events at the Calendar Bureau of the Imperial Board of Astronomy (Qintianjian 欽天監), which was a department

8 It is important to note that in the early seventeenth century, astronomy and astrology in both Europe and China remained interrelated and were only beginning to split into separate disciplines.

of the Board of Rites, whose candidates were chosen via competitive examinations in the Chinese civil service.

In 1669, the young Kangxi emperor 康熙帝 (1654–1722) arrested the last corrupt regent and ascended to the throne. There existed during this period great suspicion between the ethnic Han Chinese and the Manchus who had recently conquered their country. As the ruler of an alien dynasty, and in order to maintain control of a huge country and bureaucracy, the Kangxi emperor followed a policy that paid scrupulous attention to Chinese administrative tradition and practice in every regard. In the same year, Kangxi was informed that serious errors had been found in the calendar for 1670, which had been drawn up by Yang Guangxian 楊光先 (1597–1669), an official of the Astronomical Board and a determined critic of the Jesuits (in fact, an extra month had been inserted to correct for accumulated errors since the last calendar revision, a not uncommon practice in either China or Europe). When informed of this error, Kangxi became alarmed that public documents as critical as the nation's calendar (which had been approved and promulgated in his name) might be in error and he realized that this could have disastrous consequences. Verbiest suggested that the errors be corrected, and the extra month discarded.

This was an extremely serious matter for the newly established Qing government. In the Manchu tradition of public competitions, Kangxi ordered a test to compare the merits of European and Chinese astronomy. The test was designed to predict three events: the length of the shadow thrown by a gnomon of a given height at noon of a certain day; the absolute and relative positions of the sun and the planets on a given date; and the exact time of an anticipated lunar eclipse. It was decided that Yang Guangxian and Verbiest should each use their mathematical skills to determine the answers and that the contest should be held at the Bureau of Astronomy in the presence of senior-ranking government ministers and officials from the observatory. In the event, Verbiest correctly predicted the results in all three tests and was immediately installed as Head of the Mathematical Board and Director of the Observatory.[9]

9 In reality, the competition and appointment were more complicated. The memorial petitioning a promotion for Verbiest to the rank of Bureau Vice-Director [*jianfu* 監副] was accepted by the Kangxi emperor on April 1, 1669, but because of the inherent conflict of accepting an official rank prohibited by his Jesuit vows, Verbiest tried to decline the title. In an unprecedented bureaucratic move, Kangxi appointed him in charge of bureau affairs without the formal rank, describing the position as *zhili lifa* 治理曆法 [praefectus]. Under this rubric, the *shixianke* 時憲科 (or *like* 曆科) office was responsible for the production and distribution of annual calendars through various bureaus and departments. Before the remanufacturing of the astronomical instruments in Beijing was completed, the actual calendar corrections came from the translation and conversion of European mathematical tables of ephemerides such as those of Andrea Argoli (1570–1657). Calendars were state documents and were

Among the tasks assigned to Verbiest was the manufacture of new astronomical instruments for the Imperial Observatory.[10] He directed the palace workshops in the production of instruments modeled on the designs of the Danish astronomer Tycho Brahe, taken from his book *Astronomiæ Instauratæ Mechanica* (coincidentally published in 1602, the same year as Ricci's celebrated world map). The choice of Tycho's model was fortuitous. Tycho's instrument designs were very similar to Chinese patterns and easily adaptable, and the exceptional accuracy they were capable of represented a major achievement in astronomical science in the era before the widespread adoption of the telescope. The instruments in Beijing are thus modeled on those of Tycho's famous observatory of Uraniborg and built on Tychonic precedent, but refashioned in China with Chinese adaptations and decorations. Verbiest produced a sixteen-volume blockprint book in Chinese detailing the general construction of the new astronomical instruments called the *Xinzhi Lingtai yixiang zhi* 新制靈臺儀象誌. The last two volumes contain illustrations of the construction, tools, and mechanical techniques involved. A Latin edition of this, called *Astronomia Europaea*, was published in Europe in 1687. These provide a record of the steps used to create the six large instruments installed at the observatory, or *guanxiangtai* 觀象台.

Verbiest succeeded his predecessor, Fr. Johann Adam Schall von Bell, S.J. as director of the Astronomical-Mathematical Bureau in Beijing, and became a renowned and accomplished mathematician, astronomer, engineer, linguist, and diplomat, establishing a tradition of Jesuit directorship of the Bureau that lasted over a century. In his role as director, Verbiest served as editor of the national calendar and official almanac, corrected star charts and tables of eclipses, and designed and supervised the recasting and installation of the astronomical instruments at the ancient observatory in Beijing which had stood since the 14th century. In 1673 he become the personal tutor of the twenty-year-old Kangxi emperor in science and mathematics, translating the first six books of Euclid's *Elements* into Manchu and explaining to the young monarch European languages, astronomy, and scientific principles. During this time, Verbiest also collected and collated his own works on geography and

specific to each department. The *minli* 民曆 [Calendarium vulgare] or *huangli* 黃曆 was printed for common distribution and concerned only the calendrical portion itself while eliding the *buzhu* 補注 [supplement], which contained astrological content that the Jesuits were careful to avoid, for obvious reasons. The *Shangweili* 上位曆 [Calendarium planetarum] was restricted to the court and official ministries in Beijing. Finally, the office produced a manuscript-only version for the emperor's exclusive use.

10 Analysis of Verbiest's drawings reveal that a redesign of the observatory instruments had been planned for as early as 1664, but were interrupted by factional disputes.

the geographical works of his Jesuit predecessors to produce a much-enlarged version of previous world maps. In 1674 he presented the emperor with the completed *Kunyu quantu* 坤輿全圖 [Complete Map of the Earth].

European astronomical mathematics, geometry, and calculating instruments were a good fit for traditional Chinese practice. These techniques were accepted for their immediate and obvious practicality, but also because of common ideas about space and time. Both European and Chinese scientists agreed that space and time were measurable; that eclipses, solstices, equinoxes, comets, and other celestial phenomena were mathematically predictable; that predictions hinged on the agreement between the result of predictive computations and observation; that distances and directions could be accurately scaled; and that predictive systems could be perfected, i.e., one could reduce the margin of error between theoretical predictions and real observations. These techniques and concepts applied equally to cartography and architecture. Because of this common understanding, many astronomical treatises written by Jesuit scientists during the late Ming and early Qing Dynasties survive today in Chinese sources. To name just a few, these include the *Tianwen lüe* 天文略 [Epitome of Questions on the Heavens, 1615] by Manuel Dias, which describe Galileo's invention of the telescope and the observations he reported; the *Yuanjing shuo* 遠鏡說 [Explanation of the Telescope, 1626] by Schall, containing the first Chinese account of the Tychonic world system; the *Cetian yueshuo* 測天約說 [Brief Explanation of the Measurement of the Heavens, 1628], which further discussed the system; Giacomo Rho's *Celiang quanyi* 測量全義 [Full Meaning of Mensuration, 1631], devoted to Tycho's astronomical instruments; celestial atlases such as the *Chidao nan-bei liang zongxing tu* 赤道南北兩總星圖 [General Star Map of the Northern and Southern Hemispheres Divided by the Equator, 1634]; Jean-François Foucquet's *Lifa wenda* 曆法問答 [Dialog on Calendar Astronomy, 1712–1716], which introduced Copernican theory and elliptical orbits; and once the Catholic Church's formal ban on discussion of heliocentricism ended in 1757, Michel Benoist 蔣友仁 (1715–1774) quickly produced an accurate account of Copernican theory, adding it to the text legends of his revised world map, the *Kunyu quantu* 坤輿全圖 of 1761. Sino-Western scientific exchange not only contributed to the revision of Chinese methods, but also stimulated Chinese scholars to look back upon their own scientific tradition, leading to a new appreciation of previous scholarship.

Astronomy, mathematics, cartography, and the creation of calendars and almanacs were closely linked disciplines of great importance to Chinese civilization and to the Jesuit mission in particular. The logic and artistry in these disciplines connected the celestial and terrestrial realms through the person of the emperor, who embodied ultimate responsibility for harmony between heaven and earth. In traditional Chinese doctrine the emperor wielded the "Mandate of Heaven" only so long as his earthly obligations were correctly and precisely fulfilled.

China's cartographic traditions and geographical works of the "world map" type often had a close relationship with painting, calligraphy, and the visual arts. Some were pictorial representations of what the Chinese refer to as *tianxia* 天下, or "All Under Heaven"—the "Earthly Realm" with China itself, *Zhongguo* 中國, as the "Central Kingdom." While mention is often made of China being placed at the "center" of Jesuit maps, one can see this is not quite the case. In Matteo Ricci's seminal 1602 world map, the *Kunyu wanguo quantu* 坤輿萬國全圖, Ricci removed the hemispheres and formed a single surface where the Pacific Ocean is central, with China placed correctly within it. Ricci's 1602 map included sections on the polar regions, diagrams explaining astronomical phenomena, and voluminous text on the attributes and qualities of the many nations represented, but otherwise lacked illustrations of the flora and fauna described in the text.[11] Later versions of the map included exotic animals from various continents filling in the empty areas in the Antarctic and the southern regions called "Magellanica." Though not added by Ricci himself, the animal illustrations greatly animated these later printings and were an important part of later Jesuit maps, in particular the focus of this study, the 1674 *Kunyu quantu* 坤輿全圖 of Ferdinand Verbiest.

While the *Kunyu quantu* is quite different from earlier Jesuit world maps produced in China for a Chinese audience, such as Ricci's, Aleni's, or Sambiasi's,[12] it is built on their overall model. Verbiest's map is much larger and its geographical information was collected from published texts rather than completely original descriptions. Verbiest re-introduced the hemispheres elided from earlier maps, the form of which denies any "central place," and China is thus correctly located and shown to be one country among many "under heaven."

In the Chinese landscape painting tradition, humans are microscopic in comparison to forests, mountains, rivers, and oceans. World maps by Jesuit cartographers in China fit into this tradition of vast distances filled with strange lands, peoples, and customs. The Jesuits also combined elements of divine

11 For an examination of the source of illustrations added to Ricci's (and other Jesuit) maps, see Cheng Fangyi, "Pleasing the Emperor: Revisiting the Figured Chinese Manuscript of Matteo Ricci's Maps," *Journal of Jesuit Studies* 6, no. 1 (2019): 31–43.

12 Here I am distinguishing between Jesuit maps written in Chinese for a Chinese/East Asian audience and those Jesuit maps produced for European audiences in Latin or Italian, such as those by Ruggieri and Martini, who returned to Europe after many years in China.

omnipotence with the Renaissance tradition of the graphical representation of the universe as an example of the comprehensibility of nature through mathematics, measurement, and logic. That universe was also reflected in the human body[13] itself, a kind of universal map reflecting in microcosm that which is macrocosmic. These maps, furthermore, served the Chinese concept of the "Investigation of Things and the Fathoming of Principles" [*gewu qiongli* 格物窮理].[14]

In creating a new world map, Verbiest was following in the scientific footsteps of his Jesuit predecessors of the old China mission. Matteo Ricci had adopted a careful method of accommodation and adaptation of Christianity and Western learning that sought acceptance by the official "literati" class. This method required a thorough understanding of Chinese language, history, science, and classical literature transmitted through a medium of mutual interest. Ricci had seen first-hand that science, mathematics, astronomy, and cartography could provide such a medium, and Jesuit expertise in these fields became a touchstone of cultural exchange during the 16th–18th centuries.

Like Ricci's *Kunyu wanguo quantu*, Aleni's *Kunyu tushuo* 坤輿圖說 and Sambiasi's *Kunyu quantu* 坤輿全圖, the title of the Verbiest map, *Kunyu quantu* 坤輿全圖, reflects a careful adaptation of Chinese classical terminology that would resonate with literate Chinese of the seventeenth century. The prosaic English words "earth" or "world" are barely adequate as a translation of the term *kunyu* 坤輿 as they fail to reflect the classical Chinese learning that Ricci and Verbiest brought into their work. Kunyu 坤輿 is a term taken from the ancient Chinese *Yijing* 易經 [Book of Changes] denoting the female principle aboard the carriage that bears all things from which the earth is formed, forming the Chinese realm of "All Under Heaven." Perhaps the word that best translates the depth of meaning of this phrase would be "Creation" in the Biblical or literary sense. Linguistically the Jesuits were treading a fine line between respect for traditional Chinese concepts balanced with the need to differentiate something that was an innovative development in a technical specialty. Other terms could have been chosen for these map titles, but doing so would have caused a disconnect with the hallowed past. Another distinction

13 See Chen Hui-Hung 陳慧宏, "The Human Body as a Universe: Understanding Heaven by Visualization and Sensibility in Jesuit Cartography in China," *The Catholic Historical Review* 93, no. 3 (July, 2007): 517–552.

14 See Nicolas Standaert, "The investigation of Things and the Fathoming of Principles (格物窮理/Gewu Qiongli) in the Seventeenth-Century Contact Between Jesuits and Chinese Scholars," in *Ferdinand Verbiest (1623–1688): Jesuit Missionary, Scientist, Engineer and Diplomat*, ed. John W. Witek (Nettetal: Steyler Verlag, 1994): 396–399.

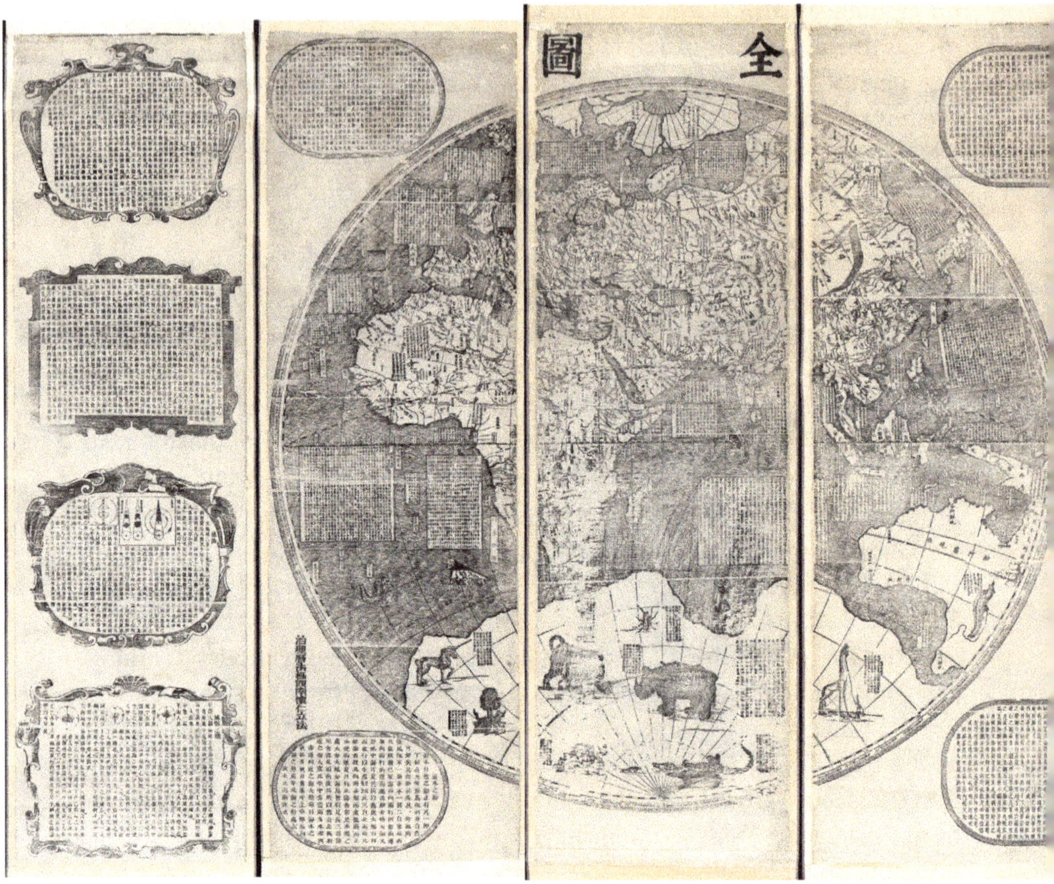

FIGURE 7.1 *Kunyu quantu* 坤輿全圖, Ferdinand Verbiest, 1674. Library of Congress.

concerning Jesuit maps is that the earlier Ming versions were openly printed and widely distributed in a relatively free atmosphere. Despite the fact that so few remain, they were widely studied and discussed in scholarly circles. After the Manchu conquest, however, censorship emerged, and the strategic nature of maps created a serious reconsideration of their sensitive contents. Also unlike the Jesuit maps produced during the Ming, Verbiest's map was the product of the government printing establishment and was not circulated widely. It seems to have been Verbiest's initiative to create it as a teaching aid, and because of its enormous size, it required special large block-printing expertise. It also coincided with Verbiest's commission to re-cast and reconstruct the astronomical instruments for the Imperial observatory, which were all completed by 1674.

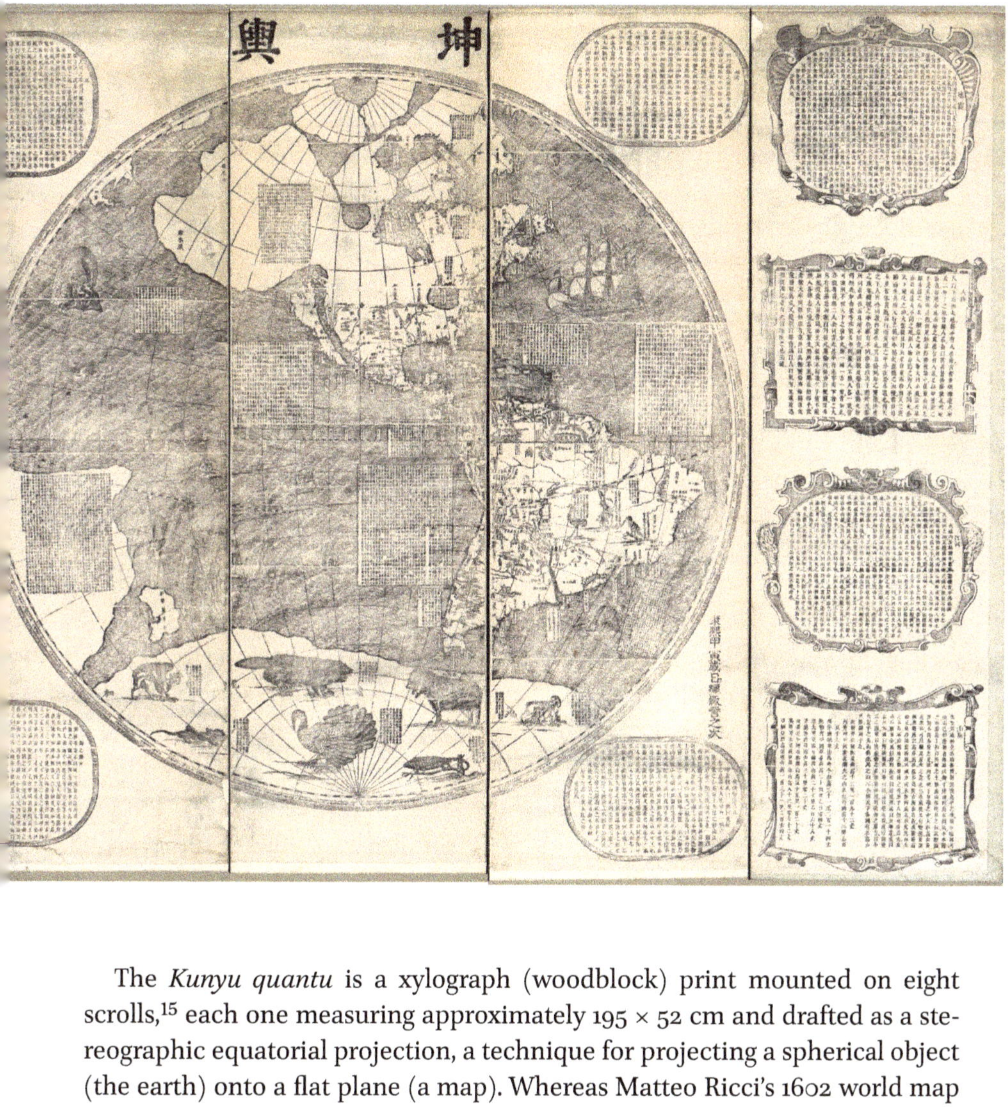

The *Kunyu quantu* is a xylograph (woodblock) print mounted on eight scrolls,[15] each one measuring approximately 195 × 52 cm and drafted as a stereographic equatorial projection, a technique for projecting a spherical object (the earth) onto a flat plane (a map). Whereas Matteo Ricci's 1602 world map removed the continental landmasses from their hemispheres (thereby freeing a large area in the center for descriptive text), Verbiest restored them.

15 For an excellent analysis of the material and construction of this map, see H. Creasy, H. Metcalf, and N. Pearce, "History and conservation of *Kunyu Quantu* (A Map of the Whole World) by Ferdinand Verbiest, 1674," in *Printed on Paper: The Techniques, History, and Conservation of Printed Media*, ed. J. Colbourne and R.F. Snyder (Newcastle, UK: Arts and Social Sciences Academic Press, Northumbria University, 2007): 33–41.

The hemispheres are each 1.5 square meters and comprise the inner six scrolls while the outermost pair display eight of the fourteen descriptive boxes known as "cartouches," each of which contain information in various categories. The title appears in large characters, two characters each on panels three and six, that read from right to left. The map produced for the Kangxi emperor was in color.

Only six complete 1674 maps plus five incomplete examples survive. The examples at the Hebei University Library in Baoding, China, and the Kobe Municipal Library[16] in Japan are in color, with the remaining (including the Library of Congress example on which this examination is based) are monochrome. This map was presented to the Library of Congress in 1930 by Arthur William Hummel (1884–1975), missionary, sinologist, former head of the Asian Division of the Library of Congress, and first president of the Association for Asian Studies.

In a reversal of European practice, the western hemisphere is positioned on the right and eastern hemisphere on the left. Meridians are marked east-to-west in degrees.[17] The *Kunyu quantu* was the first Jesuit map to plot the prime meridian through Beijing, although the name "Beijing 北京" does not actually appear—instead the name used is *Shuntian fu* 順天府, a Ming administrative term indicating the prefectural district which included Beijing and was at the time still in common use. At an angle across both hemispheres the ecliptic is traced describing the apparent path of the sun. The inner ring of each hemisphere is marked for latitude while the middle ring marked the duration of the longest summer day and longest winter night for eighteen zones from the equator to the pole. Verbiest's name and title appear prominently in the clear space opposite the unicorn on the Euro-African hemisphere: "治理曆法極西南懷仁立法" [Enacted by the Calendar Administrator from the Far West Nan Huairen (Ferdinand Verbiest)]. Opposite the hemisphere of the Americas, the date is shown as *Kangxi jiayin* 康熙甲寅日躔娵之次 [1674]. Five continents are shown: Asia, Europe, Libya (i.e., Africa), the Americas, and Magellanica (a term that here refers to the Philippines, Southeast Asia, and Australia). The map provides toponyms for 1264 locations including mountains, seas, and rivers, and most non-Chinese place names are reproduced phonetically in Chinese characters based on names used by Mercator and Blaeu. In most cases, Verbiest follows Ricci's nomenclature, whose earlier world map was the first to show the Americas, the Arctic and Antarctic, and the western coasts of Europe and Africa

16 The Kobe Museum color copy of the Verbiest map is available online in a high-resolution zoomable format: https://artsandculture.google.com/asset/map-of-the-world-ferdinad-verbiest/2AEndoCIl64opQ?hl=en (viewed 10.8.2020).

17 The 360-degree circle was an innovation Verbiest also used to great effect on reconstructed astronomical instruments for the Imperial Observatory.

on a Chinese map. America is thus rendered *Yamolijia* 亞墨利加, California is *Jialifu'erniya* 加里伏爾尼亞, and Santa Catalina becomes *Sheng Jiadalina* 聖加大利納. North America has relatively sparse information compared to South America, which has much more detailed and reliable data. Mexico (New Spain) appears as *Xin Yixiabaniya* 新以西把尼亞 [Nueva España], the Yucatan is rendered *Yujiadan* 宇加單, and in the Caribbean we find Cuba [*Guba dao* 古把島] and the Bahamas [*Bahama dao* 巴哈麻島], names still used on modern Chinese maps. Not all places are rendered phonetically, however. Florida and the American southeast, for example, is descriptively called *Huadi* 花地, "Land of Flowers" as the name "Florida" implies.

Verbiest added interesting details throughout the map. For example, Ricci's 1602 map notes the existence of feral horses in the Americas, but in the cartouche near San Miguel Island off California, Verbiest succinctly adds that "… a century ago there were no horses [in the Americas]," indicating his awareness that the modern horse was introduced to the New World by the Spanish. Verbiest's pictorial menagerie is printed across the expanse of Antarctica arranged by hemisphere, and includes the rhinoceros, chameleon, salamander, alligator, giraffe, lion, crocodile, beaver, hyena,[18] ostrich, unicorn, American turkey, and various snakes and arachnids. In the sea appear whales, seahorses, and monsters of various types. Geographical works written in Chinese by the Jesuits Alfonso Vagnone, Giulio Aleni, Francsco Sambiasi, and Sabbatino de Ursis were primary sources for the text and images. Other images, such as the rhinoceros, derive from Konrad Gessner's zoological treatise, *Historia Animalium*, produced between 1551 and 1586, the origins of which can be traced back to Albrecht Dürer's famous woodcut of 1515. Text and images for the unicorn were taken from Verbiest's own geography, the *Kunyu tushuo* 坤輿圖說 [Illustrated Explanations of the Earth], completed the previous year.[19]

The depiction of fabulous creatures on maps, gazetteers and travelogues was a time-honored tradition both in China and in Europe. European mapmakers with only hazy knowledge of the lands of the East often depicted them filled with fabulous flora and fauna, inserted the presumed location of the iron walls of Gog and Magog, or the realm of the mysterious Prester John. Chinese maps and geographies did likewise with the unknown regions of Central Asia and the West, most famously in the 4th century BCE *Shanhaijing* 山海經 [Classic of Mountains and Seas], which is filled with fantastic peoples, beasts,

18 Unfortunately, badly obscured on the Library of Congress example.

19 For an examination of the use of exotic animals in Jesuit *mappamundi*, see Eugenio Menegon, "New Knowledge of Strange Things: Exotic Animals from the West [存廣異聞：西方異獸]," *Gujin lunheng* 古今論衡 [*Disquisitions on the Past and the Present*] 15 (October, 2006): 39–48.

and dubious geographical information. While it is a bit surprising that one so dedicated to scientific accuracy as Verbiest included mythical animals among his menagerie, these creatures were thought likely to exist and had been documented over the centuries.

The first cartouche (upper right) describes the Western (Aristotelian) conception of the Four Elements. Others describe geocentrism, the earth's sphericity, natural and meteorological phenomena, the poles, the formation of clouds and rain, the movement of the air and wind, the tides and ocean currents, earthquakes, rivers, and mountains. In the cartouche "On Humankind," Verbiest relates the commonality of human nature (against superficial appearances) and its spiritual component using the Confucian term *lingxing* [靈性] (*ren suo gong tong zhe ji lingxing ye* 人所公同者。即靈性也) while mentioning the Creator [主宰天下者] (*ci qi zhong you zhu zai tian xia zhe* 此其中有主宰天下者). This is the only direct reference to God on the map. Verbiest also employs a number of nautical and navigational anecdotes to make his point.

When Western science contradicted a Chinese concept, Verbiest offered illustrated arguments that appealed to reason. In the cartouche "*Diti zhi yuan*" 地體之圜 [The Sphericity (Spherical Form) of the Earth], Verbiest reflects that the Chinese conception of a round heaven and square earth do not refer to their actual shapes but to their cultural and epistemic properties. Using three small diagrams based on Sabbatino De Ursis's *Biaodu shuo* 表度說 [The Gnomon Explained, 1614] he shows why people in different locations on the globe must experience time zone differentials, reasoning that if the earth was flat the sun and stars would appear uniformly everywhere. The observable fact that some stars and constellations are visible in some latitudes and not others is proof of the sphericity of the earth.

Several European ships are illustrated and some of the information provided is quite detailed. The panel directly under the three-masted Dutch East Indiaman describes the layout and dimensions of European ships: their length, beam and deck arrangement; the functions of ballast and its effect on shiphandling; storage of sails, goods, and provisions; the role of the crew, ship's master, and navigators; and the necessity of accurate charts.

The cartouche to its immediate left (under the prow of the ship) describes South America and the area called "New Spain" including the islands of the Caribbean with mention of its climate and flora (including deadly plants). Latin American animals drawn over what is now Brazil include the anaconda and the rhea, a large flightless bird related to the ostrich. A sea monster just south of the Brazilian coast spouts water and appears to be taken directly from Ortelius's map of Iceland. These details provide the viewer with a lively sense of movement and animation. Western sources for the Verbiest map include Johannes Blaeu's *Nova Theatrum Orbis Terrarum* (1635), Rumold

Mercator's *Orbis Terrae Descriptio* (1579), and Abraham Ortelius's *Theatrum Orbis Terrarum* (1612). Blaeu's works included the discoveries of Abel Janszoon Tasman (1603–1659) and William Barrentz (1550–1597), and Verbiest also had Spanish and Portuguese sources from the Americas. One obvious and amusing mistake concerns the state of California, which Ricci had mapped correctly but which became an island on Verbiest's map, a common error of the period that was widely repeated on European maps from 1622 until the late 18th century.[20]

FIGURE 7.2 *Kunyu quantu* 坤輿全圖, Ferdinand Verbiest, 1674. Detail depicting California as an island. Library of Congress.

20 The Glen McLaughlan Map Collection of *California as an Island* at Stanford University has over 700 examples of such maps. It fell to the Jesuits Juan de Ugarte (1721), Ferdinand Konščak (1746), and Wenceslaus Linck (1766) exploring Baja California to survey the region, but the matter was not settled until the expeditions of Juan Bautista de Anza traveling between Sonora and the west coast of California in 1774–1776, a century after Verbiest printed his map in China.

Chinese sources for this map (including earlier works by Ricci, Aleni, Sambiasi, and Martini) are uncertain but include Luo Hongxian's 羅洪先 (1504–1564) *Guang yu tu* 廣輿圖 [Enlarged Terrestrial Atlas] of 1579. The outline of China itself is well formed and includes the names of cities and provinces such as Shandong, Bei Zhili, Hangzhou Fu, Zhejiang, Yunnan, Wuchang Fu, and Guangdong. The Great Wall is indicated running through Manchuria almost to Korea. North of the Wall lies the Gobi Desert stretching up into Mongolia. Borders of individual countries and provinces are unmarked. A large, canted text box placed to the north of Japan gives details on Chinese civilization, and in the "Great Qing Ocean" we find Taiwan and the Philippines, with Luzon prominently featured. Japan presents a curious case; while Verbiest copied Ricci's pithy description of the country, he did not use the original Japanese place names (as Ricci had done and which were easily found) but instead transliterated them phonetically from Western sources.

Whether mythic or real, some of Verbiest's descriptions were eventually incorporated into important Chinese texts. For example, a merman and mermaid appear off the south coast of Australia and the text states: "In the Great East Sea there are sea creatures called *xileng* 西楞 (perhaps derived from Siren) whose upper body resembles a man or woman while the lower half is [like a large] fishtail. Their bones can stanch bleeding; the female's [bones] are even more efficacious." This information was considered important enough for medical purposes that the *xileng* was included in the expanded edition of the famous medical encyclopedia and pharmacopeia *Bencao gangmu shiyi* 本草綱目拾遺 [Supplement to the Compendium of Materia Medica].

The *Kunyu quantu* stands as a monument of scientific study and physical beauty, reflecting the blending and adaptation of European late Renaissance and early Enlightenment mapmaking practices with traditional Chinese cartographic and cosmological practice.

1 Addendum

The following is a translation of several cartouches on the *Kunyu quantu*. These items are presented to illustrate just a few of the wide variety of concepts recorded on the map surface.

1.1 *The Sphericity of the Earth*[21]

On the Library of Congress example of this map, this cartouche is found on the right panel second from the bottom. Other examples have it mounted on

21 Proving that the earth was round was a continuing theme on China Jesuit maps beginning with Ricci's 1602 version. It is clear this concept was considered fundamental to the

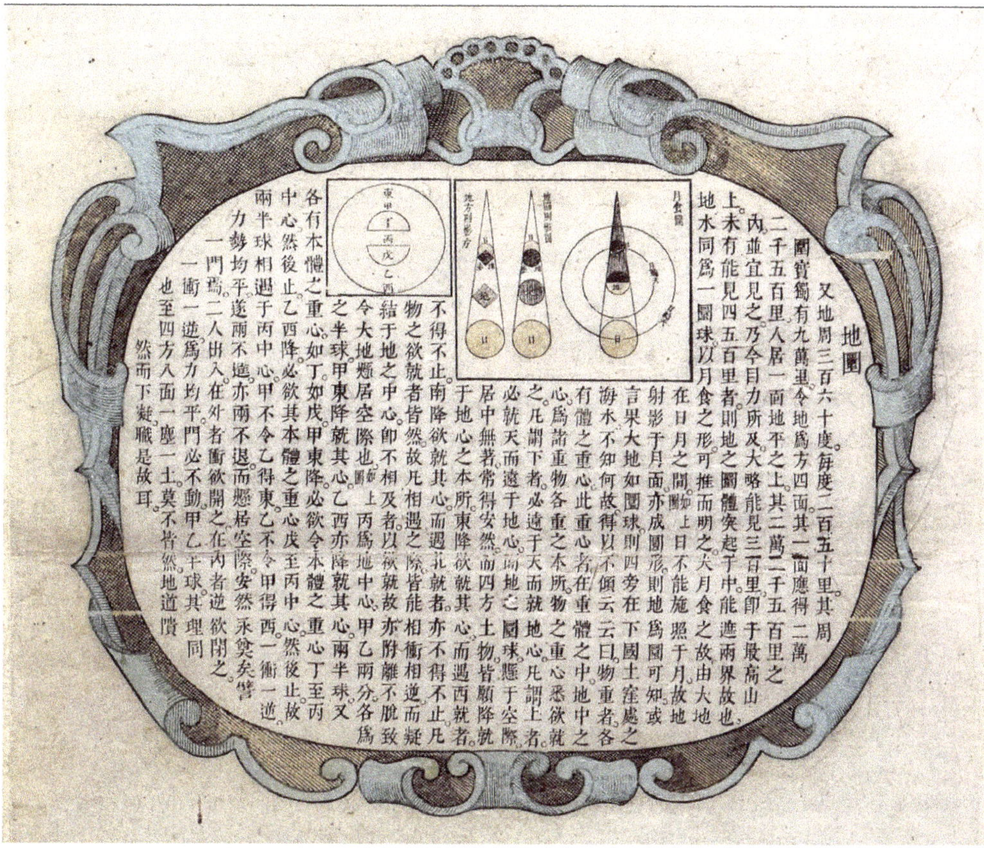

FIGURE 7.3 Cartouche from the *Kunyu quantu* 坤輿全圖. Detail describing the Sphericity of the Earth. Kobe Municipal Library.

the left panel; in some examples the end panels are missing altogether. Also, Verbiest uses the *li* 里, a Chinese unit of distance which varied over time, usually about one-third of an English mile, standardized during the Kangxi reign as equivalent to 500 meters. Since the earth's circumference is 24,901 miles, using the Chinese unit overstated the actual size by 5000 miles (90,000 *li* = 30,000 miles.)

•••

To repeat, the earth has a circle of 360 degrees with each degree measuring 250 *li*. Its circumference is 90,000 *li*. If the earth were square, each side would have 22,500 *li*. A person living on the side above the horizon should be able

geography constructed upon it. There are actually two cartouches on the Verbiest map on this topic.

to see things within the range of 22,500 *li*. But now what one's eyes can see is only about 300 *li*. Even on the top of the highest mountain, there has never been anyone who could see four or five hundred miles away. This is because the round body of the earth protrudes in the middle and blocks the two ends. The earth and the water are on the same round sphere. This can be explained and proven [known] by the shape of a lunar eclipse. The reason for the lunar eclipse lies in the earth being between the sun and the moon (see Figure 7.3). The sunlight does not shine (directly) on the moon, therefore the projection of the earth on the surface of the moon also has a round shape. This is how we know that the earth is round. If someone says, "If the earth is surely round like a sphere, then the countries below on its four sides and the sea water in the depressions are not able to drop down," and so forth. I say, the heavy objects have their own center of gravity. This center of gravity is within the body of a heavy object. The center of the earth is the original place of all different heavy objects. All the centers of gravity of the objects want to move towards it. All that is called "below" must be far from heaven but close to the center of the earth; all that is called "above" must be close to the heaven but far from the center of the earth. Thus, the round sphere of the earth suspended in the midair, having no attachment and always at ease. All things from the four directions want to fall close to their original position at the center.

If a thing from the east wants to fall close to its center, it will meet another thing from the west falling close to the center and it has to stop. If a thing from the south wants to fall close to its center, it will meet another thing from the north falling close to the center and it also has to stop. All things that want to be close to the center are the same at this point. So, whenever they meet together, they all may attack and block each other so as to condense at the middle of the earth. Even those that are not in touch will attach without going away because of the tendency to be close to the center. This leads to the suspension of the earth in midair. (As in the picture): C is the center of the earth. A and B divide it in half, each being a hemisphere of the earth. A falls down from the east to the center, B also falls down to the center from the west. In both hemispheres, there are two centers of gravity for each body, like E and F. When A falls from the east, it must have its own center of gravity, which is E; it goes to the center C and then stops. When B falls from the west, it must have its own center of gravity, which is F; it goes to the center C and then stops. Therefore, the two hemispheres meet at the C center; A would not allow B to go to the east while B would not allow A to go to the west. One attacks and the other blocks, neither will advance nor retreat because of their equal force and power. So, the earth is suspended in midair and is lodged there peacefully forever. It is just like a door where two people exit and enter. The outsider wants to attack to open it, while the insider wants to block to close it. One attacks and the other blocks, yet the door is unmovable because the two use equal force. The A and B hemispheres are the same in principle. Neither the four directions and eight sides nor even a tiny particle of dust and soil may behave differently. Only because of this, the way of the earth can be peaceful and fall down to condense.

1.2 *On the Element Qi* (*of the Aethers*) *or, "The Movement of the Air"*

In ancient times there were those people who thought that because the air had no [discernable] color, it should not belong to the Five Elements and did not in fact exist. This argument is completely in error and can be proven using six [arguments]. First, the spaces within the [regions] of heaven would be empty without air. How could the earth then be suspended there in the middle? How can the ten thousand (myriad) things be born? How could the sun, the moon, and the stars send their light through to the outside? Possibly they nourish the ten thousand (myriad) things through their "hidden virtue?" Since objects can support and protect each other only by means of a connected whole, such vacuous emptiness would be untenable (a taboo) and thus is to be avoided. Second, [without air] birds could not fly with nothing to support them. To fly is to use wings to ride on the air, just as a person floating on the water uses his hands. Third, if a person runs quickly [when the wind is not blowing], he will still feel that something ahead [of him] is touching his body. What else could it be if not the air? Fourth, when a person snaps a whip into the sky, there must be a sound produced, the same as with all "shooting" actions. Any sound produced comes from the striking between two things. If there is no air in the sky, there would be nothing else to produce the sound. Fifth, [let us suppose] there are two doors facing each other in a room. If one is opened or closed, the other one will also be [drawn open or pressed closed]. Moreover, if a man walks quickly into the room, the window paper and any lightweight objects hanging in the room will also move. If not because of the air, what else could it be? Finally, there is no wind in a quiet room, but one sees dust rolling up and down in the shadows. Why is this called "wild horse?" It perforce must be the air that makes this happen. Isn't this enough to prove the existence of the air following the above six [arguments]?

Regarding the constant shifting of the air, it is due to the differences between large reactions and small reactions, and is not hard to understand. The air is a real thing and absolutely not removable (i.e., it is essential and cannot be dismissed). For one thing, it is needed for breathing. For another, it can transport the light, image, and sound of both humans and animals. Yet another, it can maintain the nature of fire, water, and other elements. Whenever air is absent, humans and animals would suffocate and the fire in their hearts as well as their vital energy will be extinguished. In addition, the light radiating from the sky, the images of shaped objects, and the sounds traveling from many bodies would have nothing to depend on, and no means to reach where they are supposed to go, and no way to [benefit] what they contain inside their bodies.

If one says that the air is non-existent because it has no visible color, then will he say that all invisible things—wind, sound, smell, ghosts, and the souls of humans—do not exist? When our external eyes cannot perceive, one then must rely on our "internal eyes" of principle (reason) to attain. Air is thick and has three regions—above, middle, and below. The area above is nearest the fire (the sun) and should be warm at all times. The area below is nearest water and earth, which are illuminated by the sun and warm us sufficiently (for life). The area in the middle is far from the heaven above and the earth below. Therefore it is cold. In what way can we delineate the boundaries of these three regions? (Using) the highest mountain [for an example]. The area above the summit is the "top" (stratos) region, without any wind and rain. The air is very thin and people cannot live here. The area below the summit is the middle region. This is where the rain and snow gather and form. Below this is called the lower region. There are also differences between the warm region and cold region, the thick region and the thin region. The regions around the two poles, because of their distance from the sun [are cold], while warmer areas above and below are thin, but the cold area in the middle is dense. However, in the regions around the equator, because of their closeness to the sun, the warm areas above and below are thick, but the cold area in the middle is thin. From this we know that the air has different areas (strata).

1.3 *On the Sequence of the Four Elements and Their Form*

The Four Elements are not random or disordered, indicating perforce a certain sequence between them. Thus an element will remain in a peaceful (stable) state if it stays at the correct position, but will release [energy, force] if not in the correct position. When it is forced and has exhausted its energy, it will return to its original location. Where is the original location? Earth is above and Water is next to it; Fire is above and Air is next to it. This is a fixed order.

There are three reasons for the order: First is the measure of weight (heaviness vs. lightness). Heavy materials tend to be in a low position; lighter materials tend to be in a high position. By this we separate what is above from what is below, what is heavy from what is light. Moreover, there is a further division. Since there is a lower part of what is above and a higher part of what is below, by this we divide them into four elements. Water is lighter than Earth, while Air is heavier than Fire. Thus, Water is above Earth and Air is below Fire. But what we say that Water is heavy and Air is light is due to a comparison with other elements. In general, Water is said to be light when compared with Earth, but heavy when compared with Fire and Air. Air is said to be heavy when compared with Fire, but light when compared with Water and Earth. From this we know that Water must flow downward but not upward and Air must flow upward but not downward.

The second may be called harmony. The elements will function closely with each other if in a compatible situation; (but) they will remain far from each other if in an incompatible situation. For example, dryness and coldness form Earth, humidity and ice form Water. Earth and Water are compatible because of the (situation) of coldness. They are thus close to each other. Humidity and warmth form Air, while humidity and ice form Water. Water and Air are compatible because of [the situation] of humidity. Therefore they are also close to each other. Dryness and warmth form Fire, while humidity and warmth form Air. Air and Fire are compatible because of heat. They are also close to each other. If the elements are incompatible and opposite, they will be far from each other. For example, Water is cold and wet while Fire is hot and dry. The two [conditions]are just the opposite, so they are far apart. The reason that Earth and Fire are compatible by dryness but far apart from each other is because, although they are formed of compatible condition, they have quite different [measurements, or] "weights." Therefore by balancing these two principles one can determine the [natural] order of the Four Elements.

The third [method] is by visual examination. The order of the Four Elements can be readily examined before one's eyes. Fire originates as an ember and rises often at a sharp angle upward [from below]. In the West it is called a "fire shape (i.e., teardrop flame)," because it would not rest at a low point but strive upward to a very high point. When Air mixes with Earth or Water by accident, it is not able to rest (maintain stability) but will move upward. When (air) enters into Earth it makes earthquakes and landslides; when in Water it creates marshes and ponds. If one tried to force a ball under water, it will suddenly emerge. If Water is boiling in a field full of air, it must be with force and is not able to calm down. When the outside force is exhausted, it will return to its original state. Similarly, what makes rainfall is due to the sun evaporating the humid earth to form clouds. Scattered clouds belong to the Air [element], and thus are light and can float. (But) dense clouds belong to Water, and as such are "heavy" [and falling down]. To fall is to return to their original state. When Earth enters Water, it must go downward and settle as it reaches the bottom. The Four Elements must therefore be [formed] in a circular sequence. There are two reasons for this: First, the entire universe is itself spherical. The sphere consists of five large components—Heaven, Fire, Air, Water, and Earth. Since the shape of Heaven is round, it is certain that the Four Elements are all circular in shape; second, all of the Four Elements are tangential under the moon. If they have other shapes, then above the shape of Fire there might be either square or sharp-pointed but not circular shapes. It must be that they are not able to be tangential under the moon. Then there could be an empty area, which is not compatible with the nature of [physical] materials. Since it is circular above the Four Elements, it should be likewise below them. If there is any other shape, then what is surrounding the Earth will not be circular. Since the Earth is round, one may know that Water and Air connected with Earth will also be circular in shape. Nothing will survive unless it is circular. If it is square, it will easily fall apart and perish. For this reason, not only are Heaven and Earth and the Four Elements circular, but also none of the limbs of humans and animals as well as the plants, trees, and fruits are not round. Just like a drop of water must turn to round shape, it is how things would concentrate and survive rather than fall apart and perish.

1.4 *On Earthquakes*

An inquiry was raised about the cause of earthquakes, a topic of much discussion since ancient times. Some said that the earth moved by itself due to the vital forces within it; others said that the earth moved with the waves like a ship floating on the sea; still others said that the earth has unstable pieces and when these separate from the whole and fall on the hollow underground within, and might do no more than shake the landmass and loudly rumble; yet others say there was a *jiaolong* 蛟龍 (a legendary dragon with the ability to control rain and floods) or an *aoyu* 鰲魚 (a huge mythological sea-turtle) gyrating (in the deep) under the earth that make earthquakes. This is all nonsense and not deserving of further comment. One only need use reason to discuss a few aspects and natural features as follows.

First, earthquakes arise from the burning hot air within the earth. Since the sun outside at all times shines over the earth the fire inside burns constantly. When the hot air increases it squeezes into crevices and gaps [in the rock above]. The accumulated air grows increasingly dense and when finally unable to be contained it must burst out. When it cannot find a quick means of escape it therefore moves reciprocating forward or backward, turning over ceaselessly until it finally breaks through any obstruction and makes the earth tremble and roar. Like a cannon loaded with gunpowder that when ignited will explode and penetrate all obstacles, thus producing a loud report. For those who suspect, with doubting minds, that air alone is able to shake the ground, then they must know that the force of air is fierce and invincible. The wind, for example, may originate in a slight breeze but it can accumulate enough energy to move rocks and uproot trees, knock down houses and capsize ships. When air is trapped within the earth, the rupture must be powerful. It is a natural principle that the air bursts and shakes the body of the earth. How can this be strange?

If one wants to prove this truth, there could be two items that explain it. First, earthquakes mostly occur during spring and autumn. This is because the air can be easily created during these two seasons. Secondly, earthquakes must locate at domains with dry and loose soil as well as many caves. This is because these can easily contain a lot of air. Therefore, where the mountains collapse and expose many caves inside, the earthquake around could be even denser. If there are some openings on the ground facing toward the sky, they may disperse the air contained inside and the earthquake will not happen.

The islands in the seas have many earthquakes. This is because the sea water outside and sulphur contained inside can produce hot air. When the hot air becomes powerful, the earthquake will happen. Therefore the native people always dig many wells. They want to ventilate the air and let it disperse easily. In this way they can avoid the earthquake.

Normally before or after an earthquake, there could be either a lengthy drought or fierce wind so that the air becomes turbulent. There are three reasons for this: first, all the caverns within the earth are filled with air. When new air is produced and added within, it cannot move along with the existing air. The new air [pressure] will then burst and strive to escape. This leads to the shaking of the earth. Second, when the earth is covered by cold air, it will shrink. The air automatically flows inside and crashes on the earth without any order. Third, the hot air stored in the earth, once exposed to the cold air from outside, will retreat and shrink. The more it shrinks, the more force there will be, and the more liquefied in quality it will become. The more liquefied it becomes, the greater tendency to explode (powerful expansion) it will have. Through this it shakes the body of the earth.

The duration of earthquakes depends on the situation of the air. If there is a great volume of thick air, the quake will dissipate slowly. But if the air is in much less volume and thinner, the quake will disappear quickly. The topography [of the ground] is of secondary importance. The ground will be opened easily if the earth is loose and soft. The air can hardly escape if the earth is harsh and solid. Since the tremor may last a long time, sometimes it is disrupted while other times it continues. The shaking thus looks like a long-term movement of the earth. In reality it is just one movement and cannot last very long.

1.5 *On Humankind*

Under heaven throughout the wide world, there is nowhere that people do not abide. In ancient times, most doubted that anyone could live at the equator or the North and South Poles, primarily due to the heat or frigid climates there. Yet seafarers have now traveled to every corner around the world and everywhere found inhabited lands. This is sufficient to recognize the error of the old argument.

If one cares to enlighten himself, he may read the discussion in the *Kongji gezhi* 空際格致 [Investigation into Celestial Phenomena]. From East to West, in all places equidistant from the equator, people are very similar. In regions located at very different (longer) distances from the equator, the people abiding there are also quite different. This transformation on the earth arises from the circular movement through heaven from East to West with the sun, the moon, and the five stars. The nature of the Four Elements being cold/hot and dry/wet changes accordingly. The sun, moon, and the five stars move according to the ecliptic. The equal division of the ecliptic lies at the equator. That which all people under heaven jointly share is their spiritual nature. The regulation of the five moral relationships may vary in complexity, laws may be lax or meticulous, manners may be refined or rustic, and though all of these might be different in detail, ultimately none deviate from the basic principle. All share their spiritual nature though they differ in appearance and voice. Within the same species animals are similar in appearance, but a single human individual is unique. They are recognized by their appearance as such. This applies to the entire world, to a country, to a place, to a family. This is the intent of the Sovereign of All Under Heaven. Recognition of each other is profound and related to the moral relationships and proper governance. Were individuals to look alike, husbands and wives would go unrecognized, and fathers and sons remain indistinguishable. People could not govern their impulses, and licentiousness would emerge. Even if one wished to, how could it be controlled? Is not the similarity of individual animals within the same species due to the lack of moral relations and government? Having different appearances, voices can also be used to distinguish among people. As people have different visual acuity, they may not recognize each other if they meet in the night. Thus, they have to rely on their voices to verify their identities.

1.6 *Both South and North Poles of the Globe Must Correspond to the Poles of Heaven—the Earth Is Not Apart from the Center of Heaven*

The center of the Earth is the center of all the heavens. From the principle of a lunar eclipse this is evident. These are the basic arguments in the formation of the new (reformed) calendar. The reason the southern and northern poles of the earth correspond to the poles of the heaven and not away from the center can be seen clearly from the height of the poles themselves. It is likely the ten thousand countries under heaven have investigated the local height of the two poles since ancient times. What they observed then is the same as today.

The reason for not being away from the heavenly poles lies in the effective transformation of the ten thousand things. The effect of growth and transformation of the ten thousand things in different regions under heaven all derives from the sun and the stars. They come into being by following the order of the four seasons. There are differences above or below the horizon in different countries, such that hardness and softness, dryness and humidity follow to appropriately suit the ten thousand things. Now, if the two poles of the earth are not corresponding to the poles of heaven but are away from the latter two, either above or below or to the right or left, the ten thousand countries under heaven must be shaking in chaos after that. What was originally to the north of the equator now suddenly changes to the south of the equator; what was originally at its south now changes suddenly to its north. The near becomes distant, while the distant becomes near. The heat of summer suddenly changes to the coldness of the winter.

Then the order of the four seasons becomes confused and the effect of growth and transformation will be in great chaos. The ten thousand things will be extinguished! Knowing this, the two poles of the Earth eternally correspond to the two heavenly poles and will forever remain unchanged. How can one be confused by this?

Even if the earth has some changes by accident and deviates from the heavenly poles, it must itself have the ability of rotation in order to return to its original poles in accordance to the original places directing to the south and north poles in heaven. How the earth has its own ability of rotation is not at all different from a magnet. There is nothing strange about the power of a magnet but the power to face the south and north poles. Probably because the magnet is a type of pure soil in the earth, its natural *qi* 氣 (vital force) is the same as the natural *qi* of the earth. The so-called pure soil is one of the Four Elements and no other element is being mixed with it. The shallow soil and mixed soil, because of the shine of the sun, moon, and the stars, produce the five cereals (rice, two kinds of millet, wheat, and beans), one hundred fruits, plants, and trees with the effect of nourishment. But the pure soil stays in the deepest part of the earth, like in an iron ore mine inside the mountains. Knowing of this, [all elements in] the whole body of the earth are harmonized with each other. It seems as if there is a network of arteries and veins connecting them.

I have investigated the famous mountains in the ten thousand countries under heaven, and the mines of the five metals. I observed all the large rocks and the deep mines and saw clearly that at their steepest sides, there is a net of arteries and veins at each layer. Never has it not been from the bottom to the top and not faced the south and north poles. I, Ferdinand Verbiest, came to the Middle Kingdom from the Far West and have experienced a voyage of ninety thousand miles. I freed my will and looked at the mountains with steep cliffs along the coasts. When I inspected the network of arteries and veins on both south and north sides, they in general all directed to the South and North Poles. There is another network in the middle where the angle it intersects with the horizon at that place is in accordance to the angle where the north pole at the place intersects with the horizon. The caves with mines of the five metals and stone are the same. Within this kind of network, there is often the *qi* of the magnet.

Ever have I also investigated the cartographic books and maps of the ten thousand countries under heaven, [and] all of the famous mountains and large rivers of the five continents intertwining with each other. These extend to a distance of several thousand miles, from south to north, winding and mixing together. Their outline on the earth is obvious to see. Among these [separate] networks are connections with each other, just as the arteries and veins as well as the joints and bones are connected in the human body. They crisscross and connect with each other to make the entire body.

1.7 *On Rivers*

Within the confines of the earth a great repository of water is stored. It is widely noted that mineworkers encounter [subterranean] pools or valleys with rapids, and by digging wells in many places these fountains may be reached no matter how shallow or deep the well might be. Furthermore, in some cases spring water comes out of the dry ground, sometimes forming lakes, sometimes flooding the people's homes. In this way we ascertain that these events could not occur were there not a great store of water within the earth.

Moreover, when the Creator (*Zaowuzhe* 造物者) returned the floodwater into the abyss, much remained in the earth. [And this] also formed cavities and hidden ditches everywhere, so that [His] favor of nourishment (*runze* 潤澤) could spread, just as there are veins, bones, and muscles within the human body throughout which the nourishment of blood and *qi* can be circulated. In the beginning the earth was extremely arid. It could not by itself solidify without the [nourishing moisture] of water. Neither could flowers, metal, and stone be supported for use by human beings. Therefore we know that since the creation of heaven and earth, there were already many springs, rivers, and lakes for later use.

Rivers, creeks, and springs derive from the sea. Here are four aspects that prove it: first, all the world's rivers daily flow into the sea, yet they do not spill. There must be a pathway out. If there is no pathway out but still no spillage, it would be extremely difficult to comprehend. Second, there could be no other "great origins" if large rivers did not originate from the sea. The water that

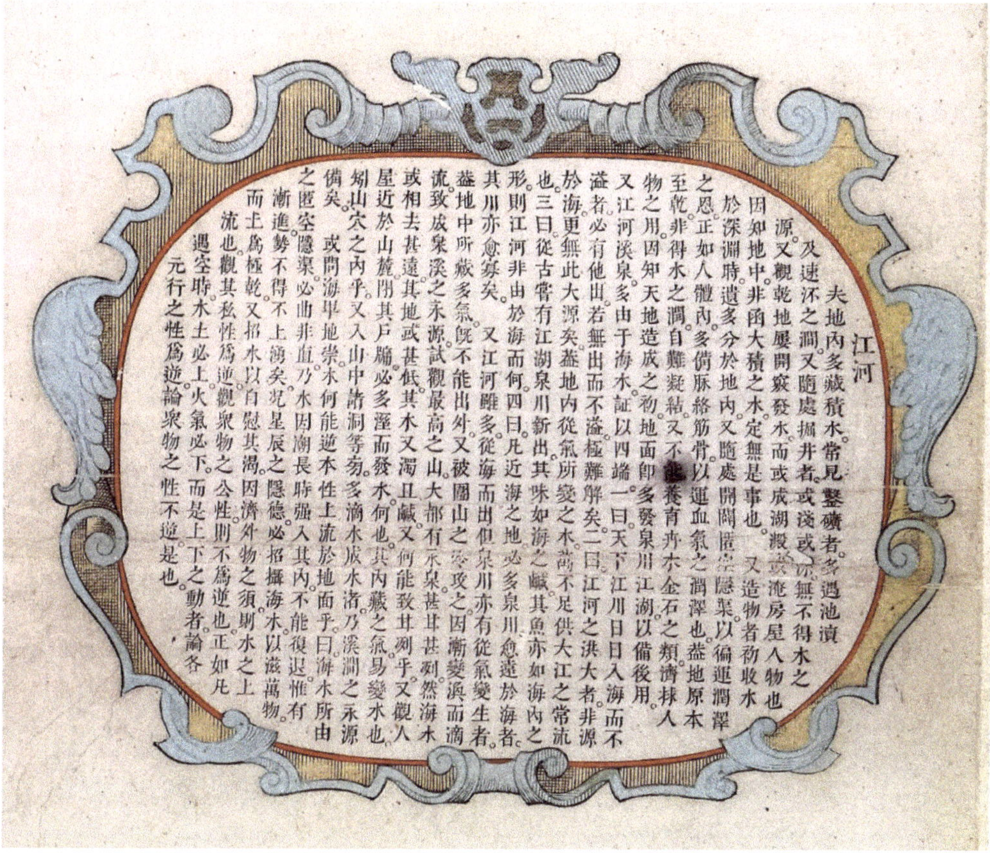

FIGURE 7.4 Cartouche from the *Kunyu quantu* 坤輿全圖. Detail on Rivers. Kobe Municipal Library.

comes out of the *qi* inside the earth is not enough to support the constant flow of the big rivers. Third, since ancient times there were newly formed rivers, lakes, and fountains, tasting similar to salty sea water and with fish similar to those in the sea. Thus, aside from the sea, where else could rivers have come from? Fourth, there are many springs and rivers in places close by the sea. The farther a place from the sea, the fewer rivers there would be.

Although the rivers mostly originate from the sea, some springs and streams may be born from the air (*qi*). A great deal of air is stored within the earth. It cannot escape and if trapped frozen by the cold surrounding mountains, it therefore gradually transforms (condenses) into dripping water (subterranean streams) and becomes the source of springs and creeks. When looking at the highest mountains we often find permanent springs, very sweet and cool, while the sea is far away. At times when the location is quite low and the water is turbid and salty, how can it be sweet and cool? Moreover, when people build their houses near the foot of a mountain, it must be very humid and damp. Why does this happen? The air hidden within [the mountain] can easily condense into water, not to mention that from inside of caves within. Further, in those mountain caves, where there is an excess of dripping (i.e., subterranean) water to make small ponds and islets, these become permanent sources of storage for canyon springs.

Someone inquires "The sea is low but the earth is high, how can water flow up to the ground against its nature?" [I] respond the empty cavities and ditches that seawater pass through must be meandering and serpentine instead of straight. When the tide rises to flood, the water is pressurized under force. Since it cannot return [to its source], it must proceed gradually and overflow onto the ground. Also because of the 'hidden virtue' of the stars, they must attract seawater to nourish the ten thousand things. But since it is extremely dry above the surface, they must [first] pull the water to satisfy their thirst to support the needs of other things. This is why water flows upward. When we observe its particular nature, it could seem rebellious. But if we observe its outward nature, it is not rebellious at all. Whenever there are openings, water and earth must go upward, while fire and air must go downward. These movements up and down are against the nature of the Four Elements. But when we speak of the nature of various things, it is not against their nature.

Acknowledgement

I wish to express my profound thanks to Prof. Song Gang 宋剛 of the University of Hong Kong for his generous help with the translation of Verbiest's often difficult Chinese neologisms and concepts in the descriptions.

Bibliography

Chapman, Allen. "Tycho Brahe: Instrument Designer, Observer, and Mechanician." *Journal of the British Astronomical Association* 99, no. 2 (1989): 70–77.

Chen Hui-Hung 陳慧宏. "The Human Body as a Universe: Understanding Heaven by Visualization and Sensibility in Jesuit Cartography in China." *The Catholic Historical Review* 93, no. 3 (July 2007): 517–552.

Cheng Fangyi. "Pleasing the Emperor: Revisiting the Figured Chinese Manuscript of Matteo Ricci's Maps." *Journal of Jesuit Studies* 6, no. 1 (2019): 31–43. doi: https://doi.org/10.1163/22141332-00601003.

Creasy, H., H. Metcalf, and N. Pearce. "History and Conservation of *Kunyu quantu* (A Map of the Whole World) by Ferdinand Verbiest, 1674" In *Printed on Paper: The Techniques, History, and Conservation of Printed Media*, edited by J. Colbourne and R.F. Snyder, 33–41. Newcastle upon Tyne, UK: Arts and Social Sciences Academic Press, Northumbria University, 2009.

Donne, John. "Hymne to God, My God in My Sicknesse." In *John Donne, Selected Poetry*, edited by John Carey. Oxford World Classics, 210. Oxford: Oxford University Press, 1998.

Menegon, Eugenio. "New Knowledge of Strange Things: Exotic Animals from the West" 存廣異聞：西方異獸. *Gujin lunheng* 古今論衡 [*Disquisitions on the Past and the Present*] 15 (October 2006): 39–48.

Niayesh, Ladan. "All Flat Maps, and I am One: Cartographic References in the Poems of John Donne." *Études Épistémè*, no. 10 (automne 2006). https://doi.org/10.4000/episteme.955.

Schütte, Josef Franz, *Valignano's Mission Principles for Japan*. St. Louis, Institute of Jesuit Sources, 1980–1985.

Standaert, Nicolas. "The Investigation of Things and the Fathoming of Principles (格物穷理/*Gewu Qiongli*) in the Seventeenth-Century Contact Between Jesuits and Chinese Scholars." In *Ferdinand Verbiest (1623–1688): Jesuit Missionary, Scientist, Engineer and Diplomat*, edited by John W. Witek, 396–399. Nettetal: Steyler Verlag, 1994.

Verbiest, Ferdinad. *Map of the World, 1674*. Online in a high-resolution zoomable format, Kobe Museum. https://artsandculture.google.com/asset/map-of-the-world-ferdinad-verbiest/2AEndoCIl640pQ?hl=en (viewed 10.8.2020).

PART 3

Reverberations of Ricci's Maps in East Asia

CHAPTER 8

Representing an Ideal World Order of the Past: The Cultural Function of the Jesuit World Maps in Eighteenth-Century Korean Government

Lim Jongtae

1 Introduction

The Chosŏn (1392–1910) government commissioned two reproductions of Jesuit world maps in the eighteenth century, both of which are currently housed at Seoul National University. The first one, in the care of the University Museum, is Matteo Ricci's *Kunyu wanguo quantu* 坤輿萬國全圖 [The Complete Map of Ten Thousand Countries in the World]. It was produced in 1708, on the king's order, by the dynasty's Royal Astronomy Bureau. This copy reproduced Ricci's original as faithfully as possible, preserving its original form consisting of large scrolls. Seen from numerous pictures of European sailing vessels and exotic animals featured in the map, the original Ricci map is not the famous Beijing edition of the same title, printed in 1602. Researchers believed that the original was a copy of the one presented to the Ming Wanli 萬曆 emperor in 1608 (Figure 8.1).[1] The other one, preserved in the Kyujanggak Institute for Korean Studies, was presumably drawn in around 1790, based on Giulio Aleni's *Wanguo quantu* 萬國全圖 [The Complete Map of Ten Thousand Countries]. Although there is no clue, the Chosŏn reproducers used the version included in Aleni's world geography text, *Zhifang waiji* 職方外紀 [Records of Regions beyond the Jurisdiction of the Imperial Geographer, 1623], which was in a wide circulation in Korea during the eighteenth century. Unlike the reproduction of Ricci's map, this remake of Aleni's map is included in an atlas of three volumes,

1 O Sanghak, *Chosŏn sidae segye chido-wa segye insik* 朝鮮時代世界地圖와世界認識 [World Maps and Worldviews in the Chosŏn Period] (P'aju: Ch'angbi, 2011), 180–184. A photo of a much better copy of the same reproduction of Ricci's map is also preserved in Seoul National University—this one in the Kyujanggak Institute for Korean Studies. This photo was taken in 1932 at the Pongsŏnsa Temple, a Buddhist temple in the eastern suburb of Seoul where the map had been preserved. This map was later destroyed during the Korean War (1950–1953). As for this Pongsŏnsa copy, see O Sanghak, *Chosŏn sidae*, 180.

© LIM JONGTAE, 2024 | DOI:10.1163/9789004684782_010
This is an open access chapter distributed under the terms of the CC BY-NC-ND 4.0 license.

FIGURE 8.1 *Kunyu wanguo quantu* 坤輿萬國全, 1708. Seoul National University Museum, Korea.

entitled *Yŏji to* 輿地圖 [Terrestrial Maps] (Figure 8.2).[2] As far as I know, these two maps are all that have so far been identified as official reproductions of Jesuit maps from seventeenth- and eighteenth-century Korea.

This article takes these two eighteenth-century government reproductions of the Jesuit maps as a window through which to examine the ways in which the government of a premodern East Asian dynasty viewed and used world maps of foreign origin. Despite their rarity, the very presence of these fine reproductions demonstrates that the Chosŏn government and its officials were not blind opponents of the new Western maps. On the contrary, the government, as early as from the early seventeenth century, played the role of a patron of Jesuit maps in the late Chosŏn period. It encouraged its tributary

2 Although there is no written evidence showing that this atlas was a government compilation, researchers believe it to be so on the basis of the excellent drawing skill on display and the use of high-quality paper and pigment. See O Sanghak, *Chosŏn sidae*, 186–187. The date of its compilation is estimated to be between 1789 and 1795. See the bibliographical note to this atlas provided by the Kyujganggak Institute for Korean Studies: https://kyu.snu.ac.kr/ (accessed on Oct. 11, 2017).

envoys to China to bring the new Jesuit maps to Korea and by later commissioning, albeit occasionally, these fine reproductions of them accorded to the suspicious Western maps an official endorsement. Why, then, did the Chosŏn government pay such positive attention to Western maps? What kind of values, whether political, cultural, or scientific, did the government officials find in them?

One possible answer might be that government officials appreciated the Jesuit maps' scientific value, which enabled a more objective description of the world than their traditional counterparts. This is a plausible hypothesis, given the contemporaneous advancement of cartography in Korea, evidenced in a series of accurate maps of Chosŏn that were produced or sponsored by the government in the eighteenth and nineteenth centuries. The Jesuit map might have been welcomed by Korean map makers as a better example of the precision and accuracy that could be applied to the production of maps of the Chosŏn kingdom. The only problem in this line of reasoning is that there is no direct connection between the two phenomena: the production of high quality, national maps in the eighteenth and nineteenth centuries was made

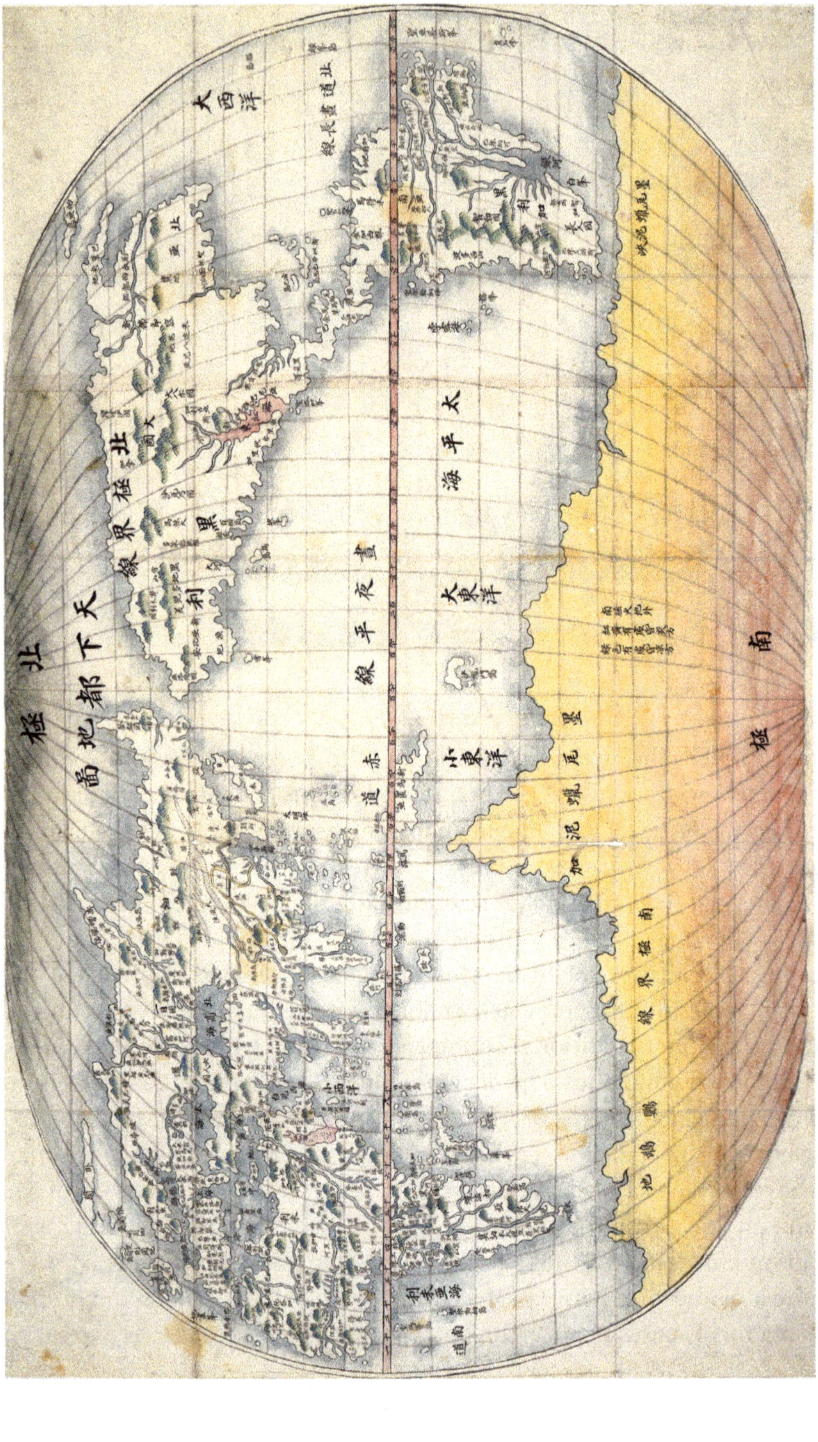

FIGURE 8.2 "Chŏnha to chi do" 天下都地圖. A Reproduction of Aleni's Map, ca. 1790. Kyujanggak Institute for Korean Studies, Seoul National University, Korea.

possible mostly by applying traditional techniques, such as the square grid system, rather than the Western methods of triangulation or astronomical observation.[3]

What further complicates the issue is that the Jesuit maps were not the only kind that the Chosŏn government produced as its official world maps. An example of the alternative kinds of world maps produced by the government is shown in Figure 8.3. This map, entitled "Ch'ŏnha to" 天下圖 [A Map of All under Heaven], was included in another government atlas from the mid-eighteenth century, *Haedong chido* 海東地圖 [Maps of the Eastern Kingdom].[4] Dubbed by Joseph Needham as an example of East Asian "religious cosmography," this so-called Korean "wheel map" presented a highly mythological world picture, drawn from the *Shanhai jing* 山海經 [The Classic of Mountains and Seas], the famous geographical-cum-mythological text of ancient China.[5]

This strange coexistence of two seemingly incommensurable pieces in the list of the dynasty's official world maps discourages us from taking the alleged scientific merits of the Jesuit maps as the sole explanation for the Chosŏn government's patronage of them. It rather prompts us to examine the official reproductions of the Jesuit maps in a much broader context, one in which the Chosŏn government produced and used various kinds of maps for various purposes. To this end, I will first locate the reproduction of the Jesuit maps in a longer historical process in which the Chosŏn government introduced and then consumed world maps from China. And, in the second part, I will narrow down the focus to the specific contexts surrounding the reproduction of the two Jesuit maps in the eighteenth century. Two kinds of context will be considered: first, a larger political and ideological environment that informed the official reproduction of the Jesuit maps, and second, the context of the late Chosŏn maps, in which the Jesuit maps were located. In particular, I will examine the specific relationships the reproduction of Aleni's map established with other maps included in the same government atlas, *Yŏji to*. This approach will lead us to a rather surprising conclusion: the Jesuit maps appealed to the Chosŏn officials not so much for their technical potential for "scientific" cartography as for their historical origin in Ming China.

---

3 Lim Jongtae, "Matteo Ricci's World Maps in Late Joseon Dynasty," *The Korean Journal for the History of Science* 33, no. 2 (2011): 277–296.

4 *Haedong chido* 海東地圖 [Maps of the Eastern Kingdom] (Seoul: Kyujanggak Institute for Korean Studies, 1995), 1:2.

5 Joseph Needham, *Science and Civilisation in China* (Cambridge: Cambridge University Press, 1959) 3: 583–590. See also O Sanghak, "Circular World Maps of the Joseon Dynasty: Their Characteristics and Worldview," *Korea Journal* (Spring 2008): 8–45.

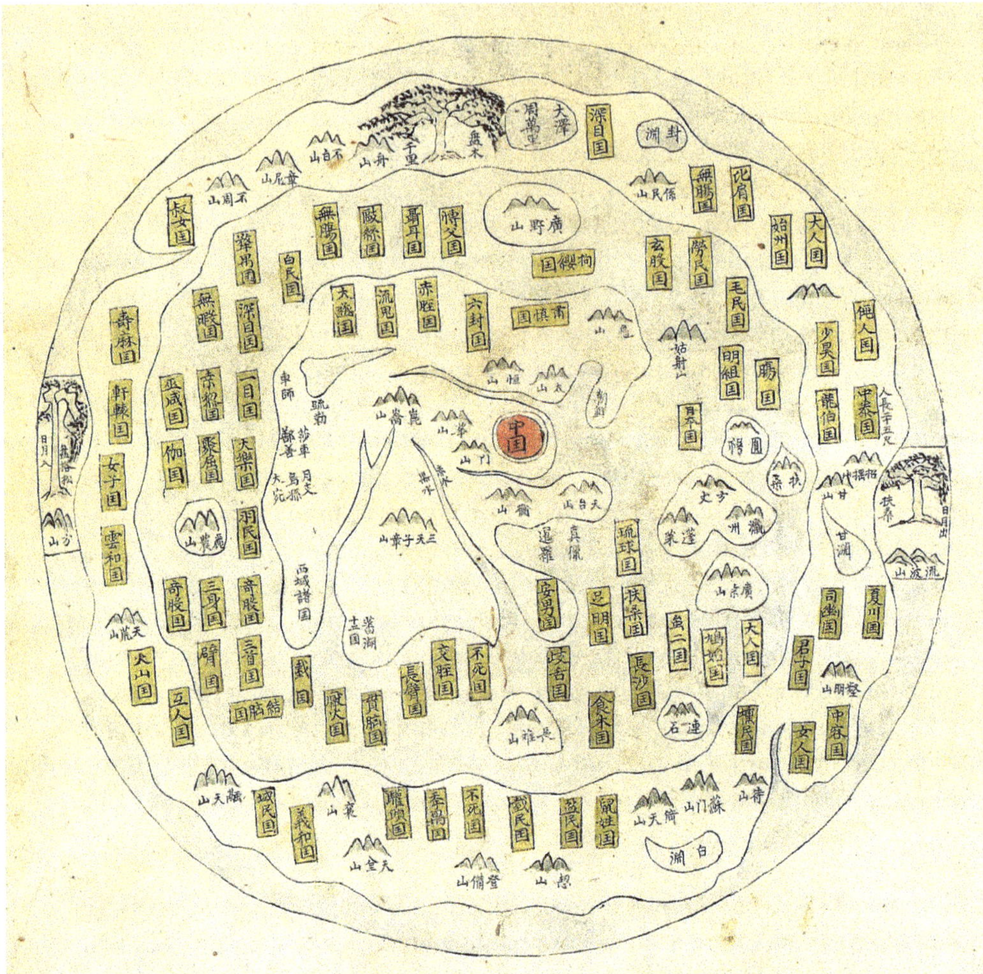

FIGURE 8.3 "Ch'ŏnha do" 天下圖 in *Haedong chido* 海東地圖 (ca. 1750). Kyujanggak Institute for Korean Studies, Seoul National University, Korea.

2 World Maps in the Chosŏn Government

In most of its premodern history, Korea was not so much a creator of world maps on its own as a consumer of those produced in China.[6] For this reason,

6 As for general surveys of world maps in premodern Korea, see O Sanghak, *Chosŏn sidae*; Gari Ledyard, "Cartography in Korea," in *The History of Cartography*, vol. 2, bk. 2, *Cartography in the Traditional East and Southeast Asian Societies*, eds. J.B. Harley and David Woodward

the world maps that circulated in premodern Korea were not greatly distinguishable from those in China, either in their appearances or in their underlying cartographical methods. But Korea's age-old position as a consumer of Chinese world maps accorded to these world maps a significantly different cultural and political meaning in Korea that was not usually assumed in China. And, in this regard, the Jesuit world maps in Korea were not especially different. To the Koreans, they were another kind of world map from China, even though they were well aware that the maps were produced by Western priests. In fact, importing world maps of non-Chinese origin from China was not entirely new: the oldest extant world map of Korea, dated 1402, also partly originated in the Western world.

As is well known, this map, entitled *Honil kangni yŏktae kukto chi to* 混一疆理歷代國都之圖 ["Map of Integrated Lands and Regions of Historical Countries and Capitals"], hereafter *Kangni to*, was a synthesis of two Mongol maps (Figure 8.4).[7] This map's vast coverage of the known world, including Europe and Africa, indicates that the original Mongol maps incorporated elements of knowledge from the Islamic world in the course of vibrant cultural exchanges between the East and the West under Mongol rule.[8] This map also illustrates that premodern Koreans were able to access geographical information concerning the wider world only through the geographical knowledge possessed by the Chinese empires.

Korea's dependence on China in regards to world geography no doubt reflected its relatively limited range of international relations. During the Chosŏn period, Korea's foreign relations, whether political or commercial, were largely confined to its direct neighbors: China, Japan, Ryukyu, and the Jurchen peoples on the northern border. But there is an additional factor, which might be called ideological, that amplified Korea's reliance on China's world geography: the idea of the so-called tributary system that specified a hierarchical relationship between the rulers of Korea and China. According to this, Chosŏn kings were supposed to assume a subordinate position to Chinese emperors, who, under the mandate of heaven, ruled "all under heaven" (*tianxia*

(Chicago: The University of Chicago Press, 1994), 235–345; O Sanghak, *Han'guk chŏnt'ong chirihak sa* 韓國傳統地理學史 [A History of Geographical Learning in Traditional Korea] (P'aju: Tŭlnyŏk, 2015).

7 As there are several systems of romanization for Korean, the *Kangni to* is also sometimes referred to as the *Gangnido*, as in Chapter 9 of this volume, or *Kangnido*, as in Chapter 6.

8 Gary Ledyard, "Cartography in Korea," 244–249; Miya Noriko, *Mongoru teikokuga unda sekaizu* モンゴル帝国が生んだ世界図 [A World Map Produced by the Mongol Empire] (Tokyo: Nihon keizai sinbunsha, 2007); O Sanghak, *Chosŏn sidae*, 91–133.

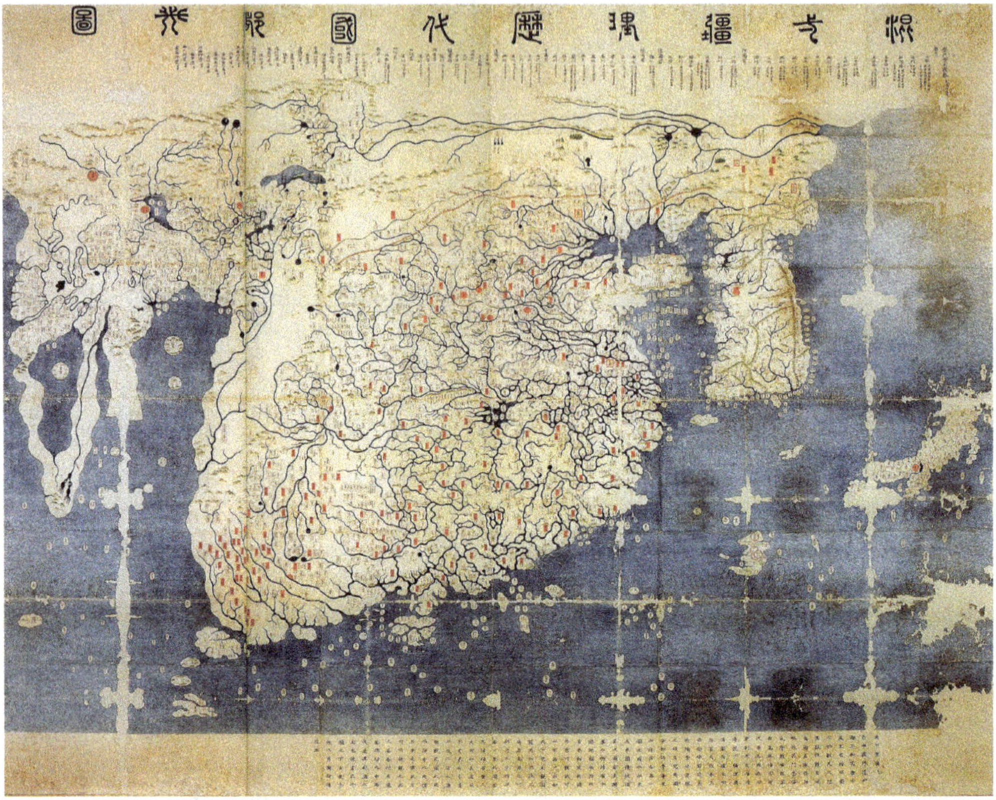

FIGURE 8.4 A later copy of the *Honil kangni yŏktae kukto chi do* (originally produced 1402). Honkōji Tokiwa Museum of Historical Materials, Nagasaki, Japan.

天下).[9] In this scheme, "*tianxia*" was simply outside the purview of a Chosŏn king and thus could not be a legitimate object of his political gaze. Although this factor never prevented the Chosŏn kings and literati officials from producing and consuming world maps, it no doubt discouraged the Koreans from being creators of their own world maps.

9 For a brief account of the idea of the tributary system, see J.K. Fairbank, "A Preliminary Framework," in *The Chinese World Order: Traditional China's Foreign Relations*, ed. J.K. Fairbank (Cambridge, MA: Harvard University Press, 1968), 1–19. As for the Sino-Korean relationship in the Chosŏn period, see Donald N. Clark, "Sino-Korean Tributary Relations under the Ming," in *The Cambridge History of China, Volume 8: the Ming Dynasty, 1368–1644, Part II*, ed. Denis Twitchett and Frederick W. Mote (Cambridge: Cambridge University Press, 1998), 272–300; Lim Jongtae, "Tributary Relations between the Chosŏn and Ch'ing Courts to 1800," in *The Cambridge History of China, Vol. 9: The Ch'ing Empire to 1800. Part II*, ed. Willard J. Peterson (Cambridge: Cambridge University Press, 2016), 146–196.

The 1402 Korean world map, *Kangni to*, illustrates this point very well. Although previous research has argued that the State Council of the Chosŏn Dynasty compiled this map to highlight the Chosŏn king's power,[10] this is quite unlikely. King T'aejong 太宗 (r. 1400–1418), whose accession to the throne two years earlier was morally problematic, had sought to consolidate his weak political legitimacy by receiving the Ming emperor's endorsement of his kingship as a tributary ruler. And, in 1401, just a year before the production of the *Kangni to* map, King T'aejong had become the first Chosŏn king to earn the Ming emperor's investiture. As can be seen from the map, the official map makers fully acknowledged the central and dominant position of the Ming Empire, while limiting the range of Korea's agency to the eastern part of the world. This map thus represented the marginal, yet still significant, position of the Chosŏn Dynasty in a world then dominated by the Ming Empire. This case illustrates that Chosŏn officials needed a Chinese representation of the world before they could inscribe into it the political and cultural position of their state. This function of a world map might be viewed as political, but it was so largely in a symbolic sense.[11]

A more directly political function, relating to the legitimacy of royal power and the practical matters of state governance, was ascribed to national maps. In the same year that the *Kangni to* world map was made, the Chosŏn government also produced a national map, entitled *Pon'guk chido* 本國地圖 [A Map of Our State], as a special gift to celebrate King T'aejong's birthday.[12] By presenting the national map to the throne, government officials wanted to emphasize the king's sovereignty over the kingdom. In addition to playing this symbolic function, the national maps, together with the geographical gazetteers of the kingdom, were also expected to provide a rich repository of practical information to be used in various fields of statecraft, such as taxation, national defense, and the moral edification of the people.

This stark contrast between the roles of world maps and national maps in the Chosŏn government explains the different trajectories that the two fields—world geography and national geography—would take later in the Chosŏn period. Due to its obvious practical importance, national geography received far more attention from the government, as seen in the government's

10 See, for example, Miya Noriko, *Mongoru teikokuga*, 218–235; O Sanghak, *Chosŏn sidae*, 95.

11 Lim Jongtae, "Ŭijŏngbu-ŭi segye chido, Honil kangni yŏktae kukto chi to" 議政府의 世界地圖, 混一疆理歷代國都之圖 [Honil kangni yŏktae kukto chi to as a World Map by the State Council of the Chosŏn Dynasty], *Munhwa yŏksa chiri* 文化歷史地理 27, no. 1 (2015): 1–14.

12 *Chosŏn wangjo sillok* 朝鮮王朝實錄 [Veritable Records of the Chosŏn Dynasty], the 16th day, 5th month of the 2nd year (1402) in King T'aejong's reign.

continuous efforts to produce and update the maps and the geographical treatises of the kingdom. This government support was one of the major forces that drove the remarkable progress of Chosŏn cartography in the field of national maps in the eighteenth and nineteenth centuries.[13] In comparison, world maps, which were mainly symbolic to kings and officials, were of a secondary importance. Although the Chosŏn government did not entirely neglect the field of world geography, the government's support was less frequent and mainly consisted in purchasing world maps from China instead of making its own.

Due to their assigned purpose of representing Korea's symbolic position in the world, the occasional interest the Chosŏn government showed in world maps was provoked only in the periods when kings and elite officials had reasons to feel an acute need to readjust Korea's (symbolic) relationship with the outer world. Notable examples were the periods after major rearrangements of the regional political order, such as dynastic changes in China. In those periods, the Chosŏn government paid unusual attention to purchasing maps from China, and even to reproducing those maps to represent Korea's new position in the newly emerging order in East Asia.

We can identify three of those special moments through the end of the eighteenth century. The first one occurred during the early decades of the fifteenth century, when the new Chosŏn Dynasty made a strenuous effort to establish a favorable relationship with the new Ming Empire. This period witnessed the production of the above-mentioned *Kangni to* map. The second moment took place in the early sixteenth century, when the Chosŏn king Chungjong 中宗 (r. 1506–1544) showed an unusual interest in purchasing world maps from China. This royal initiative, according to recent research, gave rise to a world map, entitled *Honil yŏktae kukto kangni chido* 混一歷代國都疆理地圖, a modified reproduction of a Ming map, "Yudi tu" 輿地圖, with Yang Ziqi's 楊子器 postscript.[14] We do not know the exact reason for the king's interest, but it might have been a part of his larger political agenda of enhancing his weak

13 The Chosŏn government's investment in the field of national geography is well documented in O Sanghak, *Han'guk chŏnt'ong chirihak sa*. As for the advancement of cartography in late Chosŏn period, see Yang Pogyŏng, "Taedong yŏjido-rŭl mandŭlgi kkaji" 大東輿地圖를만들기까지 [History of Cartography in Korea up to the "Taedong yŏjido"], *Hanguksa simin kangjwa* 韓國史市民講座 16 (1995): 84–121.

14 Son Koŭn, *16segi Cho-Myŏng woegyo-wa chiri chisik kyoryu: 'Honil yŏktae kukto kangni chi doo'rŭl chungsimŭro* 16世紀朝明外交와地理知識交流:'混一歷代國都疆理地圖'를中心으로 [Diplomatic Relations and Geographical Knowledge Exchange between Chosŏn and the Ming in the Early Sixteenth Century: Focusing on the "Honil yŏktae kukto kangni chido"] (master's thesis, Korea University, 2017).

political legitimacy by strengthening his status as a tributary ruler of the Ming emperor. Researchers agree that King Chungjong's reign was a crucial moment through which the relationship between Chosŏn and Ming was dramatically consolidated, to such an extent that Chosŏn now assumed the position of a filial son with regard to the Ming empire.[15] The king might have wanted the Ming's world maps as a visual means of supporting his political drive based on a Sino-centric world view.

The third and final period in which the Chosŏn government showed an unusual interest in world maps was the early eighteenth century, half a century after the Ming-Qing dynastic change in Beijing in 1644. In this period, the Chosŏn government reproduced Matteo Ricci's world map in 1708, one of the two government reproductions of Jesuit maps that I mentioned at the beginning of this paper. What specific relationship, then, can we find there, between the historic event of the Ming-Qing dynastic change in China and the Chosŏn government's reproduction of a Western world map?[16]

3 Ming Loyalism and the Chosŏn Government's Reproductions of the Jesuit Maps in the Eighteenth Century

The 1708 reproduction of Ricci's map was, as far as I know, the first case in which the Chosŏn government officially expressed a significant interest in reproducing and disseminating Jesuit world maps. Before that, we cannot find any record of Jesuit maps in the government's chronicles, although Jesuit maps had been known to Chosŏn society since the early seventeenth century.

This long silence about Jesuit maps in the official dynastic records does not mean that the government was entirely indifferent to them. During the seventeenth century, the Chosŏn government quietly played its role, particularly in introducing Jesuit maps from China. The main conduits were Chosŏn's

15 See, for example, Kye Sŭngbŏm, *Chungjong-ŭi sidae* 中宗의 時代 [King Chungjong's Era] (Seoul: Yŏksa pip'yŏng sa, 2014).

16 If we extend the scope of this paper into the nineteenth century, we may add the fourth instance of the Chosŏn government's interest in world maps from China. This renewed interest was kindled by a series of military crises in China caused by Western imperial powers in the mid-nineteenth century. In this period, the Chosŏn government and officials made efforts to introduce the new world geographical treatises from China, such as *Haiguo tuzhi* 海國圖志 [Illustrated Treatise on the Maritime Kingdoms], as sources to learn the situation and the way to cope with the Western threat. This period also witnessed a renewed Korean interest in the old Jesuit world maps, as shown in the 1860 official reproduction of Verbiest's world map, *Kunyu quantu* 坤輿全圖 [Complete Map of the Earth]. See, O Sanghak, *Chosŏn sidae*, 188.

tributary missions to China. The official envoys were generally encouraged to bring back valuable items from China, whether they were of strategic importance or had cultural value. Maps of various kinds were among the most favored items for the envoys to present to the court as their achievements in the mission. From the early seventeenth century, Jesuit maps began to be included, as shown in a few remaining private records by literati officials.

Yi Sugwang 李睟光 (1563–1628) was thus able to see Matteo Ricci's *Kunyu wanguo quantu* in 1603, only one year after its production in Beijing. This copy was presented to the Chosŏn court by the envoys upon their return from China. Three decades afterward, another returning envoy, Chŏng Tuwon 鄭斗源 (1581–?), presented to the king Ricci's world map and Aleni's world geography text, *Zhifang waiji* 職方外紀, together with many other books and artifacts concerning Western learning. He had acquired these items from a Portuguese Jesuit, João Rodrigues (1561–1633), whom he had met in Shandong during his mission. Not all the Jesuit maps acquired during the mission were to be presented to the court. In some cases, the envoys purchased Jesuit maps for their own private collections. A notable example was the copy of Matteo Ricci's *Liangyi xuanlan tu* 兩儀玄覽圖 [Map of the Profound View of the Heaven and Earth, 1603], which is preserved in the Korean Christian Museum in Seoul. Through these sporadic efforts by envoys during the seventeenth century, all the major Jesuit maps, including Verbiest's *Kunyu quantu* 坤輿全圖 [Complete Map of the Earth], were introduced to Korea and then spread gradually into the wider literati society.[17]

4 1708 Reproduction of Ricci's Map

The reproduction of Ricci's map in 1708 was not simply an outcome of this ever-growing popularity of Jesuit maps in Chosŏn society. The government initiative in this reproduction has several unusual aspects.

First of all, the court discussion that initiated the reproduction of Ricci's map in the fifth month of 1708 was recorded in the official chronicles of the dynasty, *Sŭngjŏngwon ilgi* 承政院日記 [The Daily Records of the Royal Secretariat].[18] Curiously, it is one of only a few records about the official production of a world map of any kind that we can identify in the official chronicles for the entire Chosŏn period. This implies that the government accorded unusual importance to this project.

17 Lim Jongtae, "Matteo Ricci's World Maps in Late Joseon Dynasty," 278–286.

18 *Sŭngjŏngwon ilgi*, the 12th day, 5th month of the 34th year (1708) in King Sukchong's reign.

More strikingly, no one paid any attention to the Western origin of the map and its author in the court discussion of this issue, although they correctly named it *Kunyu wanguo quantu* (Konyŏ man'guk chŏndo, in Korean). To King Sukchong 肅宗 (r. 1674–1720) and his officials, it was simply "a [map] made in China" (*tangbon* 唐本). In his order to make a reproduction of the map, the king explained its provenance as follows:

> *Konyŏ man'guk chŏndo* [...] is a [map] made in China. It is said to have been brought [to us] during the Wanli 萬曆 (1572–1620) and Tianqi 天啓 (1620–1627) reign periods [of the Ming Dynasty]. It has since then been preserved in the Kanghwa Prefecture. [...] This map is indeed extremely detailed [in representing the world]. But being a Chinese edition [made of Chinese paper], which deteriorates easily, this map also has many damaged places. If a copy is not made immediately, I am afraid it would not able to be handed down to posterity.[19]

Notably, the king regarded the Jesuit map as part of the cultural heritage of the Chosŏn Dynasty, intended to be handed down to posterity. The only expressed reasons for this were, first, its high level of detail in representing the world, and second, its origin in Ming China. In other words, Ricci's map was viewed simply as a detailed world map from Ming China, and was thus regarded as deserving of reproduction by the Chosŏn government. If there is little possibility that the king and the officials were entirely ignorant of the map's Western origin,[20] then King Sukchong's refashioning of the Jesuit map into the Ming's legacy should be viewed as an attempt to give a new cultural identity to the Jesuit map.

Indeed, the reproduction of Ricci's map in 1708 signaled a drastic change in the Chosŏn government's perception of Western maps. Up until that time, Jesuit world maps had remained, at best, passing exotica from the remote Western countries. In the early seventeenth century, Yi Sugwang placed his remarks on Ricci's map in a chapter of his encyclopedia devoted to the marvels and strange things of remote foreign lands. At times, Jesuit maps were even criticized for their underlying idea of a round earth. In spite of the growing popularity of Jesuit maps, they failed to establish themselves as reliable

19 *Sŭngjŏngwon ilgi*, the 12th day, 5th month, of the 34th year (1708) in King Sukchong's reign.

20 As will soon be discussed in footnote 22, however, the officials who reproduced *Kunyu wanguo quantu* in 1708 misidentified its author as Adam Schall von Bell (Tang Ruowang 湯若望), the maker of a star map which was reproduced by the Chosŏn government together with Matteo Ricci's map.

representations of the world and thus legitimate objects deserving literati officials' serious interest.[21] In 1708, however, Western maps suddenly received official recognition as a noble legacy of the Ming Dynasty, to which the government paid unusual attention.

Why did this sudden change in the government's attitude toward Jesuit maps occur at this specific moment in the first decade of the eighteenth century? Extant documentary evidence suggests that it was an outcome of the overlap of two seemingly unrelated events: the government's heightened interest in Jesuit astronomy, on the one hand, and the government's ideological program to establish the Chosŏn Dynasty as a legitimate heir of the fallen Ming Dynasty, on the other.

First, in the first decade of the eighteenth century, the Chosŏn government gave its full support to the Astronomy Bureau's project of learning the *Shixian li* 時憲曆 system, the official calendrical system of the Qing Dynasty which was presented by the Jesuits and then promulgated upon the Qing's occupation of Beijing in 1644. Although the Chosŏn Dynasty adopted the *Shixian li* system as early as in 1653, its implementation of it was only partial, limited to the issuance of annual civil almanacs. A full-scale adoption of the Jesuit methods, including those for calculating planetary movements and lunar and solar eclipses, did not materialize during the second half of the seventeenth century in spite of successive proposals by several interested officials. This general lack of government interest in the *Shixian li* system might partly be explained by an anti-Qing ideology, which had been deeply entrenched in the literati officials' minds since the Qing's invasion of Chosŏn in 1637 and then of the Ming in 1644. To them, the Jesuit calendrical system promulgated by the Qing symbolized the barbarian world order. It was not until 1705, when the Chosŏn government found embarrassing discrepancies between the calendars of Chosŏn and Qing for that year, that it began supporting the Astronomy Bureau's proposal for a full implementation of Western astronomy.[22] The reproduction of Ricci's map in 1708 was also an extension of this government patronage of Jesuit astronomy. In fact, Ricci's map was reproduced as one of a pair of charts, the other being Adam Schall von Bell's star map.[23] Naturally, the king ordered his Bureau of Astronomy to take charge of the reproduction of the two maps.

21 Lim Jongtae, "Matteo Ricci's World Maps in Late Joseon Dynasty," 279.

22 Lim Jongtae, "Learning Western Astronomy from China: Another Look at the Introduction of the *Shixian li* Calendrical System into Late Joseon Korea," *The Korean Journal for the History of Science* 34, no. 2 (2012): 219–221.

23 The reproduced star map is now lost. As mentioned, both charts were considered to have been made by Adam Schall, as seen from the preface to the two charts by the Chief State Councilor Ch'oe Sŏkchŏng, who was in charge of the project. His preface is found in the

Second, the government's unusual support of Jesuit science coincided with the government-led surge of Ming loyalism in the aftermath of the sixtieth anniversary of the Ming Dynasty's fall in Beijing in 1644. In 1704, the government built an altar in the Royal Palace, at which the Chosŏn king would perform rituals to honor the Ming's Wanli 萬曆 emperor, the alleged savior of the Chosŏn Dynasty who had sent rescue forces to fight Japanese invaders in the 1590s. This movement was an attempt to consolidate Chosŏn's cultural identity as the sole inheritor of the Ming's Confucian legacy in a world then dominated by what the Chosŏn literati officials viewed as barbarian Manchus.[24]

This fervent Ming loyalism in the first decade of the eighteenth century determined the ideological tone of the government's support of Jesuit science. Proponents used Ming loyalism to legitimize this science. Although Western astronomy now served the barbarian emperors, they claimed it had in fact originated in the Ming period, during the Chongzhen calendar reform of the 1630s; it could therefore be viewed as a noble legacy of the Ming that the Chosŏn state, the sole inheritor of Ming culture, had to preserve and hand down to posterity.[25]

The key figure who forged a successful connection between Jesuit science and Ming loyalism was the then Chief State Councilor, Ch'oe Sŏkchŏng 崔錫鼎 (1645–1715). Being a renowned literati official of the period, he was also famous for his unusual interest in mathematical sciences. As a competent mathematician, Ch'oe used his political power to support the Astronomy Bureau's project to learn Jesuit astronomy. It was also Ch'oe who, as Chief State Councilor,

---

left margin of the Ricci reproduced map in Figure 8.1, and is also included in the collection of Ch'oe's literary works. See Ch'oe Sŏkchŏng, "Sŏyang kŏnsang konyŏdo yibyŏng ch'ongsŏ" 西洋乾象坤輿圖二屛總序 [Preface to the Celestial and Terrestrial Maps of the West], in *Myŏnggok sŏnsaeng munjip* 明谷先生文集 [The Collected Works of Master Myŏnggok], vol. 8 (Seoul: Kyŏng'in munhwasa, 1997), 172–173. In this preface, Ch'oe states that both charts were made by Adam Schall in the early years of the Chongzhen 崇禎 reign period (1627–1644) of late Ming China. In other words, Ch'oe viewed these charts to be important legacies of the Chongzhen calendar reform—a historic event in which the Chinese government attempted to adopt Western astronomy as its official calendrical system. As is well known, the outcome of the reform was later presented by Adam Schall to the Manchu conqueror, in 1644, and thus was finally implemented as the *Shixian li* system.

24 Hŏ T'aeyong, *Chosŏn hugi chunghwaron-kwa yŏksa insik* 朝鮮後期中華論과歷史認識 (Seoul: Ak'anet, 2009); Jongtae Lim, "Tributary Relations," 164–170.

25 Lim Jongtae, "Learning Western Astronomy," 197–217; Kim Seulki, "Sukchong-dae Kwansanggam-ŭi sihŏnnyŏk haksŭp: ŭryunyŏn yŏksŏ sagŏn-gwa kŭ-e taehan Kwansaggam-ŭi taeŭng-ŭl chungsim-ŭro" 肅宗代觀象監의時憲曆學習：乙酉年曆書事件과그에對한觀象監의對應을中心으로 (master's thesis, Seoul National University, 2016).

supervised the 1708 reproduction of the Western world map and star map. In his famous preface to those maps, written in the left margin of Ricci's map (Figure 8.1), Ch'oe put the two Jesuit works into the ideological frame of contemporary Ming loyalism.

Curiously, in his preface, Ch'oe did not give full credence to the veracity of Ricci's world map. Although highly praising Schall's star map for its portraying the "true shape of the heavenly images," Ch'oe denounced the Jesuit map's representation of the world as being exaggerated, groundless, and disturbing. He seems to have been particularly offended by its locating China in the northern part of the world, not at its center. If so, why did the government under his leadership, reproduce this problematic world map? It was because even though the Western representation of the world was not completely true, it had not been proven entirely false either. For this reason, Ch'oe suggested, it would be wise to preserve it temporarily and thereby to "broaden [the store of] strange knowledge" (*kwang imun* 廣異聞).[26]

At the end of his preface, however, Ch'oe abruptly changed his hitherto critical tone, by describing the moving moments when he first saw the two maps. He was particularly touched, he confessed, at finding in them such phrases as "the *mujin* year (1628) of the Chongzhen reign period" (*Sungjŏng mujin, Chongzhen wuchen* in Chinese, 崇禎戊辰) and "the United Realm of the Great Ming" (*Tae Myŏng ilt'ong, Da Ming yitong* in Chinese, 大明一統). Ch'oe even claimed that the Jesuit world map and star map reminded him of "the world-ordering of Yu the Great" and "the calendar of the Zhou dynasty" (*wubong churyŏk* 禹封周曆), respectively—in other words, the sacred institutions of geography and astronomy created by the sage rulers of Chinese antiquity. In spite of all the criticism he had raised concerning the Jesuit world map, the four-character phrase "*Tae Myŏng ilt'ong*" in the map finally persuaded him to give to this otherwise suspicious map full ideological legitimacy.[27]

In sum, the Chosŏn government did not reproduce the Jesuit world map because the king and officials fully trusted its description of the world. Its credibility and precision were not their main concern. What mattered to them was, above all, the fact that the map had its origin in Ming China and therefore reminded them of the ideal Confucian world order of the past.

---

26 Ch'oe Sŏkchŏng, "Sŏyang kŏnsang konyŏdo," 172–173.

27 Ch'oe Sŏkchŏng, "Sŏyang kŏnsang konyŏdo," 172–173.

5 Aleni's World Map in a Government Atlas

The second example of official reproductions of Jesuit maps appeared in the last decade of the eighteenth century. As mentioned, this map, entitled *Ch'ŏnha to chido* 天下都地圖 [The Complete Map of All under Heaven], was a remake based on Aleni's *Wanguo quantu* and was included in a government atlas compiled around 1790. This three-volume atlas, entitled *Yŏji to*, does not have a preface or any other document showing how it was made, by whom, or for what purposes. But this atlas still can tell us many things about the function and meaning that the compilers ascribed to the Jesuit map, if we pay attention to the peculiar positioning of the Jesuit map in the atlas.

Atlases became a popular form of making and consuming maps in Korea in the seventeenth century. Produced by private map makers and the government, the late Chosŏn atlases reflected the growing popularity of maps in Chosŏn society.[28] They are usually collections of existing maps: rather than creating maps on their own, the atlas compilers mostly copied existing maps, often with some modifications to them. Compilers' agency can thus be found mostly in their ways of selecting and arranging maps in their atlases. Except for the fact that the result of their compilation had to be in book form, there was no fixed set of editorial standards for the late Chosŏn atlas compilers to abide by, such as the number and the kinds of maps included, the order of their arrangement, and so on. Many late Chosŏn atlases contained only maps of the Chosŏn kingdom, but there are a number of atlases that included, at least in part, maps other than those of the provinces and prefectures of Chosŏn. This part of the atlases, which might be called the volumes of "miscellaneous maps," consisted of a set of heterogeneous maps. It usually included a complete map of Chosŏn, a map of its capital city Seoul, a world map, and maps of such foreign places as China, Japan, and Ryukyu. By including this volume of "miscellaneous maps," the atlas came to convey a comprehensive picture of the world, of which the Chosŏn Dynasty was a part.

The *Yŏji to* atlas was one of those comprehensive atlases, consisting of 31 individual maps compiled into three volumes. A special feature of this atlas is that it gives much more space than usual to the "miscellaneous maps." Except for Volume 2, which consists of eight maps of Chosŏn's eight Provinces, the other two volumes are devoted to the "miscellaneous maps." While Volume 3 contains 16 maps of China's capital area and provinces, Volume 1 consists of the following seven "miscellaneous maps":

28 O Sanghak, *Han'guk chŏnt'ong chirihak sa*, 242–243.

1. World map, entitled "Ch'ŏnha to chido" [Aleni's world map (Figure 8.2)].
2. Map of a traditional style, *tianxia tu* (untitled) (Figure 8.5). As will be discussed, this is not simply a map of East Asia. It is also a kind of world map and belongs to a tradition that can be traced back to the *Huayi tu* 華夷圖 [Map of Chinese and Foreign Lands, 1121], inscribed in a Song Dynasty stone stele.[29]
3. Map of the route of tribute missions, from Ŭiju to Beijing (untitled) (Figure 8.6).
4. Map of Beijing, with a long title, *Beijing ducheng san jie liu shi wu tan ba miao ditu* 北京都城三街六市五壇八廟全圖 [the Complete Map of the Three Streets, Six Markets, Five Altars, and Eight Shrines in the Capital City Beijing].
5. Map of Seoul (untitled) (Figure 8.7).
6. Map of Chosŏn, Japan, and Ryukyu (untitled) (Figure 8.8).
7. A map of Chosŏn, entitled *Wo guo zong tu* 我國總圖 [Complete Map of Our State].

At first, we can surmise that one of the main aims of the official atlas makers in compiling this volume was to portray Chosŏn's relationship with the outside world. Perhaps this was the main reason that three of the seven maps (Maps 2, 3, and 4) were devoted to the most important aspects of Chosŏn's foreign relations, that is, its tributary relationship with China. While Map 2 (a *tianxia tu*) seemed to represent Chosŏn's position in an idealized Sino-centric world, or *tianxia*. Map 3 (Chosŏn's route for tributary missions to Beijing) portrayed the real, contemporary diplomatic relationship with China. In contrast, the map of Chosŏn, Japan, and Ryukyu (Map 6) represented the other, less important, aspect of Chosŏn's diplomatic relationships, those with other of China's tributary states. What kind of role, then, did the compilers expect the reproduction of Aleni's map to play in this atlas? Was it just there to give a whole world picture in which the Chosŏn dynasty interacted with its neighbors, as in our modern atlases?

A closer examination shows that the relationships among individual maps in this atlas diverges from that in modern atlases in a significant respect. Although the *Yŏji to* atlas as a whole conveys a comprehensive picture of the world, consisting of Chosŏn and the surrounding world, this picture is a highly syncretic one. To our modern eyes, it appears that the compiler failed to maintain editorial consistency in selecting and arranging maps: each individual

29 On this Song Dynasty map, see Cordell D.K. Yee, "Cartography in China," in *History of Cartography*, 46–48.

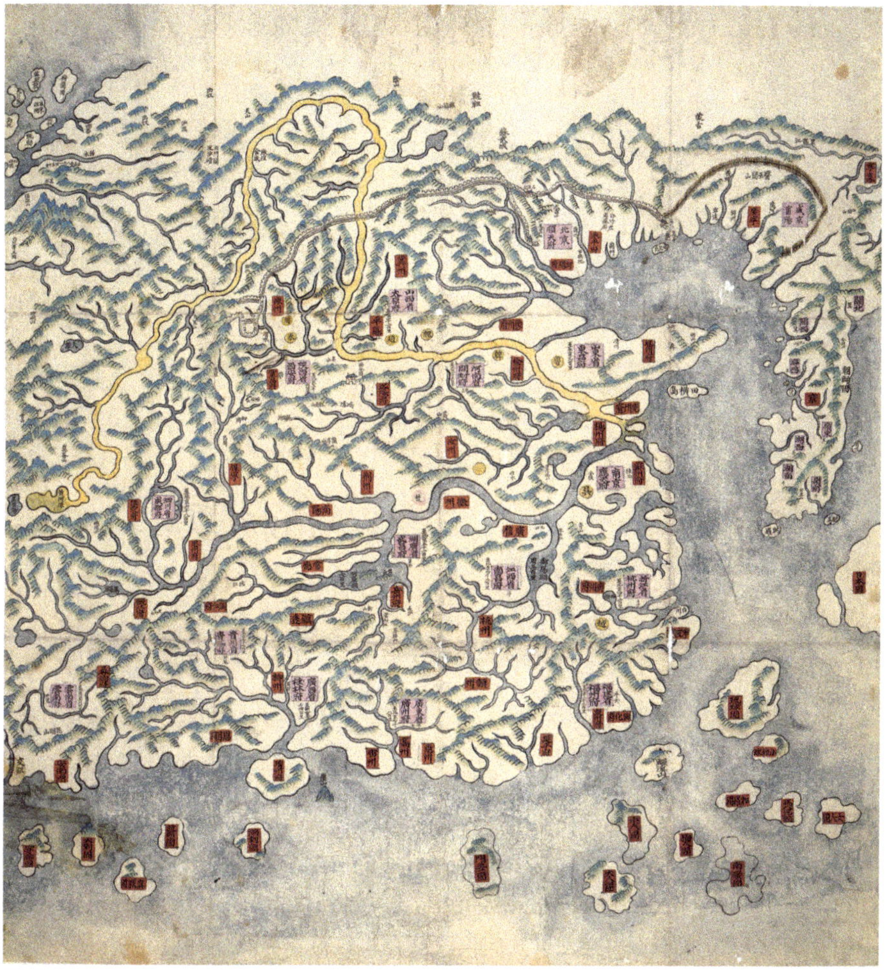

FIGURE 8.5 Traditional-style *tianxia tu* (map 2 of vol. 1), in *Yŏji to*, Kyujanggak Institute for Korean Studies, Seoul National University, Korea.

map of the atlas cannot be viewed as a part of an organic whole. The atlas looks like a mixture of heterogeneous maps, in the following two senses.

First, the atlas was not homogeneous in terms of the cartographic styles and methods applied in individual maps. We can identify several different styles in the atlas, and this is one reason that these maps had varying degrees of precision or accuracy. On the one hand, there are maps whose makers seem not to have cared much about cartographic precision and geographical accuracy. The traditional style *Tianxia tu* map (map 2 of vol. 1) and the maps of Ryukyu and Japan (map 6 of vol. 1) might belong to this category. On the other hand,

FIGURE 8.6 Map of the Route of Tributary Mission (map 3 of vol. 1), in *Yŏji to*, Kyujanggak Institute for Korean Studies, Seoul National University, Korea.

there are cases in which the makers applied certain mathematical principles to attain higher geographical accuracy. To this group might belong Aleni's world map, a map of Seoul (map 5 of vol. 1), a complete map of Chosŏn (map 7 of vol. 1), and the maps of Chosŏn's provinces in Volume 2 (Figure 8.9).

Even these maps of higher precision and accuracy show a remarkable difference in their specific methods. While Aleni's map used a geometrical projection of European origin, the precision of the maps of Seoul and the provinces of Korea was the outcome of applying a different method of indigenous origin. Researchers view these maps of Korea in the atlas as belonging to a tradition of late Chosŏn cartography, stemming from Chŏng Sanggi 鄭尙驥 (1668–1752), a legendary cartographer in the early eighteenth century who with his masterpiece *Tongguk chido* 東國地圖 [Map of the Eastern State] remarkably enhanced the accuracy of the maps of the Chosŏn kingdom. A cartographic tool Chŏng applied to his map was the "one hundred-*li* scale" (*paengni ch'ŏk*

FIGURE 8.7 Map of Seoul (map 5 of vol. 1), in *Yŏji to*, Kyujanggak Institute for Korean Studies, Seoul National University, Korea.

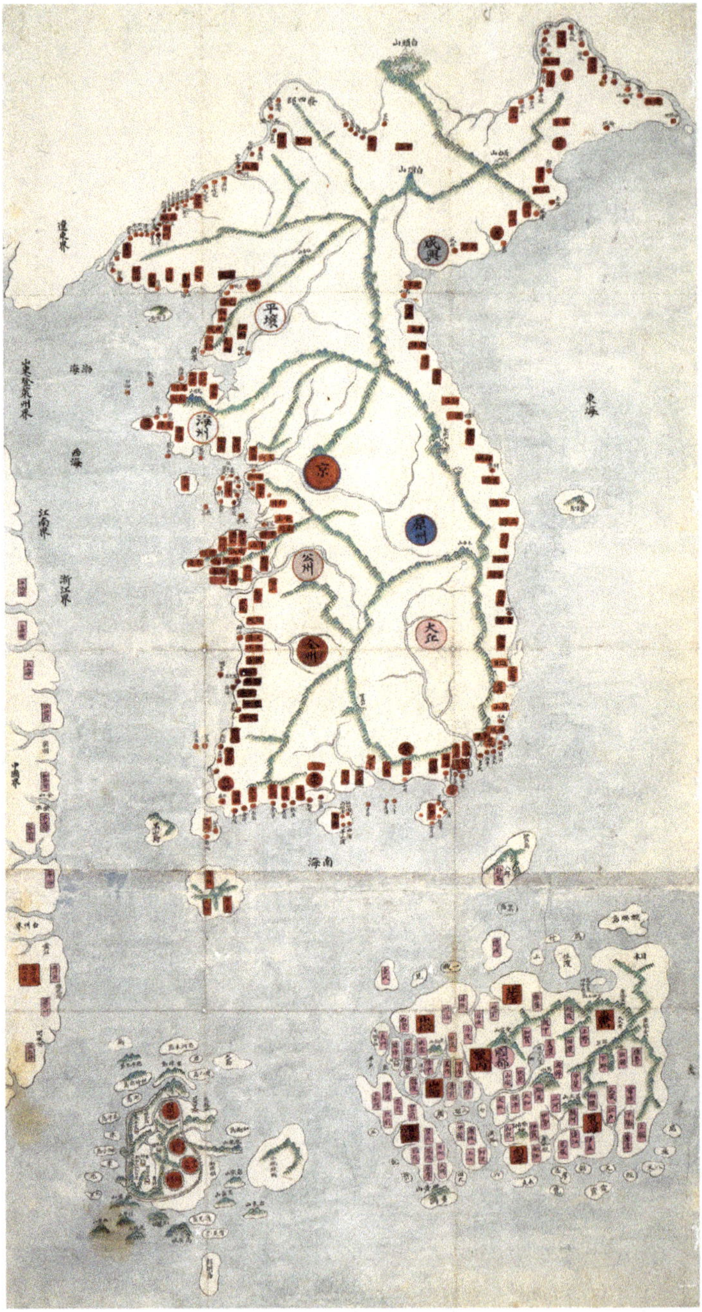

FIGURE 8.8 Map of Chosŏn, Japan, and Ryukyu (map 6 of vol. 1), in *Yŏji to*, Kyujanggak Institute for Korean Studies, Seoul National University, Korea.

FIGURE 8.9 Map of Kyŏnggi Province (map 1 of vol. 2), in *Yŏji to*, Kyujanggak Institute for Korean Studies, Seoul National University, Korea.

百里尺), which enabled him to represent geographical reality in a fixed mathematical proportion.[30]

The second aspect of heterogeneity to be found in the atlas is related to the temporality that each map represents. Rather strikingly, the maps in the atlas

30 O Sanghak, *Han'guk chŏnt'ong chirihak sa*, 246–257.

did not simply portray different aspects of the same contemporaneous world. On the contrary, these maps can be divided into two groups, each representing the worlds of two different temporal strata—the past and the present. First, there are maps the aim of which was to represent the contemporary world. This category includes maps in volumes 2 and 3, which portray the Provinces of Chosŏn and Qing China; in volume 1, the maps of Seoul, Beijing, and the tributary route connecting the two cities also belong to this category. They are mostly the maps of higher precision and accuracy that I mentioned above.

There are also, however, maps of the world of the past. The traditional style *tianxia tu* map in volume 1 (Figure 8.5) is a case in point. The China portrayed in this map is basically China of the Ming period, which had two capital cities (Beijing and Nanjing) and thirteen provinces. This map fails to incorporate the major administrative changes of the early Qing period, such as the establishment of Jiangnan 江南 and Gansu 甘肅 Provinces. The preservation of the old Ming geographical scheme does not necessarily reflect the official compiler's ignorance. The maps of China's provinces in volume 3 demonstrate that the official compilers knew, although only partially, the new administrative divisions of the Qing Dynasty. This volume includes, for example, a map of Jiangnan Province, in which the Ming's old southern capital Nanjing is now identified as Jiangning Prefecture 江寧府, the city's new name given by the Qing.

This suggests that the compiler's selection of this *tianxia tu* map using an old geographical scheme was not intended to represent the geopolitical reality of the contemporary world, a world then ruled by the "barbarian" Manchu dynasty. On the contrary, this map was included perhaps as a reminder of a bygone world.

It should be noted here that there is nothing special about the inclusion of this *tianxia tu* map in a late Chosŏn atlas. Going back to the *huayi tu* of the Song, or more recently to the *Gujin xingsheng zhi tu* 古今形勝之圖 [Map of Advantageous Terrain, Past and Present, 1555] of Ming China,[31] this kind of cultural representation of a China-centered *tianxia* was quite popular among the literati in the late Chosŏn period, and many atlases of the period had a similar map as one of their world maps. Confucian literati could have found in it a

31 On "*huayi tu*," "*Gujin xingsheng zhi tu*," and other maps of similar kinds, see Richard J. Smith, "Mapping China's World: Cultural Cartography in Late Imperial Times," in *Landscape, Culture, and Power in Chinese Society*, ed. Wen-Hsin Yeh (Berkeley: Institute of East Asian Studies, University of California, Berkeley, 1998), 62–71.

visual guide to Chinese civilization, a spatial representation of its history from the time of the ancient sages to the recent past of the Ming Dynasty.[32]

For this reason, the "*tianxia*" as portrayed in this map was not that of any specific moment of the past, say, that of the Ming period. It was rather an overlay of multiple temporal layers. The thirteen provinces of the Ming Dynasty thus coexisted with several different geographical schemes of the past, such as that of the feudal states of the ancient Zhou Dynasty. The outer part of the map was populated by tributaries and other barbarians from different historical moments and different literary sources, even including several monstrous peoples featured in the *Shanhai jing*. In this representation, the Ming Dynasty's thirteen provinces and two capitals were given a special role as a cosmic reference frame: they were associated with their corresponding stars, or lunar lodges, according to the principle of field allocation (*fenye* 分野).[33] This map as a whole, therefore, offered a spatial summary of the entire history of Chinese civilization, or a visual synopsis of the classical world order in which the Middle Kingdom (*Zhongguo* 中國) had ruled all under heaven (*tianxia*).

Despite the overall similarity of the *tianxia tu* map in the *Yŏji to* atlas with those made in China, it gives a significant twist that might have been given to the original Chinese map by the anonymous Korean copiers: Chosŏn's peculiar positioning in this China-centered world. Although Chosŏn was only one of China's many tributary states, this map gave it a more significant position. Chosŏn's provinces were marked in the same style as that of the Ming's provinces, and, moreover, Chosŏn was the only country outside of China that enjoyed the privilege of being associated with stars: three lunar lodges (the Dou 斗, Xu 虛, and Ji 箕 lodges) were respectively allocated to the southern, middle, and northern parts of Chosŏn. In a sense, Chosŏn was portrayed as an eastern extension of the Middle Kingdom. This map thus visualizes the dominant worldview of late Chosŏn literati officials, according to which Chosŏn was the only legitimate heir of the Middle Kingdom lineage after the fall of the Ming Dynasty.

Now, let us turn our attention to the reproduction of Aleni's world map, which appears first in volume 1 of the atlas and is followed by the

---

32 On *Gujin xingsheng zhi tu* and other similar maps as history maps, see Richard J. Smith, "Mapping China's World," 68–71. Special thanks to Dr. Chu Pingyi for pointing out the possibility that the *Tianxia tu* map was highly popular and that the inclusion of this kind of map into atlases was also a widespread practice in late Chosŏn map making.

33 As for a brief account of the field allocation system, see John B. Henderson, "Chinese Cosmographical Thought: The High Intellectual Tradition," in *The History of Cartography*, vol. 2, bk. 2, *Cartography in the Traditional East and Southeast Asian Societies*, ed. J.B. Harley and David Woodward, (Chicago: The University of Chicago Press, 1994), 208–210.

above-mentioned *tianxia tu* map (Figure 8.2). It is not difficult to notice that this map also preserved the old geographical scheme, that of the Ming period. Like Ricci's map reproduced in 1708, this map also has the four-character announcement, "the United Realm of the Great Ming," a clear sign showing the late Ming provenance of the original Jesuit map. Although it includes Jiangnan Province, established in the Qing period, the administrative scheme is basically that of the Ming period; in addition, in the northeastern part of China, we can see the characters "*nüzhen*" 女眞 [Jurchen], the old name of the "barbarian" people who would later build the Qing Dynasty.

The inclusion of Aleni's map in this atlas reflects its popularity in late Chosŏn literati society. Its inclusion in Aleni's widely read world geography text, *Zhifang waiji* 職方外紀 [Records of Regions beyond the Jurisdiction of the Imperial Geographer, 1623] partly explains its popularity. But as with the official reproduction of Ricci's map in 1708, the official compilers of the *Yŏji to* atlas might also have taken into consideration its sacred origin in the late Ming period. In my view, the Jesuit maps' historical provenance might explain, at least partly, why there was no official reproduction of Verbiest's map, *Kunyu quantu*, in eighteenth-century Chosŏn. Unlike Ricci's and Aleni's maps, Verbiest's map might have obviously been viewed as an official world map of the Qing Dynasty, made by the Jesuit in his capacity as an official astronomer of the Qing emperor.[34]

One remaining question would be why the compilers of the *Yŏji to* atlas needed Aleni's map, if the above-mentioned *Tianxia tu* was given the task of representing the past. It is certainly not the case that the two maps were complementary to each other, with one representing the whole world and the other showing East Asia. For the "Tianxia tu" map belongs to a radically different tradition from the Jesuit maps, both in their cartographic methods and in the geographical information inscribed on them. In other words, we cannot acquire the *Tianxia tu* map simply by zooming in on the East Asian part of Aleni's map. Due to the lack of documentary evidence, we do not know the compilers' intentions in juxtaposing two radically different geographical representations of the world.

34 An official reproduction of Verbiest map was made as late as in 1860 (see footnote 15). By this time, Chosŏn government officials had changed, to a considerable degree, their previously negative view of the Qing and had begun to regard it as a legitimate Chinese dynasty. On the changing attitudes of the Chosŏn literati toward the Qing, see Lim Jongtae, "Tributary Relations," 193–196.

Seen from the perspective of conventional practice in East Asian cartography, however, this mixture of apparently conflicting representations of the world was quite common. For example, a late Ming encyclopedia, Zhang Huang's 章潢 (1527–1608) *Tushu bian* 圖書編 [Compilation of Diagrams and Writings, 1613], contains three different kinds of world maps. In addition to a cultural representation of China-centered *tianxia*, which is similar to the above-mentioned *Gujin xingsheng zhi tu*, it also includes what Confucian literati generally viewed negatively as not-so-trustworthy representations of the world: a Buddhist map and three different Jesuit maps. Though the latter kinds of maps were not wholly trustworthy, Zhang Huang pointed out they each offered some useful lessons: the Buddhist map helped one to appreciate the vastness of the world, while the Jesuit maps showed a method to portray the whole world in a map.[35] This flexible and generous attitude toward dubious representations of the world was exactly the same as that expressed by Ch'oe Sŏkchŏng in his preface to Ricci's world map in 1708.

In the case of the *Yŏji to* atlas, we may add one more factor that might have lessened the compilers' sense of contradiction in juxtaposing the two different world maps. To them, neither map was meant to provide a faithful representation of the real, contemporary world, with the help of which government officials might set strategies for diplomacy or national defense. In spite of many fundamental differences, the two maps were commensurate in their shared aim of representing an ideal world of the past, particularly that of the Ming period. From the standpoint of this shared aim, the two maps' major differences in cartographic methods and geographical information seemed to lose their importance. This association of Aleni's map with the less accurate map *Tianxia tu* suggests that my earlier attempt to classify maps in the atlas on the basis of their degrees of precision and accuracy fails to fully reflect the actor's category. To the compiler of the atlas, Aleni's map was not an example of precise cartography to be compared with those fine maps of the Chosŏn kingdom. The Jesuit map was made to represent a totally different temporal stratum, in which precision and geographical accuracy lost their significance.

Therefore, the Jesuit maps as a representation of the past could be replaced with other types of world maps possessing a similar function. The so-called

35 Lim Jongtae, *17, 18 segi Chungguk-kwa Chosŏn-ŭi sŏgu chirihak ihae: Chigu-wa tasŏt taeryuk-ŭi wuhwa* 17, 18世紀中國과朝鮮의西歐地理學理解：地球와다섯大陸의寓話 [Allegories of the Round Earth and the Five Continents: Western Geographical Knowledge in Seventeenth- and Eighteenth-century China and Korea] (P'aju: Ch'angbi, 2012), 257–261.

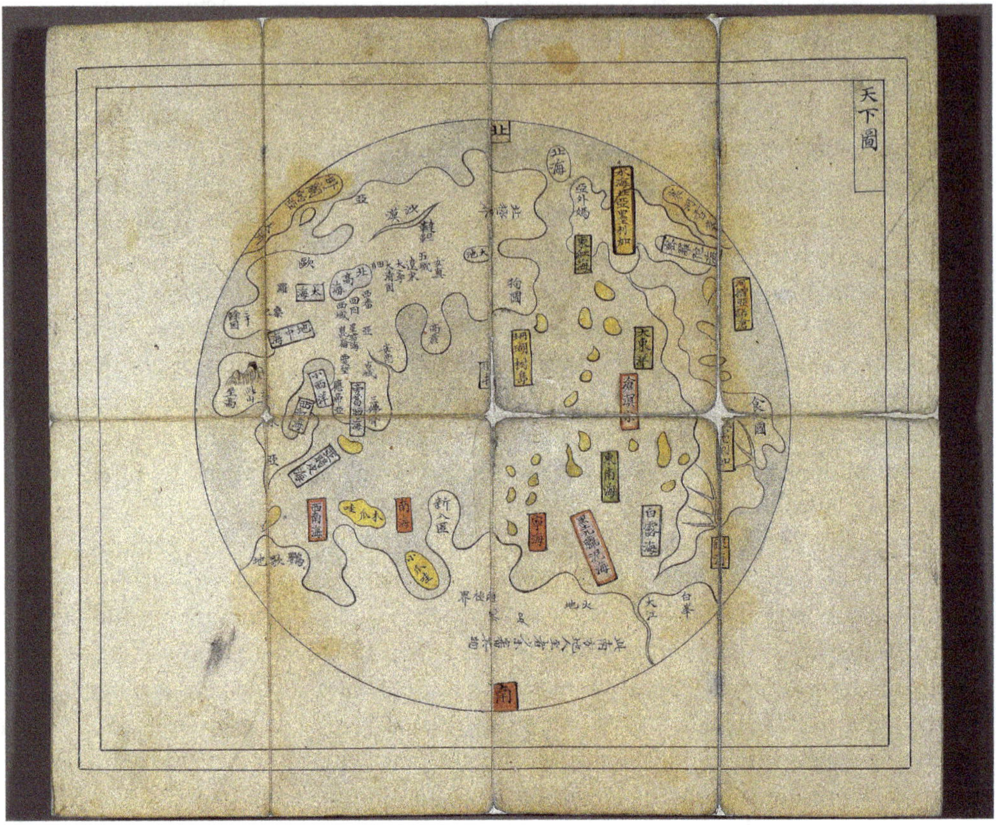

FIGURE 8.10 "Ch'ŏnha to" in *Kwangyŏ to* Atlas. Early nineteenth century, Kyujanggak Institute for Korean Studies, Seoul National University, Korea.

Korean wheel map, or the "*Shanhai jing* map," was a notable alternative, as seen in the above-mentioned *Ch'ŏnha to* map included in the mid eighteenth-century government atlas (Figure 8.3). We have numerous other instances in which late Chosŏn atlases, whether compiled officially or privately, chose the *Shanhai jing* maps as their world maps.

Were there two competing schools with different preferences in their world map choice? We simply do not know. But a final example of the reproduction of Ricci's map suggests a different possibility. This map, entitled *Ch'ŏnha to* 天下圖 [A Map of All under Heaven], was from an early nineteenth-century atlas, *Kwangyŏ to* 廣輿圖 [An Enlarged Map of the World], a privately compiled atlas based on an earlier government one (Figure 8.10). This *Ch'ŏnha to* map was a reproduction of Ricci's *Shanhai yudi quantu* 山海輿地全圖 [A Complete Map of Mountains and Seas in the World] that had been included in the early seventeenth-century Chinese encyclopedia *Sancai tuhui* 三才圖會 [Collection of Diagrams concerning Heaven, Earth, and Men]. What I would like to emphasize here is the striking similarity between this version of Ricci's map and the so-called Korean wheel maps. Perhaps the similarity was more conspicuous to late Chosŏn people. In their mind, the differences in the contours of the continents and the place names between the two maps might be largely irrelevant. After all, the two world maps were not about the real, contemporary world. They described the world of other temporal realms, whether they be the recent historic past or the imaginary world recorded in ancient Chinese documents.

Bibliography

Ch'oe Sŏkchŏng. "Sŏyang kŏnsang konyŏdo yibyŏng ch'ongsŏ" 西洋乾象坤輿圖二屏總 [Preface to the Celestial and Terrestrial Maps of the West]. In *Myŏnggok sŏnsaeng munjip* 明谷先生文集 [The Collected Works of Master Myŏnggok] 8: 172–173. Seoul: Kyŏng'in munhwasa, 1997.

Chosŏn wangjo sillok 朝鮮王朝實錄 [Veritable Records of the Chosŏn Dynasty], the 16th day, 5th month of the 2nd year (1402) of King T'aejong's reign.

Clark, Donald N. "Sino-Korean Tributary Relations under the Ming." In *The Cambridge History of China, Volume 8: the Ming Dynasty, 1368–1644, Part II*, edited by Denis Twitchett and Frederick W. Mote. Cambridge: Cambridge University Press, 1998.

Fairbank, J.K. *The Chinese World Order: Traditional China's Foreign Relations*. Cambridge, MA: Harvard University Press, 1968.

Haedong chido 海東地圖 [Maps of the Eastern Kingdom]. Seoul: Kyujanggak Institute for Korean Studies, 1995.

Henderson, John B. "Chinese Cosmographical Thought: The High Intellectual Tradition," edited by J.B. Harley and David Woodward, vol. 2, bk. 2, *Cartography in the Traditional East and Southeast Asian Societies*. Chicago: The University of Chicago Press, 1994.

Hŏ T'aeyong. *Chosŏn hugi chunghwaron-kwa yŏksa insik* 朝鮮後期中華論과歷史認識 [Sino-centric Ideology and Historiography in Late Chosŏn Period Korea]. Seoul: Ak'anet, 2009.

Kim Seulki. *Sukchong-dae Kwansanggam-ŭi sihŏnnyŏk haksŭp: ŭryunyŏn yŏksŏ sagŏn-gwa kŭ-e taehan Kwansaggam-ŭi taeŭng-ŭl chungsim-ŭro* 肅宗代觀象監의時憲曆學習:乙酉年曆書事件과그에對한觀象監의對應을中心으로 [Learning the *Shixian li* Astronomical System in King Sukchong's Era in the Chosŏn Dynasty: A Focus on the Royal Astronomy Bureau's Response to the Calendar Scandal in 1705]. Master's thesis, Seoul National University, 2016.

Kye Sŭngbŏm, *Chungjong-ŭi sidae* 中宗의時代 [King Chungjong's Era in the Chosŏn Dynasty Period in Korea]. Seoul: Yŏksa pip'yŏng sa, 2014.

Ledyard, Gari. "Cartography in Korea." In *History of Cartography*, edited by J.B. Harley and David Woodward, vol. 2, bk. 2, *Cartography in the Traditional East and Southeast Asian Societies*. Chicago: The University of Chicago Press, 1994.

Lim Jongtae. *17, 18 segi Chungguk-kwa Chosŏn-ŭi sŏgu chirihak ihae: Chigu-wa tasŏt taeryuk-ŭi wuhwa* 17, 18世紀中國과朝鮮의西歐地理學理解:地球와다섯大陸의寓話 [Allegories of the Round Earth and the Five Continents: Western Geographical Knowledge in Seventeenth- and Eighteenth-century China and Korea]. P'aju: Ch'angbi, 2012.

Lim Jongtae. "Learning Western Astronomy from China: Another Look at the Introduction of the *Shixian li* Calendrical System into Late Joseon Korea." *The Korean Journal for the History of Science* 34, no. 2 (2012): 205–225.

Lim Jongtae. "Matteo Ricci's World Maps in Late Joseon Dynasty." *The Korean Journal for the History of Science* 33, no. 2 (2011): 277–296.

Lim Jongtae. "Tributary Relations between the Chosŏn and Ch'ing Courts to 1800." In *The Cambridge History of China, vol. 9: The Ch'ing Empire to 1800. Part II*, edited by Willard J. Peterson. Cambridge: Cambridge University Press, 2016.

Lim Jongtae, "Ŭijŏngbu-ŭi segye chido, Honil kangni yŏktae kukto chi to" 議政府의世界地圖, 混一疆理歷代國都之圖 [Honil kangni yŏktae kukto chi to as a World Map by the State Council of the Chosŏn Dynasty]. *Munhwa yŏksa chiri* 文化歷史地理 27, no. 1 (2015): 1–14.

Miya Noriko. *Mongoru teikokuga unda sekaizu* モンゴル帝国が生んだ世界図 [The world map produced by the Mongol empire]. Tokyo: Nihon keizai sinbunsha, 2007.

Needham, Joseph. *Science and Civilisation in China*, Vol. 3. Cambridge: Cambridge University Press, 1959.

O Sanghak. *Chosŏn sidae segye chido-wa segye insik* 朝鮮時代世界地圖와世界認識 [Worldmaps and Worldviews in Chosŏn Period Korea]. P'aju: Ch'angbi, 2011.

O Sanghak. "Circular World Maps of the Joseon Dynasty: Their Characteristics and Worldview." *Korea Journal* (Spring 2008): 8–45.

O Sanghak. *Han'guk chŏnt'ong chirihak sa* 韓國傳統地理學史 [A History of Geography in Premodern Korea]. P'aju: Tŭlnyŏk, 2015.

Smith, Richard J. "Mapping China's World: Cultural Cartography in Late Imperial Times." In *Landscape, Culture, and Power in Chinese Society*, edited by Wen-Hsin Yeh. Berkeley: Institute of East Asian Studies, University of California, Berkeley, 1998.

Son Koŭn, "16segi Cho-Myŏng woegyo-wa chiri chisik kyoryu: 'Honil yŏktae kukto kangni chi to'rŭl chungsimŭro" 16世紀朝明外交와地理知識交流:'混一歷代國都疆理地圖'를中心으로 [Diplomatic Relations between Chosŏn and the Ming and Exchanges of Geographical Knowledge between Them: Focusing on Honil yŏktae kukto kangni chido]. Master's thesis, Korea University, 2017.

Yang Pogyŏng. "Taedong yŏjido-rŭl mandŭlgi kkaji" 大東輿地圖를만들기까지 [A History of Cartography in Korea up to the Taedong yŏjido]. *Hanguksa simin kangjwa* 韓國史市民講座 16 (1995): 84–121.

Yee, Cordell D.K. "Cartography in China." In *Cartography in the Traditional East and Southeast Asian Societies*. The History of Cartography, vol. 2, bk. 2, edited by J.B. Harley and David Woodward. Chicago: University of Chicago Press, 1994.

Yŏji to 輿地圖 [Terrestrial Maps]. Kyujanggak Institute for Korean Studies, call number: 古 4709-78-v.1-3. Its images with bibliographical notes are available at the official website of the Kyujanggak Institute: https://kyu.snu.ac.kr/.

CHAPTER 9

Entering Asia: The Repositioning of Japan

Kären Wigen

1 Introduction

In March of 1885, a hard-hitting editorial was published in the Japanese newspaper *Jiji shinpō* 時事新報, arguing that Meiji Japan should distance itself from the conservative governments of its Chinese and Korean neighbors and align itself instead with the West. The title of this essay—"Datsu-A Ron 脱亜論"—is best translated as "On Leaving Asia." Although the editorial was anonymous, it has long been assumed that it was penned by Fukuzawa Yukichi 福沢諭吉, the prolific modernizer whose face can be seen on Japan's 10,000-yen bill to this day.[1]

If one thinks about it, the metaphor of Japan "leaving Asia" made sense only if Japan had *entered* Asia in the first place. And this is not something to be taken for granted. Asia was, after all, a foreign term, introduced to the Japanese and their neighbors in the sixteenth century by European missionaries. Like their Chinese and Korean counterparts, most literate Japanese at the time felt no particular need for this alien geographical category. If some found it amusing to be told that they resided in a part of the globe called Asia, others undoubtedly found it insulting to have their entire known world recast in this way. Nonetheless, by Fukuzawa's day, the transliterated term *Ajia* had long-since taken root in the Japanese lexicon. And it took hold as part of a larger geographical taxonomy: one that included Europe, Africa, the Americas, and Magellanica (Terra Australis) as well. As a result, to understand how the Japanese came to embrace the unfamiliar concept of Asia requires looking at how they dealt with the larger meta-geographical framework of which it was a part.

The current essay takes a cartographic approach to that question. Surveying a lively field of visual evidence, it asks what the surviving corpus of Edo-era

1 Albert M. Craig, *Civilization and Enlightenment: The Early Thought of Fukuzawa Yukichi* (Cambridge: Harvard University Press, 2009). See also Pekka Korhonen, "Leaving Asia? The Meaning of Datsu-A and Japan's Modern History," *The Asia-Pacific Journal / Japan Focus* 11 (issue 50), no. 1 (December 2013): 1–18. Article ID 4083. https://apjjf.org/2014/11/50/Pekka-Korhonen/4083/article.html.

© KÄREN WIGEN, 2024 | DOI:10.1163/9789004684782_011
This is an open access chapter distributed under the terms of the CC BY-NC-ND 4.0 license.

world maps can tell us about Japanese thinkers' reception of, and responses to, the European continental scheme. There was no single answer to that question, for Tokugawa cartography was a divided enterprise. Maps of the world came in markedly different styles, reflecting the fact that the men behind them ranged from Confucian scholars to commercial artists to Buddhist priests. Likewise, their audiences varied. If some maps were marketed widely to the public, others were hung in classrooms, and still others were viewed in private salons by like-minded insiders. Yet inasmuch as they all grappled with how to locate and label the different landmasses of the globe, members of these different communities participated in a *de facto* conversation about the shape of the world and Japan's place in it. Before delving into that conversation, it is worth briefly taking stock of what they were responding to. What precisely was the continental scheme as introduced by the adventurers and missionaries who brought the term "Asia" eastward from Europe in the sixteenth century?

Before Columbus, most thinkers in the Christian and Islamic realms posited a world of three continents, a scheme inherited from the ancient Greeks.[2] Asia, the largest landmass, was typically depicted at the top of the world, with Europe and Africa below. The medieval T-O schema shown below (Figure 9.1) was the product of eager Christian intellectuals doing their best to put a biblical stamp on what was originally a pagan vision. For medieval Christians, Asia was particularly important as the site of the Holy Land, where the seminal events of the Bible had taken place; paradise was believed to lie at its farthest extreme.

Even before Christopher Columbus's pioneering trans-Atlantic voyages, the already-hybrid T-O framework was being pried apart to accommodate new geographical information about Asia and Africa. Some of the news came by way of Marco Polo, whose account of his travels along the silk routes opened up new vistas for Europeans. Additional information, especially about the Indian Ocean world, came from Islamic geographers, whose works began filtering into the libraries of Europe during the Renaissance. Both of these sources informed the landmark Fra Mauro map, made around the year 1450 by an Italian cartographer of the same name. Painted in brilliant color on parchment, and set in a wooden frame that measures more than two by two meters, this magnificent wall-map (which hangs in Venice to this day) accommodated hundreds of detailed illustrations and over 3000 descriptive texts covering the entire Eastern Hemisphere—all contained within the traditional T-O shape.

2 Martin Lewis and Kären Wigen, *The Myth of Continents: A Critique of Meta-Geography* (Berkeley: University of California Press, 1997).

FIGURE 9.1 European T-O map, from a fifteenth-century printed version of *Etymologiae* by Isidore of Seville (ca. 560–636), identifying the three known continents as populated by the descendants of Noah: Sem (Shem), Iafeth (Japheth), and Cham (Ham).

The resulting image is considered by many scholars to mark the culmination of biblically-based geography in Europe.[3]

If Fra Mauro's map represented the high-water mark for the tri-continental scheme, it also marked its end. By the time it came to Japan, the Greek conception of the world had been fully burst open by the discovery of major new

3 Piero Falchetta, *Fra Mauro's World Map, with a Commentary and Translations of the Inscriptions* (Turnhout: Brepols; Venice: Biblioteca Nazionale Marciana, 2006).

land areas unknown to the ancients: the Americas (originally conceived as one massive continent) and a conjectured Southern continent (Terra Australis or Magellanica). It was thus as one of five continents—now conventionally shown with north at the top—that Asia was introduced into Japan by the Portuguese in the mid-sixteenth century.

The person who was most responsible for making this vision of the globe legible for an East Asian audience, of course, was Matteo Ricci. Of the five world maps that Ricci is known to have made during his years in China, the large-format woodblock-print version published in Beijing in 1602—with its long explanatory preface and over 1000 place-names transliterated into Chinese—was the most detailed. It was this image that first introduced the term "Asia," transliterated phonetically into the Chinese characters 亜細亜, to a Chinese, Korean, and Japanese audience. And Ricci's successors continued in his footsteps. Having witnessed the keen cartographic interest of the Chinese literati, subsequent Jesuits devoted considerable time to writing about global geography and creating maps of their own. One of the most influential among the next generation was Giulio Aleni (1582–1649), whose "Account of countries not listed in the records office" (*Zhifang waiji* 職方外紀) appeared some twenty years after Ricci's map.[4] In Aleni's simplified version (published under the title *Wanguo quantu* 萬國全圖 [Complete Map of the Myriad Countries,]) the names of the continents were prominently featured.

With varying degrees of fidelity, Ricci's and Aleni's maps—along with their explanations and phonetic spellings for the continents—were taken up and reprinted in a variety of forms over succeeding years, starting with an influential Chinese encyclopedia of 1609 called the "Illustrated Compendium of the Three Fields of Knowledge" (*Sancai tuhui* 三才圖會). Here, the Jesuit claims about the earth were given space, but they were set alongside quite different maps and models derived from indigenous cartographical traditions. This treatment set the pattern for how European geographical ideas would be engaged in China under the subsequent Qing dynasty as well. As pointed out by Pekka Korhonen, Ricci and his map were neither condemned nor taken seriously. The Jesuit map was simply allowed to exist as one strange argument about the world.[5]

4 Giulio Aleni, *Zhifang waiji* (1623), in Laura Hostetler, *Qing Colonial Enterprise: Ethnography and Cartography in Early Modern China* (Chicago: University of Chicago Press: 2001), 56–57.

5 Pekka Korhonen, "Asia's Chinese Name," *Inter-Asia Cultural Studies* 3, no. 2 (2002): 253–270. For further evidence of late imperial agnosticism about competing theories of geography, see Matthew Mosca, *From Frontier Policy to Foreign Policy: The Question of India and the Transformation of Geopolitics in Qing China* (Stanford: Stanford University Press, 2013).

Outside China, Ricci's maps encountered different degrees of receptivity. The environment in Korea was relatively restrictive. Ricci's 1602 map was brought to the Korean royal house soon after it was published, but few people are believed to have seen it, and the original does not survive. What do survive are hand-painted manuscript copies of Ricci's and Aleni's maps, commissioned by royal order in the 1700s and kept within the court. As one Korean historian puts it, these maps might well have shocked Korean intellectuals, but few had an opportunity to see them in their original large format. Most Korean scholars accordingly encountered the European continental scheme in the grossly simplified form in which it was repackaged in the Chinese encyclopedia. Frowned on by the Confucian establishment, and accessible only in this crude caricature, Jesuit geography not surprisingly failed to get much traction with literate Koreans.[6] This was in contrast to the situation in Japan, where knock-offs of Ricci's 1602 wall-map would eventually be widely seen. More than a score of Japanese variants would be published over the next 250 years, in small formats and large, along with dozens of hand-made tracings and copies that circulated among elite scholars and literati interested in geography.

For Japanese scholars interested in global geography, the Jesuit view of the world was reconfirmed when their continental scheme was introduced a second time, by way of the Protestant Dutch. After all other Europeans were expelled from the country in the 1630s (and even as increasingly severe censorship was placed on books from the West), the V.O.C. was allowed to retain a presence in Nagasaki for two centuries more. To be sure, the Dutch who came to Japan were clerks and merchants, not intellectuals like the Jesuits. Moreover, they rotated through their posts on short-term assignments and were not encouraged to learn Japanese. Nonetheless, on the Japanese side, there slowly emerged a network of translators and scholars who took a serious interest in Dutch learning. Their work was boosted significantly in the 1720s by the eighth shogun, Tokugawa Yoshimune 徳川吉宗 (1684–1751). Persuaded that the shogunate needed to improve its intelligence about global geopolitics, Yoshimune lifted the ban on Western books (as long as they did not explicitly advocate for Christianity). The easing of censorship allowed Jesuit maps, as well, to come out of the shadows. It took another two or three generations for Dutch studies to find footholds beyond Nagasaki, but by the turn of the nineteenth century,

6 Bae Woo Sung, "Worldviews and early Cartography," in *The Artistry of Early Korean Cartography*, ed. Han Young-woo, Ahn Hwi-Joon, and Bae Woo Sung (Larkspur, CA: Tamal Vista Publications, 1999), 93–129.

world maps and geographies based on Dutch originals began finding their way into print.[7]

Given this history, it is no surprise that the standard account of Japanese reckoning with European geography has been written from the vantage-point of Deshima. Rather than focus on the Dutch Studies school, however, this essay takes a different tack, asking how mapmakers from *other* intellectual traditions responded to the European continental scheme. The answer cannot be singular, for the simple reason that Japan was not a unified intellectual field during the Tokugawa period. On the contrary, at least three distinct schools of thought coexisted during these decades. One consisted of self-identified Neo-Confucians, for whom China was the measure of civilization and the center of the world. A second was made up of Buddhist scholars, for whom not China but India was the pivot of the world. Finally, a third discursive community coalesced over time: one that increasingly centered the world map on their own homeland, an archipelago they proudly labeled "Dai-Nippon" or "Great Japan." The leading lights at the core of this movement were nationalist scholars, Shinto ideologues, and poets writing in the Japanese vernacular.[8] But graphic artists also played a role, popularizing the ideas of these thinkers. What we might call the "Our Land" (*honcho* 本朝)-centric vision found vivid expression in the work of woodblock artists employed by the Edo publishing industry during the eighteenth and nineteenth centuries. Judging from the exuberant market for printed products that conveyed a Japan-centric view of the world, we might deduce that this loose discursive community had touched the lives of millions of urban consumers during the last years of the Tokugawa shogunate—including many with only rudimentary levels of literacy. Taking up Confucian, Buddhist, and Japanist visions in turn, the remainder of this chapter will trace how each group mapped the world, and where "Asia" fit into their competing global (and cosmic) visions.

2 Neo-Confucians Map the World

The worldview that was most widely shared across the Sino-sphere before Ricci is captured in a magnificent Chinese map from the fourteenth century,

7 Grant K. Goodman, *Japan and the Dutch 1600–1853* (London: RoutledgeCurzon, 2000). See also Donald Keene, *The Japanese Discovery of Europe, 1720–1830* (Stanford: Stanford University Press, 1952) (rev. ed. 1969).

8 Susan Burns, *Before the Nation: Kokugaku and the Imagining of Community in Early Modern Japan* (Durham: Duke University Press, 2003).

the "Amalgamated map of the great Ming empire" (*Da Ming hun yi tu* 大明混一圖), ca. 1389. This is a *mappa mundi* dominated by the Chinese state, which squarely fills the frame. The empire is surrounded on three sides (though rarely from above) by a ring of lesser tributaries and barbarian lands. The presence of the Arabian Peninsula and Africa on the far left signal that this is an image of Eurasia as a whole, and not just of those areas controlled by the Chinese. As many readers will know, the ability of Ming cartographers to sketch and name such distant lands was a legacy of the Yuan dynasty, whose Mongol rulers had been in regular communication with their kin and co-religionists across a vast swath of Afro-Eurasia.[9]

The surviving artifact that best embodies the Yuan world view may be a Korean map, known as the *Gangnido*, whose title can be translated as "Map of integrated lands and regions of historical countries and capitals" [Korean: *Honil gangni yeokdae gukdo ji do* 混一疆理歷代國之圖], 1402.[10] This too is a magnificent manuscript, hand-painted on silk and measuring about 1.5 meters on each side. Evidently made for the Korean court, it was probably confiscated during Hideyoshi's 豊臣秀吉 invasions of Korea in the 1590s, and it now resides in Japan. But as suggested by the aggrandizement of Korea and the belittling of Japan (both by its small size and by its placement underneath Korea), the *Gangnido* map clearly embodies a Korean interpretation of a China-centered world.[11]

This Sino-centric paradigm was apparently more congenial to Koreans—who styled themselves as the most important tributaries of the Chinese—than it was to the Japanese, who lay outside that geopolitical sphere (having effectively turned themselves into a pariah state under Hideyoshi by attacking their continental neighbors). Nonetheless, the basic world-view it embodied was deeply ingrained in Japan as well, and Confucian cartographers played variations on this theme all the way through the Edo period. In the incarnation shown in Figure 9.2—a Japanese reprint of an earlier Chinese original—we can see that a map of China and its environs is being transformed in an experimental way into a world map by the extension of the ocean into a circle that completely surrounds the Great Ming, on the north as well as the west, south, and east sides.

9 Hyunhee Park, *Mapping the Chinese and Islamic Worlds: Cross-cultural Exchange in Premodern Asia* (Cambridge: Cambridge University Press, 2012).

10 As there are several systems of romanization for Korean, the *Gangnido* is also sometimes referred to as the *Kangni to*, as in Chapter 8 of this volume, or *Kangnido* as in Chapter 6.

11 Jerry Brotton, *A History of the World in 12 Maps* (New York: Viking, 2013), Chapter 4.

FIGURE 9.2 "Map of Nine Borders, All the Countries, Footprints, and Itineraries under the Ming Dynasty" [*Daimin kyūhen bankoku jinseki rotei zenzu* 大明九辺万国人跡路程全図] by Ou Kunho (Wang Chufu); original Chinese version 1663; Japanese version by Umemura Yahaku, Kyoto, ca. 1705. Tohoku University Library.

In the hands of a later mapmaker, this Sino-centric model took on a shape that is in some ways closer to the Korean *Gangnido* original, with two important changes (Figure 9.3). One concerns the treatment of Japan, restored here to its correct position and exaggerated in size. The other change is that the peninsula on the left no longer represents Africa, as on the *Gangnido* model, but is labeled Tenjiku, or India: a land that was of much more importance to Japanese. In fact, this colorful nineteenth-century print encapsulates the pre-Jesuit Japanese understanding of the world as made up of "three countries": India, China, and Japan.

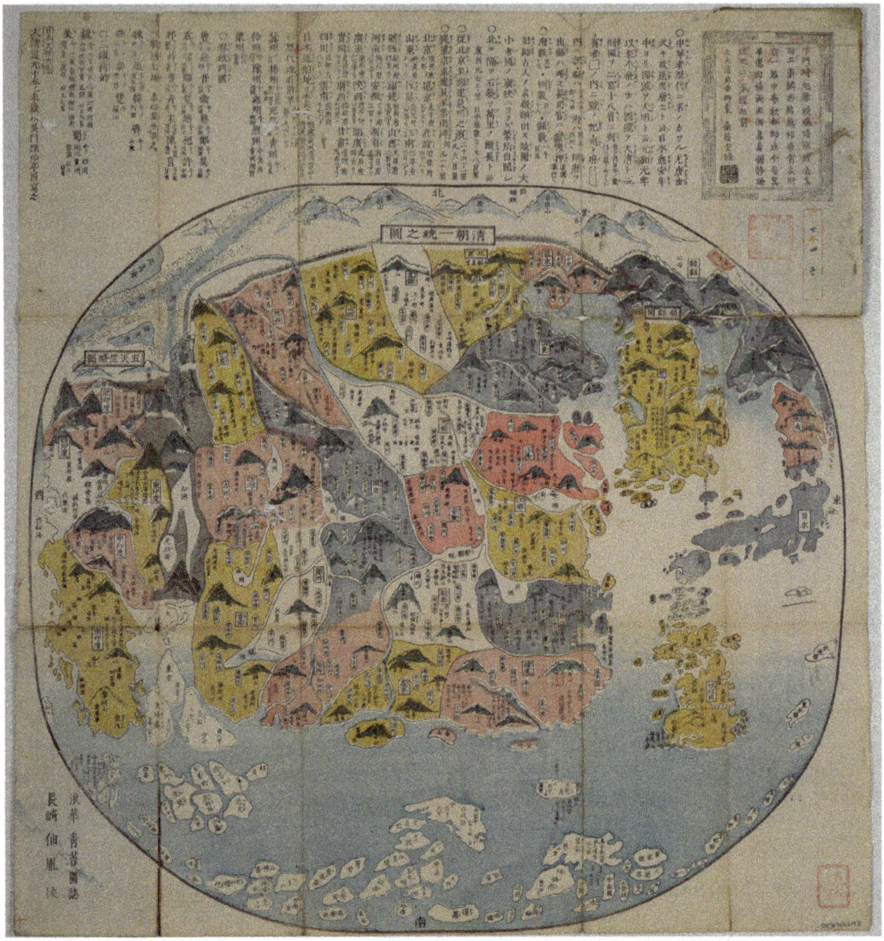

FIGURE 9.3 "Map of the Great Qing" [*Da Qing yochi zenzu* 大清輿地全図]. Woodblock print, 1835. Tsukuba University Library.

India continued to be prominent in many Buddhist visions of the world, as we will see in a moment. In the hands of Confucian scholars, by contrast, the Sinocentric mapping tradition evolved in a rather different direction: namely, toward detailed depictions of the historical Chinese heartland as "*Asia*." In other words, what had started its career as a comprehensive world map lived on in regional mapping. The cartographer who designed Figure 9.4, Nagakubo Sekisui, was a prominent and prolific geographer with a fascination for China. His crowning life-work was this massive (nearly 2 m × 2 m) rendering of "The Great Qing Empire, with longitudes and latitudes."

FIGURE 9.4 "Map of the Great Qing with Longitudes and Latitudes" [*Keiten gatchi Dai Shin kōyozu* 經天合地大清廣輿圖] by Nagakubo Sekisui. Edo, 1785. 184.2 × 198.7 cm. National Diet Library.

The same Confucian scholar also published a map of the world on the Ricci model, and just five years after his map of the Great Qing, he published yet another map, this time definitively slotting the Sino-sphere into a Jesuit frame (Figure 9.5). The name given to this map says it all: "Ajia—Tōyō," or "Oriental Asia." It seems fair to say that this is where Sino-centric cartography in Edo-era Japan ended up: as a respected, living tradition, but one that was increasingly demoted to regional status within a fundamentally European framework. The old Chinese world became the core of the new "Asia."

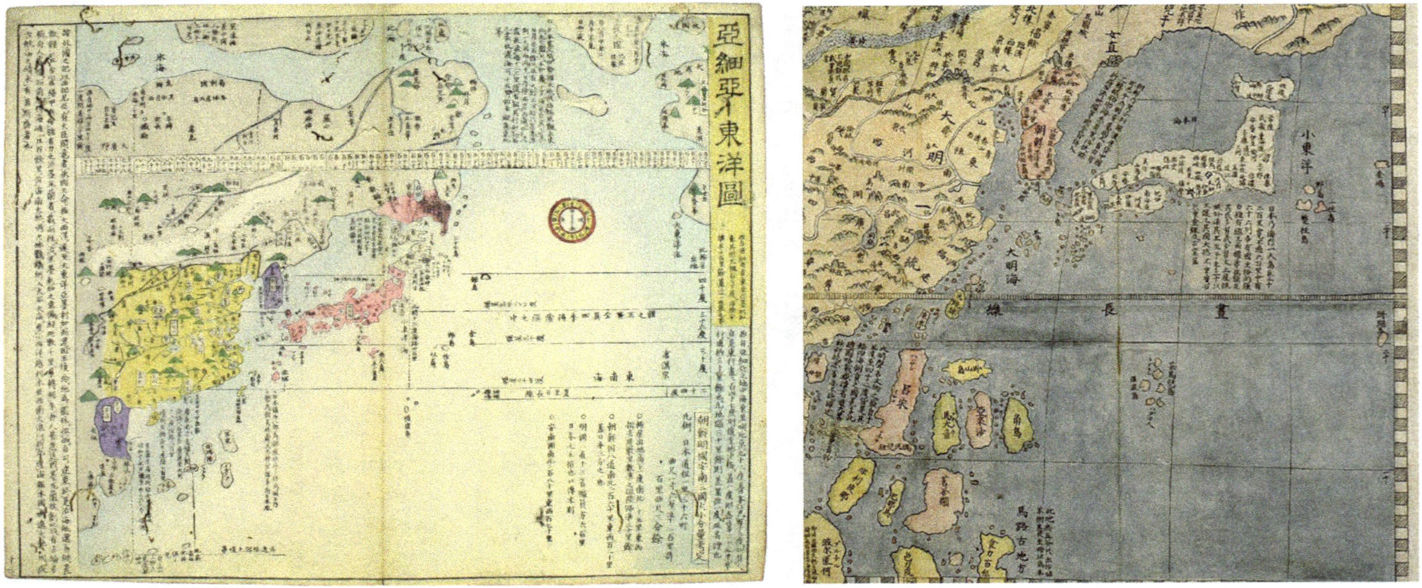

FIGURE 9.5 Left: "Map of Asia—Orient" [Ajia Tōyō zu 亞細亞東洋圖], from "Historical Atlas of the Successive Chinese Dynasties" (*Tōdo rekidai shūgun enkaku chizu* 唐土歷代州郡沿革地圖) by Nagakubo Sekisui. Edo, 1790. Right: detail from Matteo Ricci's 1602 map of the world. Tohoku University Library.

3 An India-Centered World

Buddhist cartographers took a different approach.[12] The original Buddhist conception of the world consisted of four landmasses, centered on Mount Meru (Sumeru) and surrounded by ocean on all sides. Only one of those four—the southern realm, or Jambu-dvīpa (J. Nansenbushu 南瞻部洲)—was habitable; this was the domain of all earthly geography. Since the lower half of this cosmic continent consisted of the five kingdoms of India (Gotenjiku 五天竺), traditional Buddhist maps in Japan might be called either "Gotenjiku-zu 五天竺図" or "Nansenbushu-zu 南瞻部洲図." China, Korea, and Japan were relegated to the margin of such maps, with Japan only partially appearing in the upper right corner (Figure 9.6).

World maps based on this Buddhist schema presumably developed first in China and Korea, but to my knowledge, the only place where this remained a living tradition into the early modern era was Japan. (Figure 9.6 shows one of a dozen surviving examples in Japan, this one currently owned by the Kobe City Museum.) What is notable about Buddhist cartography in Japan, besides its sheer durability, is how plastic it proved as scholar-monks grappled with European geography.[13] In 1708, about sixty years after the first printed western-style maps had begun circulating in Japan, a leading Buddhist scholar named Sōkaku 宗覚 attempted to plot a host of new places into Jambu-dvīpa, including many introduced by the Jesuits. Africa and Europe appeared on his map to the left of India, while in the far upper right corner he sketched in an unnamed landmass. (On some later maps, this would be labeled Cuba or North America).

Sōkaku's design was taken up and put into circulation in the massive, long-lived image from 1710 by the monk Hōtan 鳳潭, shown here as Figure 9.7. In Hōtan's *mappa mundi*, the tight shell around Jambu-dvīpa has begun to crack, such that new countries and even continents can be seen floating like islands in the sea around mother India. In this otherwise black-and-white version, the individual provinces of China and Japan have been distinguished in brilliant color on the right.

Hōtan's map was picked up and republished throughout the eighteenth century, becoming one of the most widely seen images of the world in Japan at the time. What is interesting for our purposes is how Buddhist cartographers beginning with Hōtan began to deploy the Jesuit continental taxonomy,

12 D. Max. Moerman, *The Japanese Buddhist World Map: Religious Vision and the Cartographic Imagination* (Honolulu: University of Hawaii Press, 2022).

13 Unno Kazutaka, "Cartography in Japan," in *The History of Cartography*, ed. J.B. Harley and David Woodward (Chicago: University of Chicago Press, 1994), 2.2: 346–477.

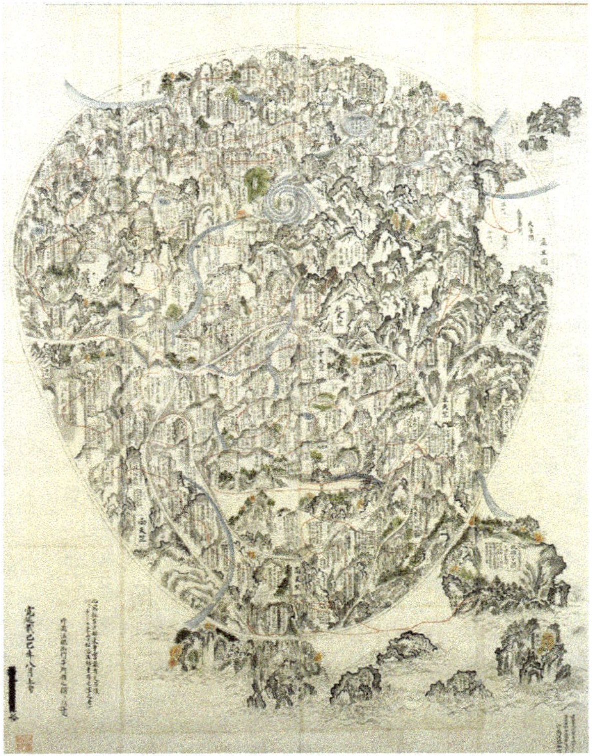

FIGURE 9.6 Left: "Map of the Five Indias" [*Gotenjiku zu* 五天竺図], 1749 (copy of 1364 original), 167 × 133 cm. Kobe City Museum. Right: English schema with explanatory labels from Unno Kazutaka, 1994, "Cartography in Japan," in *The History of Cartography* (University of Chicago Press, 1994), 2.2: 373.

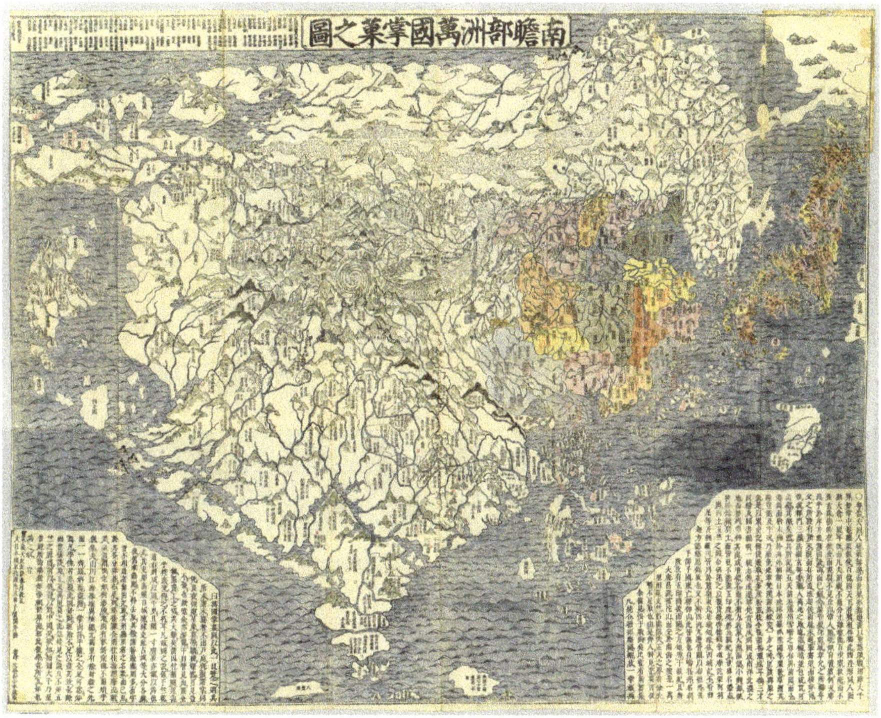

FIGURE 9.7 "Map of the Myriad Countries of Jambu-dvīpa" [*Nansenbushu bankoku shōka no zu* 南瞻部洲万国掌菓之図], by Hōtan 鳳潭. Edo, 1710; woodcut, 113.5 × 144 cm. The David Rumsey Map Center, Stanford University.

eventually including the term "Asia." Figure 9.8 below shows a crude simplification of Hōtan's map, printed circa 1744, alongside an oval world map allegedly by the same publisher. A close look at these paired images suggests one Buddhist solution to the encounter with Ricci: contracting *Jambu-dvīpa*, and slotting it into the space of "Asia" in a clearly European-derived framework. In a sense, this particular designer might be said to have given up on Buddhist cosmology—unless (as one Japanese scholar alleges) he did not fully grasp it to begin with. In either case, the framework in the second image is manifestly derived from Matteo Ricci, and Asia is labeled as such even though it retains a recognizably Buddhist shape (replete with Lake Anavatapta). While this is but one of several solutions that Buddhist cosmographers would come up with, it demonstrates the flexibility, hybridity, and compromise that characterized Edo-era Buddhist *mappae mundi* as a whole.

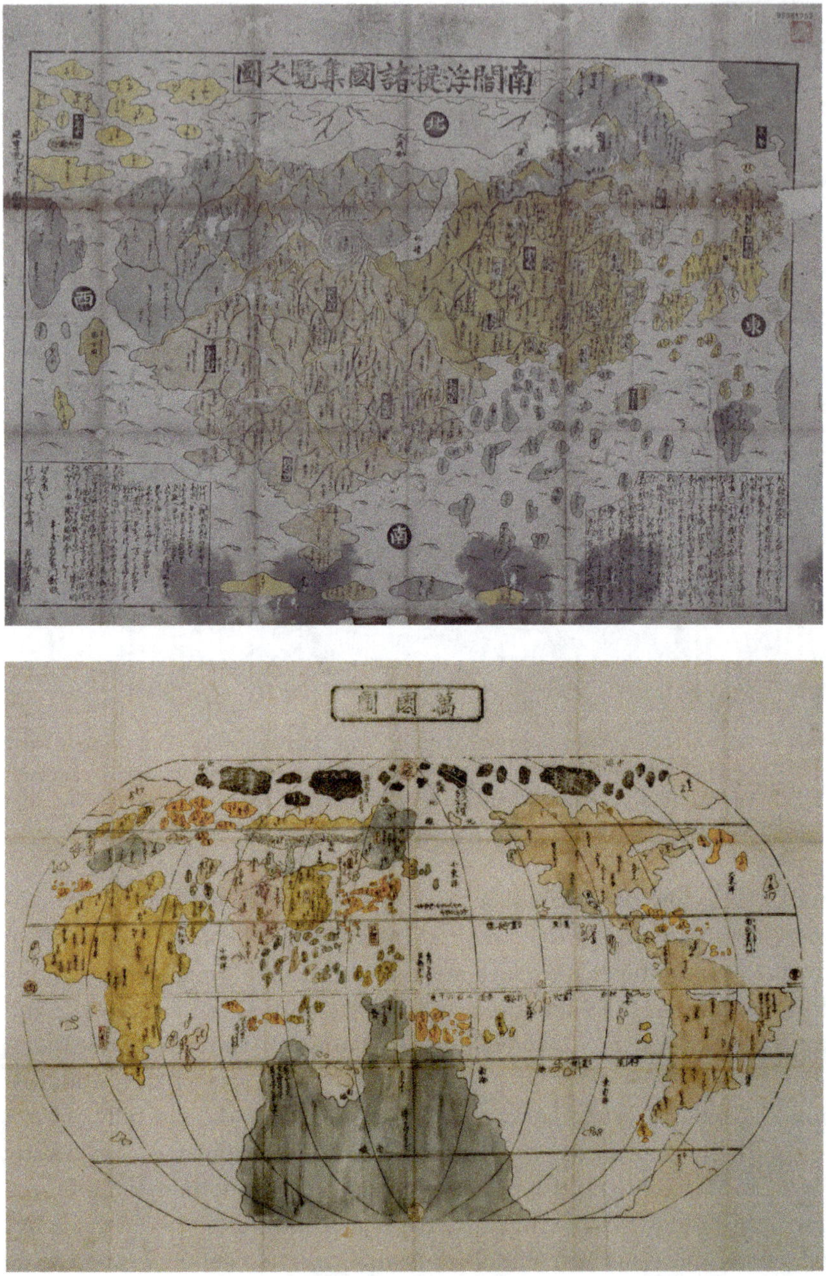

FIGURE 9.8 Upper: "Map of the Various Countries of Jambu-dvīpa" [*Nansenbushu shokoku shūran no zu* 南瞻部洲万国諸国集覧之図] by Kabō Hyōzō 花坊兵蔵, 1744. Woodcut, 43.5 × 60 cm. Kyoto University Museum. Lower: "Map of the Myriad Countries" [*Bankoku zu* 万国図], ca. 1744. Woodcut, 50 × 68 cm. Department of Geography, Graduate School of Letters, Kyoto University.

4 A Japan-Centric World

This brings us to the last discursive community: the nationalistic group for whom "Dai Nihon 大日本" or "Great Japan" occupied the center of the world. The increasing appropriation of the term "Dai" [Great] in front of "Nihon," in cartography and elsewhere, is a clue that Japan-first thinking was on the ascendant during the Edo period. Japan-centric prints may have been made more often by graphic artists than by scholarly geographers, and their cartography was often crude, but I believe we should take them seriously, nonetheless. For in size as well as diversity, this visually appealing genre expanded significantly over the course of the early modern era.

It is worth reflecting on what the continental Asian traditions offered people of this persuasion to work with. Buddhist cosmographers centered their world maps on the five Indias, with the result that the Japanese archipelago was pushed into the far-right margin. Confucians offered a different scheme, centered on the Chinese empire, but here too Japan was relegated to the edge of the frame. Given these unflattering alternatives, one can imagine the appeal of the Jesuits' radically different *mappae mundi* from Europe. In the form that Ricci gave it—an oval projection with the Pacific Ocean at the center—the Jesuit map must have been deeply attractive to many Japanese. For the first time, their own country occupied a position at the middle of the map. Even better, when the oval was rotated on its side to put East at the top, Japan ended up above China and Korea. And that is exactly the orientation that was chosen in the earliest Japanese printed world maps, all of which adopted an oval projection with East at the top.

The most enduring of these early world maps was that of Ishikawa Ryūsen 石川流宣 (Figure 9.9). In this almost totemic *mappa mundi*, the designer—a noted woodblock artist—played fast and loose with the Ricci framework, contracting the Pacific and enlarging the archipelago so that Japan ends up literally straddling the median line of the globe. It is important to note that this image does not emphasize the continents: not by shape, not by color, not by label. In fact, it might be a stretch to call this a map of the European continental scheme at all, since the only landmass so labeled is the Americas. The terms Europe and Asia do not appear anywhere on this map, and the landmasses of the old world in particular are oddly fragmented, coming across visually as a collection of islands.

But at the same time, a samurai scholar named Nishikawa Joken 西川如見 (1648–1724)—sometimes called "Japan's first scholarly geographer"—published a schematic map of his own, embedded in a work that brought to a broad Japanese audience the contents of Giulio Aleni's Chinese geographical work from 1623 (Figure 9.10). Joken's book, whose title translates as "On commercial

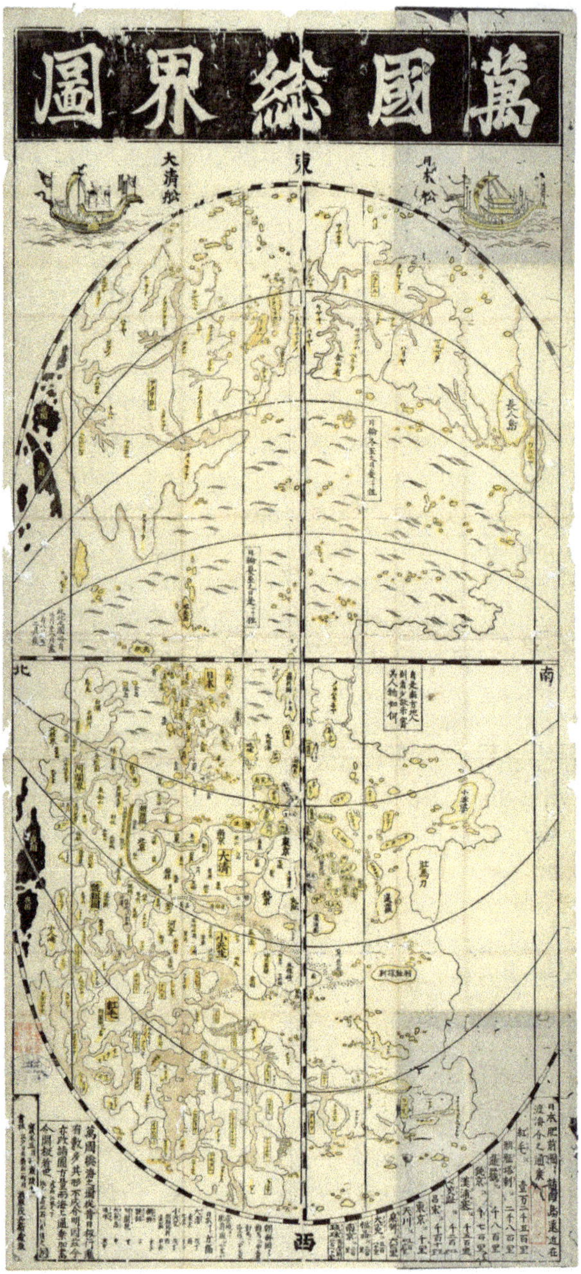

FIGURE 9.9 "Map of the Myriad Countries of the World" [*Bankoku sōkaizu*], by Ishikawa Ryūsen 石川流宣, 1708. Kobe City Museum.

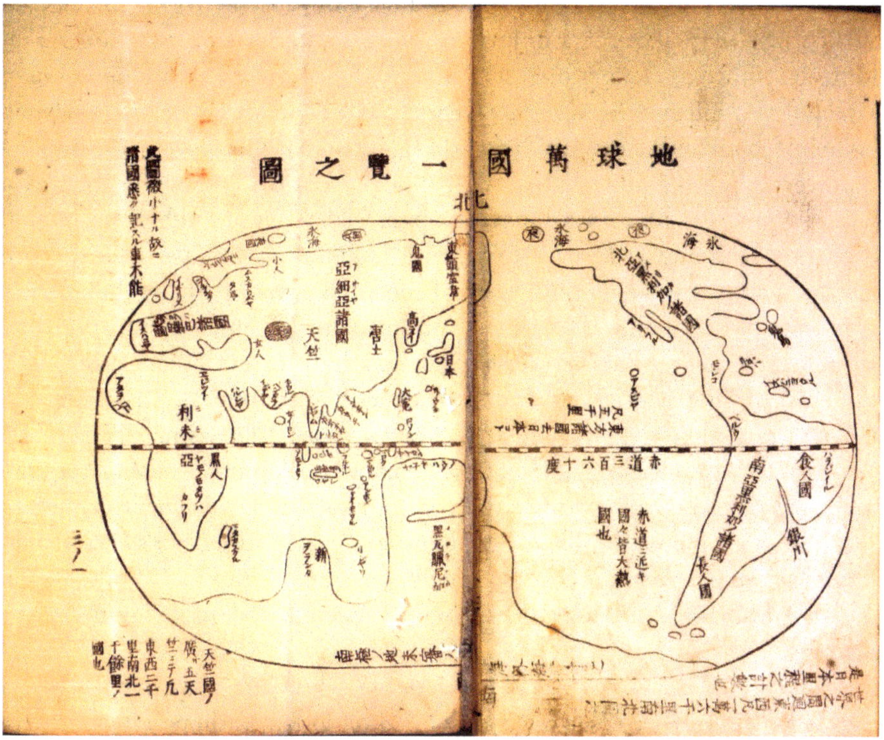

FIGURE 9.10 Nishikawa Joken 西川如見. Illustration from *Zōhō Ka'i tsūshō-kō* 增補華夷通商考 [On Commercial Relations with Chinese and Barbarians (revised and expanded version)], 1708. Waseda University Library.

relations with Chinese and barbarians" (*Ka'i tsūshō-kō* 華夷通商考) (original 1695; expanded and revised version, 1708), was one of the most widely circulated and often cited geographical texts of the entire early modern period, so it is safe to say that the accompanying map was widely seen by the literate public. And here, we do see all the continents clearly labeled with the phonetic *kanji* given to them by Matteo Ricci.

The striking similarity in the form of these two maps suggests that viewers who were exposed to texts like Joken's could readily transfer his content onto the memorable image they already carried in their head, not just from Ishikawa Ryūsen's popular map of the world but from sixty years' worth of published maps that preceded it.

Ryūsen's outdated but appealing world map dominated the Japanese print market for nearly a century. When it was finally surpassed, it was replaced by yet another oval projection based on Ricci's model: A "Complete map of the mountains and oceans of the world," published in 1788 by Nagakubo Sekisui 長久保赤水 (1717–1801). A Confucian-trained samurai from the domain of Mito, Sekisui created a national sensation with his innovative gridded map of Japan, first published in 1780. Five years later, at the urging of his publisher, he followed up with a map of the world (Figure 9.11). Both became bestsellers that were reprinted and pirated many times before the end of the Tokugawa period. But unlike his gridded map of Japan, Sekisui's "Comprehensive map of the lands, seas, and countries of the world" offered nothing particularly new. On the whole, it was a faithful copy of Ricci's 1602 wall-map, by then nearly 200 years old. The only changes Sekisui introduced were simplifications: he shrunk the image (to about one-fourth the size of Ricci's original), stripped out most of the place-names, and color-coded the continents. The result was a clear, convenient, legible, and affordable version of Ricci's masterful print. The timing of its release could not have been better. Public alarm about Japan's maritime defense was creating a vigorous market for global geography just as this map hit the bookstands in 1785. Benefitting from its author's established reputation as a cartographic authority, Sekisui's world map would prove to be a long-running hit. It would be reprinted, redacted, and widely imitated for decades.

When we compare the world maps designed by these two men (Ryūsen and Sekisui), we do not see a big leap forward. In fact, although it was made in 1785, Sekisui's world map was based on the same model as its predecessor: the classic 1602 Ricci wall map. The main differences were that Sekisui reproduced Ricci with greater fidelity, on a larger scale, and—most significant for our purposes—that he color-blocked the continents for the first time.

Over the course of 150 years, then, the Japanese had begun to slowly work their way toward a continental worldview. A visual foundation, laid by artists in the seventeenth century, was gradually filled in with scholarly content by later geographers until we reach this apotheosis—created, interestingly, by a Confucian intellectual. It is worth recalling that Sekisui was the same cartographer who made the map of "Oriental Asia" shown above. When that 1790 map is set side-by-side with Ricci's depiction of the same region (as in Figure 9.5), we can see how fully the Jesuits' continental scheme had been assimilated by the leading Japanese mapmaker of the eighteenth century.

Nor would this world-picture remain the exclusive property of the literati. Sekisui's work was hugely influential, inspiring a raft of pirated editions. In the middle decades of the nineteenth century, five decades after this pioneering geographer's demise, Japanese consumers had access to a cornucopia of maps in this style (Figure 9.12). Crude but colorful, small but cheap, the later Sekisui knock-offs boldly foregrounded the continental divisions. In the examples

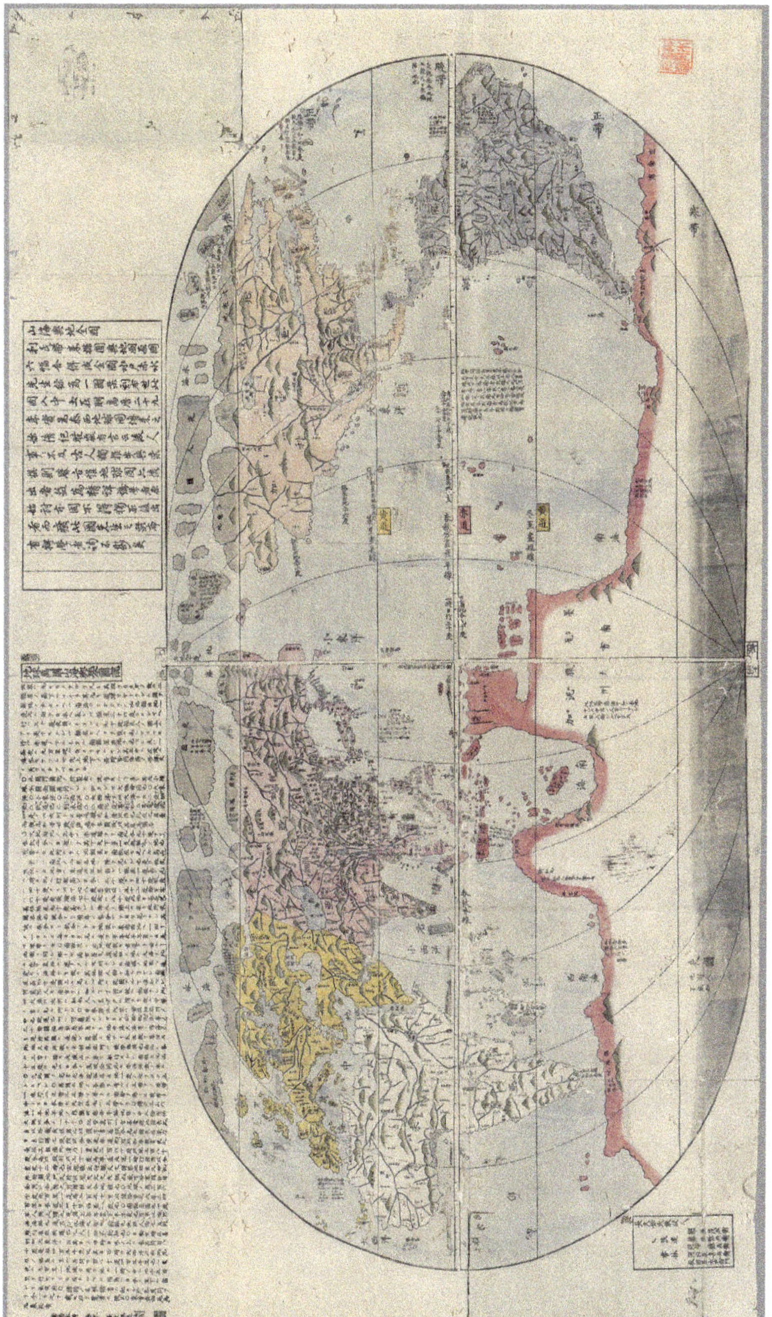

FIGURE 9.11 "Comprehensive Map of the Lands, Seas, and Countries of the World" [*Chikyū bankoku sankai yochi zenzusetsu* 地球万国山海輿地全図説] by Nagakubo Sekisui 長久保赤水. Original 1785; reprint 1834. Woodblock print, 91 × 165 cm. Yokohama City University Library.

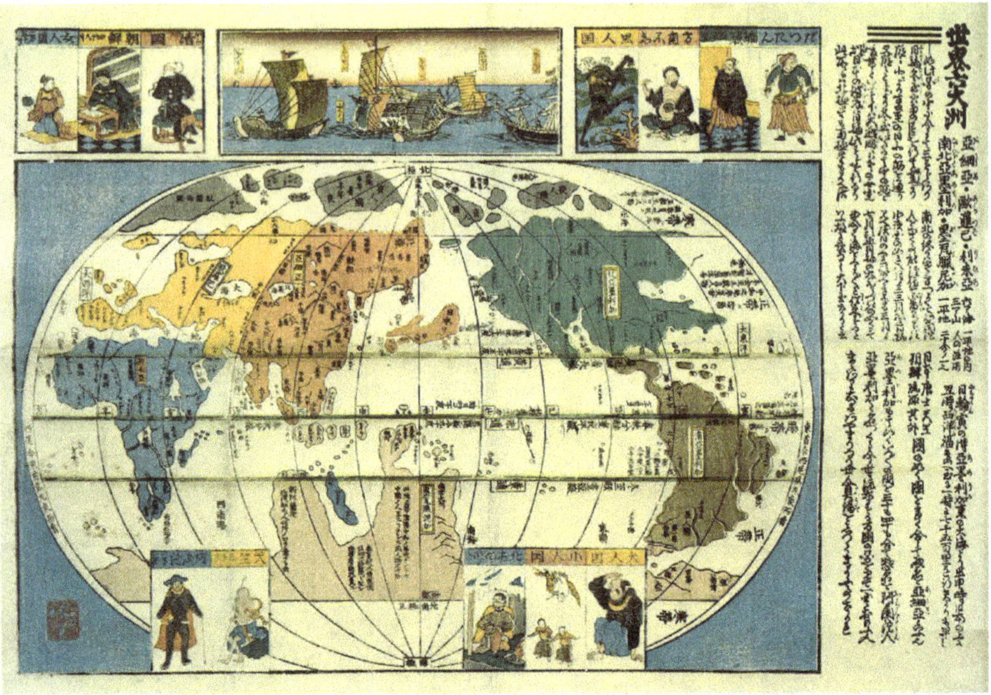

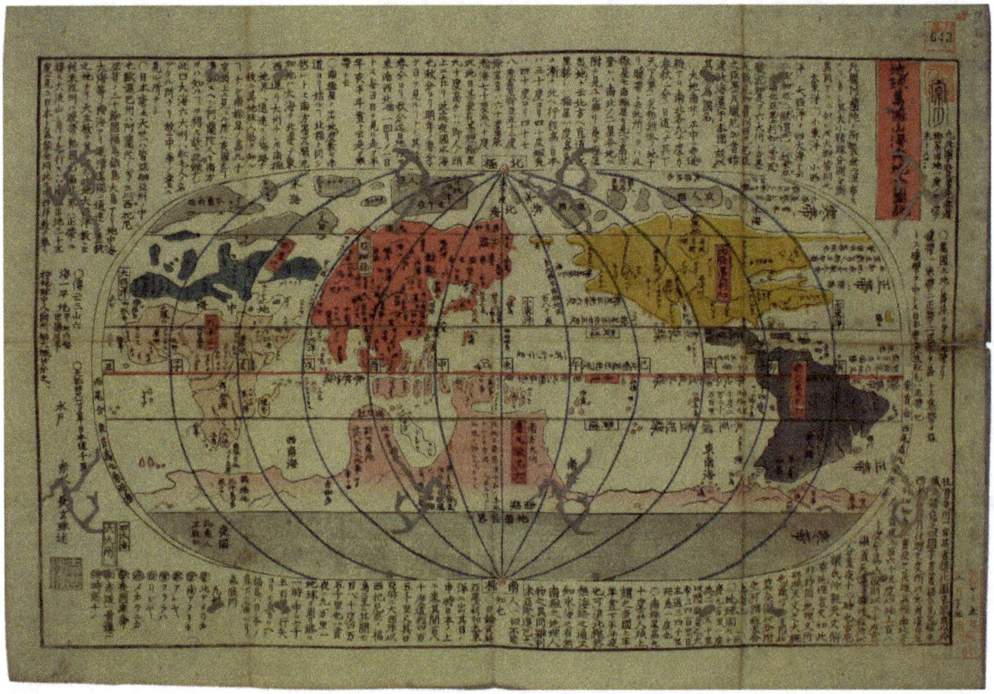

FIGURE 9.12 Anonymous world maps in the style of Nagakubo Sekisui. Woodblock prints, various sizes, n.d., mid-nineteenth century. Yokohama City University Library.

below (and many more), North and South America began to be differentiated, for a total of six continents (counting the anachronistic great Southern land mass, Terra Australis or Magellanica, which continued to be visible on all of the maps in this genre). And true to Sekisui's original color scheme, Asia was typically marked in red. As a result, it may well be that we should credit this Confucian cartographer, more than anyone else, with habituating the Japanese to *seeing* their country as part of "Ajia."

In short, by the time of the Meiji Restoration, it could be argued that Japan had already fully "entered Asia." By the 1860s, the country was awash in maps being marketed under catchy names like "Quick guide to the six continents." At a basic level, the European continental scheme was undoubtedly becoming a normal way to think about Japan's place in the world. But with the older, Jesuit-derived *mappae mundi* lingering in their minds, many people in Japan must have continued to imagine their homeland as having a special place in the world. At least, they certainly rewarded artists who found a way to imply as much.

One creative way that print-makers elaborated the Japan-centric tradition was by foregrounding Japan relative to Asia in panoramic picture-maps. Such was the novel approach of Kuwagata Keisai 鍬形蕙斎 (1764–1824), who created an immensely popular fish-eye view of Japan in the early nineteenth century. Keisai's image offered a novel perspective by exaggerating the capital region while making the northern and southern ends of the archipelago recede into the distance. The Asian continent, represented by Korea, was just visible on the horizon at the top of the print.[14]

It is no accident that the most prominent landmark on Keisai's Japan was Mount Fuji. Fuji was strikingly visible from Edo on a clear day, and had long been an object of veneration for people in the capital region. By 1800, the Fuji cult had swelled into a national phenomenon. The sacred mountain's prominence as a pilgrimage destination—and as a focus for souvenir prints of all kinds—made it a major feature of popular visual culture in the later Edo period. In effect, what Keisai had captured was an emerging vision of Japan as a divine land: one whose axis was no longer Mt. Meru, but rather Mount Fuji.

Whoever designed the commercial objects shown in figure 9.13 achieved a similar effect, albeit with less subtlety. Although it might not seem so at first glance, each of these is actually a world map. On the ceramic plate, the cloudlike forms surrounding the Japanese islands are meant to be continents. Europe, Asia, and Africa are all clearly marked on the left, the Americas on the

14 Henry D. Smith II, "Picturing Maps: The 'Rare and Wondrous' Bird's-Eye Views of Kuwagata Keisai," in *Cartographic Japan: A History in Map*, ed. Kären Wigen, Sugiimoto Fumiko, and Cary Karacas (Chicago: University of Chicago Press, 2016), 93–97.

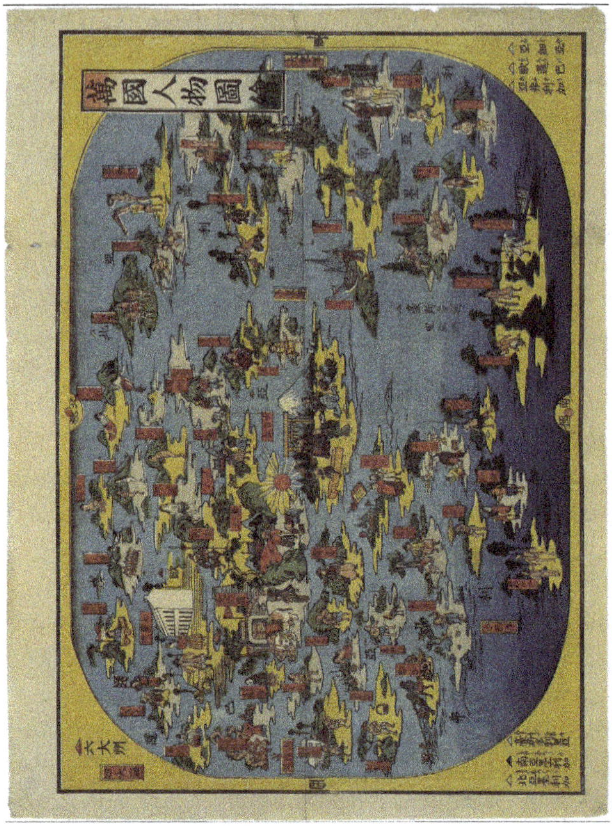

FIGURE 9.13 Left: Imari ceramic map of the world, mid-nineteenth century. Kyushu Ceramic Museum. Right: "Picture Map of Peoples of the World" [*Bankoku jinbutsu zue* 万国人物図絵]. Anonymous woodblock-print, n.d., mid-nineteenth century. Gifu Prefectural Library.

right. Likewise, the printed picture-map depicts the countries of the world as a collection of islands ringing Dai-Nihon. The names of the continents are all here, too, in large characters that wind their way between the islands; North and South America are again indicated to the right, while the names Asia, Europe, and Africa can be found to the left (even if they do not always align perfectly with the more local place names nearby).

Besides their Japan-centered continental frame, the other element that these images share is an oversized Mt. Fuji at their core. This is clearly visible in the towering snow-capped mountain on the print, but the ceramic plate as well features a simple icon in the shape of the sacred volcano. By dint of their design, this class of artifacts put Japan's sacred volcano not just at the center of the archipelago but at the heart of the world.

The popularity of images like this in the mid-nineteenth century (on plates, prints, and even board games) suggests that, in the marketplace as in the academy, the continental scheme had taken on a life of its own, albeit not exactly in the manner envisioned by the Jesuits. Creative views like these were the product of an artist's imagination. Their designers had mastered the techniques of perspective as well as the Jesuit *mappae mundi*, well enough to deploy both to their own ends. Through a combination of strategies, each managed to map the world in a way that put their own homeland in the most important position.

Perhaps the cleverest image of this kind is a cunning two-dimensional rendering of a three-dimensional globe, rotated so as to put Japan at the top. Figure 9.14 constituted the first illustration in "Separate Maps of the Countries of the World" (*Bankoku chikyū bunzu* 万国地球分図) 1856, by Utagawa Sadahide 歌川貞秀 (1807–1878). Famous for his *ukiyoe* prints and landscapes, the artist Sadahide also compiled the first and only world atlas published in Japan during the Edo era. By rotating the earth so that Japan appears to straddle Asia from above, this frontispiece image placed the archipelago in essentially the same privileged position it occupied on the earliest printed world-maps of two centuries earlier. For his readers, this would have been instantly recognizable; it was not only a familiar view but one that dignified their homeland by offering it a position of honor. And interestingly, while Sadahide followed convention by including Japan in Asia (in the separate continental map), on this initial image he separated Japan from the rest of Asia visually by giving it a distinctive red color.

Intriguingly, Fukuzawa Yukichi—the presumptive author of the infamous "Leave Asia" editorial—chose exactly the same color-patterns for the world-map in his best-selling *Sekai kunizukushi* 世界国尽 [All the countries of the world], written for children in verse (1869). Adopted as an official geography textbook in many Meiji-era schools, Fukuzawa's primer saw unprecedented

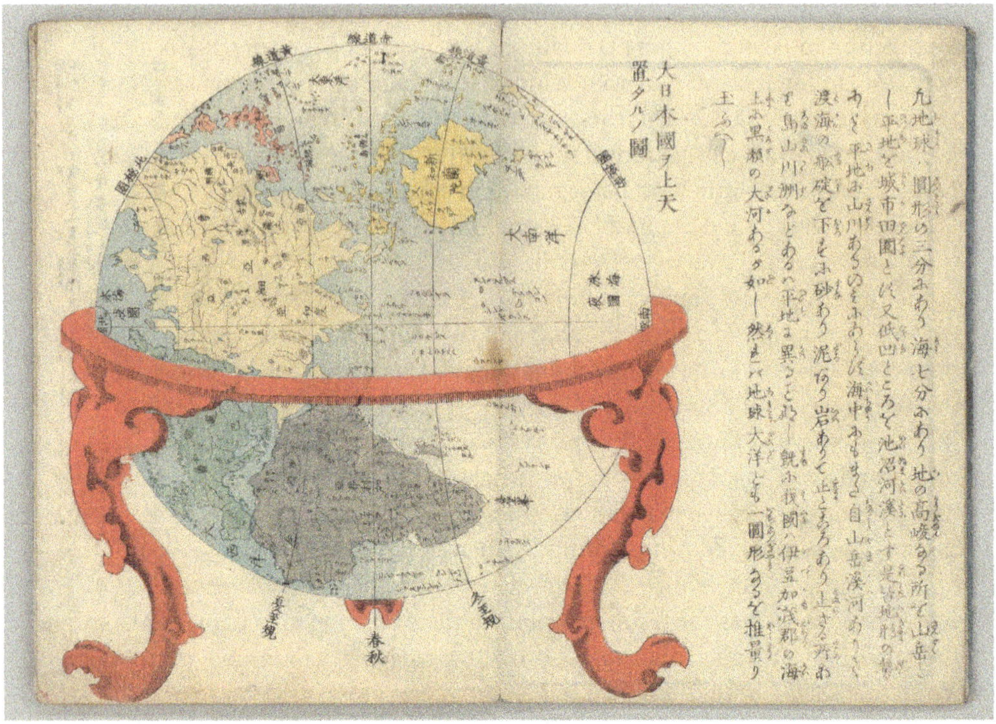

FIGURE 9.14 Utagawa Sadahide 歌川貞秀, frontispiece from *Bankoku chikyū bunzu* 万国地球分図 [Separate Maps of the Countries of the World]. Woodblock print, 1856. Yokohama City University Library.

circulation. And while cast in the dual-hemisphere projection traditionally associated with Dutch science (rather than the familiar Ricci-esque oval), Fukuzawa's map echoed Sadahide's in sending a mixed message about Japan's place in the European continental scheme. Yes, Japan was part of Asia—that much was not in dispute. But it was not *embedded* in Asia on the same terms as its neighbors.

In this way, the leading cartographers at the dawn of the modern era struck a somewhat disquieting note. On the one hand, no matter what school they belonged to, Japanese map-makers had embraced the notion of "Asia," which had had a full two centuries to sink deep roots in popular consciousness. But nationalist cartographers were finding ways to suggest a sense that "Dai-Nihon" had a special, privileged position within that region. Abetted by Japan's central location in popular Ricci-esque world maps, this budding notion would blossom in subsequent decades into dark dreams of dominating the vast region that came to be known as the Pan-East Asian Co-Prosperity Sphere. In a word, the visual evidence presented here suggests that the conceptual groundwork for Japan's ill-fated empire was quietly being laid in the visual idiom of cartography well before 1885, preparing the Japanese public to resonate with the notion of *leading* Asia—if not entirely *leaving* it—a generation before that famous editorial.

Bibliography

Aleni, Giulio. *Zhifang Waiji*. 1623.

Bae Woo Sung. "Worldviews and early Cartography." In *The Artistry of Early Korean Cartography*, edited by Han Young-woo, Ahn Hwi-Joon, and Bae Woo Sung, 93–129. Larkspur, CA: Tamal Vista Publications, 1999.

Brotton, Jerry. *A History of the World in 12 Maps*. New York: Viking, 2013.

Burns, Susan. *Before the Nation: Kokugaku and the Imagining of Community in Early Modern Japan*. Durham: Duke University Press, 2003.

Craig, Albert M. *Civilization and Enlightenment: The Early Thought of Fukuzawa Yukichi*. Cambridge: Harvard University Press, 2009.

Falchetta, Piero. *Fra Mauro's World Map, with a Commentary and Translations of the Inscriptions*. Turnhout: Brepols; Venezia: Biblioteca Nazionale Marciana, 2006.

Goodman, Grant K. *Japan and the Dutch 1600–1853*. London: Routledge Curzon, 2000.

Hostetler, Laura. *Qing Colonial Enterprise: Ethnography and Cartography in Early Modern China*. Chicago: University of Chicago Press, 2001.

Keene, Donald. *The Japanese Discovery of Europe, 1720–1830*. Revised edition. Stanford: Stanford University Press, 1969.

Korhonen, Pekka. "Asia's Chinese Name." *Inter-Asia Cultural Studies* 3, no. 2 (2002): 253–270.

Korhonen, Pekka. "Leaving Asia? The Meaning of Datsu-A and Japan's Modern History." *The Asia-Pacific Journal / Japan Focus* 11 (issue 50), no. 1 (December 2013): 1–18. Article ID 4083. https://apjjf.org/2014/11/50/Pekka-Korhonen/4083/article.html.

Lewis, Martin, and Kären Wigen. *The Myth of Continents: A Critique of Meta-Geography*. Berkeley: University of California Press, 1997.

Moerman, D. Max. *The Japanese Buddhist World Map: Religious Vision and the Cartographic Imagination*. Honolulu: University of Hawaii Press, 2022.

Mosca, Matthew. From Frontier Policy to Foreign Policy: The Question of India and the Transformation of Geopolitics in Qing China. Stanford: Stanford University Press, 2013.

Park, Hyunhee. *Mapping the Chinese and Islamic Worlds: Cross-cultural Exchange in Premodern Asia*. Cambridge: Cambridge University Press, 2012.

Smith, Henry D., II. "Picturing Maps: The 'Rare and Wondrous' Bird's-Eye Views of Kuwagata Keisai." In *Cartographic Japan: A History in Maps*, edited by Kären Wigen, Sugiimoto Fumiko, and Cary Karacas, 93–97. Chicago: University of Chicago Press, 2016.

Unno Kazutaka. "Cartography in Japan." In *Cartography in the Traditional East and Southeast Asian Societies*, edited by J.B. Harley and David Woodward, 346–477. Volume 2, book 2 of *The History of Cartography*. Chicago: University of Chicago Press, 1994.

CHAPTER 10

China's Nine-Dash Line: Cartographic Science and the Adoption of New Map Languages in the Transition from Empire to Nation State

Laura Hostetler

The coordinate system of mapping, which uses lines of latitude and longitude as accepted points of reference, is prevalent throughout the world today. This conceptual framework is essential to our understanding of the size and shape of the globe, undergirds our geopolitics, and is the basis on which our cell phones give us directions. These measures of the earth's shape and size, notated in degrees and minutes, are also correlated to our experience of time. They have become so pervasive on our world maps today that we scarcely notice their existence. Their absence, as on most traditional Chinese maps, has almost come to seem almost more remarkable. When Matteo Ricci produced his 1602 Complete Map of the Myriad Countries of the World, or *Kunyu wanguo quantu* 坤宇萬國全圖, he introduced the use of lines of latitude and longitude as cartographic symbols to China, and by extension to other parts of Asia. Although Chinese geographers had performed triangulation in specific contexts prior to the introduction of cartographic techniques from Europe, the idea of applying the method to the entire world in order to represent a usefully scaled projection of the earth on paper was new, as was the representation of lines of latitude and longitude on the resulting maps.

Earlier Chinese maps that are often cited for their scaled accuracy, such as the Song dynasty *Yu ji tu*, 1136 [Map of the Tracks of Yu] or the *Da Ming hun yi tu* [Amalgamated Map of the Ming], ca. 1390, included all of China as well as some neighboring territory, but did not attempt to map the entire world. While grids were sometimes used (see Figure 10.1), they did not represent latitude and longitudinal projections per se. Chinese world maps naturally portrayed China at the center, and also emphasized the country's importance by portraying countries around its perimeter according to their relative proximity to China and to each other, but not according to a consistent scale. In other words, the maps privileged relative locations over absolute areas.

Based on the work of Ortelius, and enhanced with observations made during his own travels, Ricci's map introduced to China a new way of viewing the world, one which emphasized China's position relative to a larger constellation

© LAURA HOSTETLER, 2024 | DOI:10.1163/9789004684782_012
This is an open access chapter distributed under the terms of the CC BY-NC-ND 4.0 license.

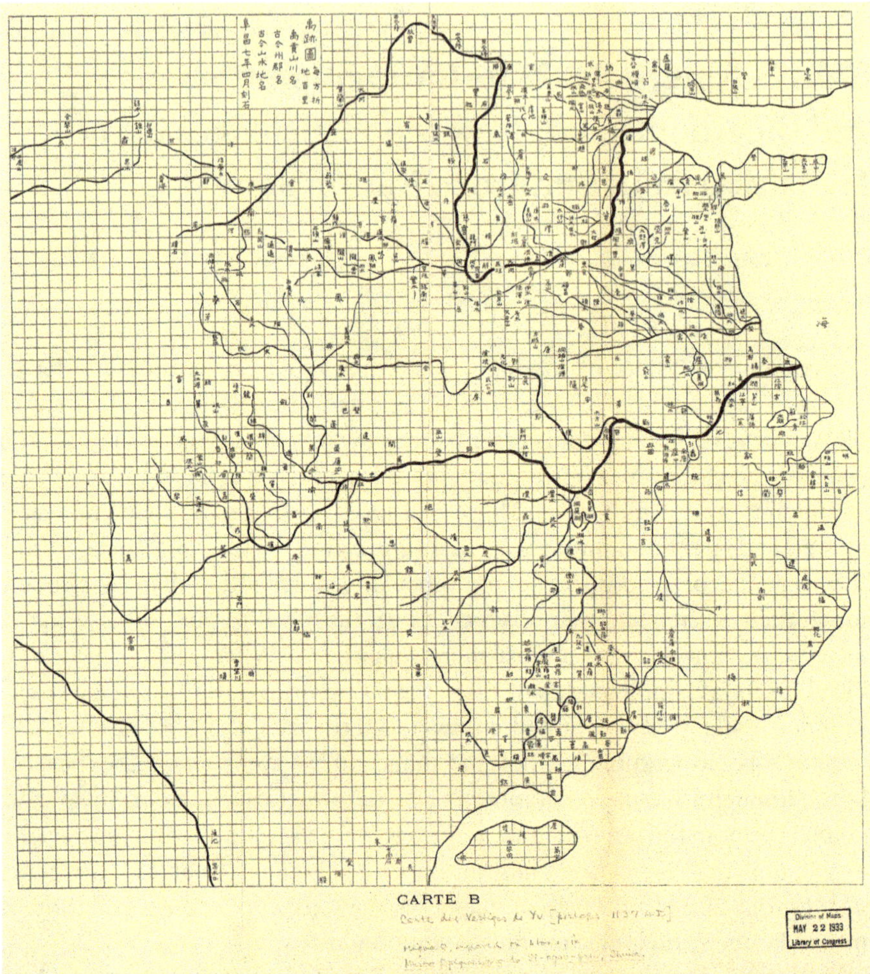

FIGURE 10.1 *Yu ji tu* 禹迹图 [Map of the Tracks of Yu]. Stone rubbing, 1903, based on Song dynasty original. Library of Congress.

of countries worldwide.[1] Yet, within three and a half centuries the coordinate system would be so thoroughly adopted that the Republic of China used it as the standard onto which to inscribe not only its territorial claims, but also its maritime aspirations.

1 Timothy Brook, *Completing the map of the world: Cartographic interaction between China and Europe* 全圖：中國與歐洲之間的地圖學互動 (Taibei: Institute of Modern History, Academia Sinica, Lecture Series no. 5. 2020), 32–33.

In 1933, the Republic of China formed a committee for the inspection of land and water maps. The effort involved cooperation and collaboration among a broad swath of government ministries including the Foreign Ministry, the Ministry of Education, the Ministry of Internal Affairs, the Marine Ministry, and the Committee on Mongolia and Tibet.[2] The goal was to claim the territorial inheritance of the Qing dynasty and to establish the new nation-state's sovereignty over these areas by marking out firm borders on paper. Extensive as they were aspirational, the claims were neither borne out by conditions on the ground nor did they dovetail precisely with the geographical expanse of the Qing. My focus here is on the unprecedented maritime claims that would be communicated in the coming years by a dashed line extending deep into the South China Sea.[3]

The historical basis of the nine-dash line sometimes gets lost in the current controversies surrounding China's maritime claims. Yet, it was clearly a product of the modern nation state's desire to establish itself and its claims on the world stage. For this reason, portrayals of the dashed line are marked on standard modern cartographic base maps that employ internationally established conventions of latitude and longitude and use accepted conventions for marking out territorial boundaries. The only, but quite significant, addition is a series of nine (or sometimes more) dashes that encompass the majority of the maritime area to China's east and southeast. The dashed portions of the line generally take on the same font, width, and color as the international borders shown for China. The contours of the line can best be described as reflecting the median point bisecting the sea between the islands claimed by China and the shorelines of adjacent countries.[4]

Although introduced into a familiar-looking base map, the nine dashes are not part of an existing cartographic nomenclature or language, nor are they correlated to specific territorial locations as determined by coordinates of longitude and latitude (despite a consistent claim to a southernmost point of 4° North). Because the nine-dash-line has no established cartographic precedents, it is difficult to engage or precisely define—and therefore limit—its claims. While the form of the maps it appears on is internationally accepted,

---

2 Li Jinming and Li Dexia, "The Dotted Line on the Chinese Map of the South China Sea: A Note," *Ocean Development and International Law* 34:3–4 (2003): 289.

3 "South China Sea" has long been used in English to denote the maritime regions south and east of China. It is important to note, however, that other countries have other names to refer to the same waters such as the East Sea (Vietnam), Luzon Sea, and West Philippine Sea. China refers to the South China Sea as *nanhai* or the southern sea, and to the East China Sea as *donghai* or the eastern sea.

4 Li and Li, "Dotted Line," 290.

in terms of *content* the nine-dash-line was absolutely new when it began to appear on maps produced in the ROC in the 1930s.

This chapter examines the nexus between cartographic practices and state formation in the context of China's transition from empire to nation state, a story that has been traced in other parts of the world as well.[5] The integration of more local and varied mapping processes into internally coherent and internationally intelligible languages was one of the projects of modernity, and inextricably linked to the formation of the modern nation state worldwide. Arguably this process began in Europe during the early modern period, a transitional time cartographically when the grandeur of courts was defined by their imperial centers rather than the specific extent of their outlying territories. When maps of France were corrected according to the newly calculated measure of longitude in 1682, Louis XIV quipped that his cartographers lost him more territory than his generals ever had. It was only coordinate mapping that had made such precision possible. Cartographic technology that defined the location and extent of states in terms of a matrix of longitudinal and latitudinal lines into which other states were also situated made it evident in a graphically new way that we all share the globe and that before long no more "undiscovered" land would exist.

At the same time, with the political transition from monarchies and empires to nation states, the definition and legitimacy of the state would be associated not with the qualities of and beliefs surrounding its ruler, but rather made manifest in its territorial integrity.[6] There would be no more room for multiple or nested sovereignties or permeable frontiers, which had been quite common in earlier times both in Europe and in Asia. As Emma Teng demonstrates, the Qing decision to annex all of Taiwan—including the aboriginal areas in which the state originally held little interest—became necessary when it became clear that if they did not, the Japanese, or another foreign power, would.[7]

---

5 See, for example Josef Konvitz, *Cartography in France, 1660–1848: Science, Engineering, and Statecraft* (Chicago: University of Chicago Press, 1987) and Thongchai Winchakul, *Siam Mapped: A History of the Geo-Body of a Nation* (Honolulu: University of Hawaii Press, 1994). A number of scholars are beginning to research this process in the Qing context: see Matthew W. Mosca, *From Frontier Policy to Foreign Policy: The Question of India and the Transformation of Geopolitics in Qing China* (Stanford University Press, 2013), and Anne-Sophie Pratte, "Mapping the Steppe: The Politics of Cartography in Qing Mongolia, 1780–1911," PhD dissertation, Harvard University 2021.

6 Laura Hostetler, "Mapping, Registering, and Ordering: Time, Space, Knowledge," in Peter Fibiger Bang, ed., *Oxford World History of Empire* (Oxford: Oxford University Press, 2020), 290, 308.

7 Emma Teng. *Taiwan's Imagined Geography: Chinese Colonial Travel Writing and Pictures, 1683–1895* (Cambridge, MA: Harvard University Asia Center, 2004), 210.

A brilliant work on the history of Thailand by Thongchai Winichakul explores the territorial aspects of geopolitics in the transition from the kingdom of Siam to the modern Thai nation.[8] Prior to the modern era Siam's frontiers were inhabited by any number of different peoples with allegiances to different states, or to no states in particular. The idea of marking off a linear border on the ground on one side of which these peoples would have to remain was utterly alien to them. Nested sovereignties in which one polity might have tributary relationships with more than one larger state were not uncommon. This situation was, in fact, typical of much of the region. For example, Korea made tributary visits to the Qing empire, but also hosted tributaries at its own court. With the introduction from Britain of the concept of coordinate mapping and notions of exclusive territorial sovereignty, however, Siam soon began to establish firmer control over the polities on its frontiers and mark out its own linear boundaries. This is a practice we can trace throughout the modern world as emergent nations re-imagined the globe and their place on it.

As elsewhere, cartographic practices in China have changed in tandem with political exigencies over the centuries—including the state's ever-changing relationship to other centers of power internationally and its shifting loci of cartographic activity domestically. The acceptance and deployment of coordinate mapping in China is part of a larger story of the global integration of space, the logic of which emergent nation states adopted for their own self-definition and preservation.[9] Whether experimented with by cartographers in the seventeenth century, adopted by the Qing court in the eighteenth, practiced by Han Chinese statecraft scholars in the nineteenth, or used to map out the territorial claims of the nation in the twentieth, the adoption of coordinate mapping came gradually to reflect China's efforts to define itself in internationally recognized terms to thereby stake out territorial claims both at home and abroad in that map discourse. The nine-dash line, however, seeks to stretch that discourse to include a new claim to maritime space.

The world map Ricci constructed was only one of a number of avenues he pursued designed to establish relationships around topics of mutual interest with Chinese literati. In gaining the respect of his Chinese peers, Ricci understood that the related art of calendrical science had even greater potential when it came to the Jesuits obtaining entrée at the court. Calendrics, including the ability to accurately predict eclipses, was directly tied to imperial legitimacy. Almost twenty years after his death, and only fifteen years before the fall

8 Thongchai, *Siam Mapped.*

9 I adopt the term "global integration of space" from Charles H. Parker, *Global Interactions in the Early Modern Age, 1400–1800* (Cambridge: Cambridge University Press, 2010).

of the Ming, a new calendrical office that was to incorporate Western methods was established.[10]

The political utility of European-style coordinate mapping would take much longer to establish at the court. Even the famed 1602 map he produced in Beijing was not an officially sanctioned version, as it was only in 1608 that the imperial court requested twelve copies be made for its princes.[11] Even then, as Cheng Fangyi argues, the Wanli emperor may have been more interested in the exotica he had figured onto exemplars of Ricci's maps.[12] Of course the maps may themselves have been understood as a kind of exotica.

Yet efforts at disseminating European mapping techniques and approaches to geographical knowledge continued after Ricci's death. As discussed in detail in Paola Demattè's chapter, Giulio Aleni's *Wanguo quantu* [Complete Map of Ten Thousand Countries] appeared in 1620 and his *Zhifang waiji* [Unofficial Records on Foreign Countries] in 1623. 1623 also saw the production of a globe by Niccolò Longobardi and Emmanuel Diaz, thought to have been held at the imperial court. Francesco Sambiasi produced a revised edition of Aleni's *Kunyu quantu* in 1647 or 1648. Thus, the geographical production and exchange begun by Matteo Ricci continued on after the Ming. Cao Junyi's world map of 1644, the *Tianxia jiubian fenye renji lucheng quantu* 天下九邊分野人跡路程全圖, contains clear indications of longitude, as did Zhang Huang's earlier (1613) *Yudi Shanhai quantu*, which was reproduced in the *Siku quanshu* compendium.[13]

Although the dynastic transition from Ming to Qing presented some complications for the Jesuits in terms of establishing their *bona fides* to the new leadership, they continued to have a presence at court with only minor interruptions. Adam Schall, who had been part of the calendrical reform effort under the Ming, would become close to the first emperor of the Qing dynasty and head up the Bureau of Astronomy, where he would be succeeded by Ferdinand Verbiest in 1666. In 1674 Ferdinand Verbiest presented his own Complete Map of the World, the *Kunyu quantu*, to the Kangxi emperor. In 1685 six mathematicians

10 Michela Fontana, *Matteo Ricci*, 234.

11 John D. Day, "The Search for the Origins of the Chinese Manuscript of Matteo Ricci's Maps," *Imago Mundi* 47, no. 1 (January 1995): 96; Michela Fontana, *Matteo Ricci: A Jesuit in the Ming Court* (Lanham, MD: Rowman & Littlefield Publishers, 2011), 221–222.

12 Cheng Fangyi, "Pleasing the Emperor: Revisiting the Figured Chinese Manuscript of Matteo Ricci's Maps," *Journal of Jesuit Studies* 6, no. 1 (March 11, 2019): 31–43.

13 Brook, Timothy, *Completing the Map of the World*, 19–21 (figure 1.1) and Song Gang, "Re-Locating the 'Middle Kingdom': A Seventeenth-Century Chinese Adaptation of Matteo Ricci's World Map" in *Mapping Asia: Cartographic Encounters between East and West*, eds. Martijn Storms, Mario Cams, Imre Josef Demhardt, and Ferjan Ormeling (Cham: Springer International Publishing, 2019), esp. 196.

were sent to China by Louis XIV, two of whom were retained in service to the court and tutored the emperor in mathematics.[14]

Examples of coordinate mapping had been present at the court for almost a century in the form of world maps when employed in the Kangxi emperor's grand survey of the Qing empire in the 1710s.[15] The resulting *Huangyu quanlan tu* 皇輿全覽圖 [Map of All Surveyed] demonstrated the imperial will to harness the value of early modern cartographic science by more precisely mapping the administrative and geographical features, as well as territorial extent, of the Qing empire. These maps, based on surveys coordinated by the imperial household department and involving both Jesuit and Manchu servants of the throne, can be understood as a major step in the global integration of space in China. The maps made their way back to Europe as well. Their appearance in Jean Baptiste Du Halde's widely translated and disseminated *Description ... de la Chine* is the most notable example.[16]

We now know that the Kangxi survey maps that were completed with Jesuit assistance in the 1710s do not represent the only Qing engagement with coordinate mapping. Later editions, substantially revised, were carried out under the Yongzheng and Qianlong emperors. Neither reached further south than the southern tip of Hainan Island at 18° North. The Yongzheng map was made in the context of contemporaneous Russian efforts to map that empire's expanding territorial claims. It covered approximately twice as much territory as the Kangxi era map by augmenting it with information gleaned from travelers' reports, other maps, and even interviews with Jesuits at the court, although Jesuit involvement in this later project was much curtailed as compared to the earlier map.[17] The Yongzheng map was also made on an entirely different projection, the emperor having stated emphatically that he did not want a map

14 Catherine Jami, *The Emperor's New Mathematics: Western Learning and Imperial Authority During the Kangxi Reign* (Oxford: Oxford University Press, 2012), especially chapters five and seven.

15 For additional background, see Theodore N. Foss, "A Western Interpretation of China: Jesuit Cartography," in *East Meets West: The Jesuits in China, 1582–1773*, 209–251 (Chicago: Loyola University Press, 1988); Laura Hostetler, *Qing Colonial Enterprise: Ethnography and Cartography in Early Modern China* (Chicago: University of Chicago Press, 2001); and Mario Cams, *Companions in Geography: East-West Collaboration in the Mapping of Qing China (c. 1685–1735)* (Leiden: Brill, 2017).

16 Jean Baptiste Du Halde, *Description ...* (Paris: Chez P.G. Le Mercier, 1735).

17 Laura Hostetler, "Imperial Competition in Eurasia: Russia and China," in *The Cambridge World History*, ed. Jerry H. Bentley, Sanjay Subrahmanyam, and Merry E. Wiesner-Hanks (Cambridge: Cambridge University Press, 2015), 6.1: 297–322. Mario Cams, "Reimagining Qing Space: Yongzheng's Eurasian Atlas (1727–8)," *Late Imperial China* 42, no. 1 (2021): 93–129.

with curved lines.[18] Longitude is, however, indicated in the form of parallel vertical lines according to degrees marked at the upper and lower margins of the map sheets. The overall effect is that the lines appearing on the map are at right angles to each other, reflecting the appearance of a grid, such as is found on the Song dynasty *Yu ji tu*, or Map of the Tracks of Yu (Figure 1.1). This projection substantially alters the appearance of land masses as one approaches the poles as compared to the Kangxi (and later Qianlong) maps. It would appear that the Yongzheng emperor was seeking to limit the appearance of European influence within his court even as he extended Qing imperial claims outward.

The map made under the Qianlong emperor ca. 1760 returned to the European-style projection used in the Kangxi era map. At the same time it depicted even more territory, reflecting not only recent conquests in Xinjiang but also lands well beyond the reach of the empire, including parts of the Indian subcontinent and even the Arabian Peninsula.[19] Again, Jesuits at the court were called on to provide what information they could offer, with Felix Da Rocha and Joseph D'Espinha commissioned to travel to Hami, Barköl, Turfan, and Ili to take geographical readings, and Michel Benoist involved in producing a copperplate edition of the map. Yet, none of these men were officially credited for their role in the production of the Qianlong map editions. Nonetheless, these later maps contributed not only to the Qing cartographic record but also made their way to Europe, where they were used in the nineteenth century as source material for maps of Central Asia.[20]

During the eighteenth century, the strength of the high Qing was manifest in the imperial inclination and ability to patronize more than one mapping language. Even as the Qianlong emperor commissioned this new map drawing on the emerging international standard of cartographic and scientific accuracy, he also continued to commission maps made according to more traditional Chinese mapping practices. The 1767 *Da Qing wannian yitong tianxia quantu* [Complete map of the Great Qing's Everlasting Unification of All Under Heaven], which continued to be revised and reprinted into the early

18 Antoine Gaubil, in Brucker, Joseph, S.J. "Correspondance Scientifique d'un Missionnaire Français a Peking Au Dix-Huitième Siècle. Le P. Antoine Gaubil, D'Après Des Documents Inédits," 172 (from a letter dated 8 October 1727 addressed to P. Gaillard). See also, Laura Hostetler, "Early Modern Mapping at the Qing Court: Survey Maps of the Kangxi, Yongzheng, and Qianlong Reign Periods," in *History in Geographical Perspective*, ed. Jeff Kyong-McClain and Du Yongtao (Lanham, MD: Lexington Books, 2013), 23.

19 Hostetler, "Early Modern Mapping," 18.

20 Hostetler, "Early Modern Mapping," 28.

nineteenth century, is a case in point (see Figure 1.2).[21] This map, which proclaimed the grandeur of the Great Qing, did so in an idiom that drew on the language of "All under Heaven," that would appeal broadly to more traditionally oriented views of universal empire. During the first half of the Qing, geographical information and mapping conventions did not need to—or rather needed *not to be*—commensurate with emerging international standards, at least domestically.

Yet, even as the Qing court was patronizing a variety of cartographic methods internally, it found it useful to adopt the emerging international idiom on the periphery of empire where frontiers with other powers had to be negotiated. Early examples can be found in the Treaty of Nerchinsk and the Treaty of Kiakhta, which were completed with the Russian empire in 1689 and 1727 respectively. Clarification of international boundaries on the frontiers introduced new mapping technologies. By the first half of the nineteenth century, geographic practices throughout China were undergoing significant change. Shifting imperatives in regard to foreign policy and national defense required that China's territories and borders be mapped in a coherent fashion that would render them both internally consistent and in sync with international mapping standards. Prior to that time, efforts to internationalize mapping technologies had been largely limited to the court, which was also responsible for foreign relations. But from circa 1820 Chinese scholars working to achieve such coherence outside the court, and who found it impossible to reconcile the multiplicity of existing conceptual systems, began to perceive the incommensurability of the geographic practices of the early Qing as a serious weakness.

In a careful study of mapping practices on China's southeast coast between 1820 and 1840, Matthew Mosca demonstrates that over a period of several decades a "different approach to foreign geography was beginning to take hold." What had become important was the commensurability of different geographic sources, and the test of this was the ability to locate them all on one map. For this reason, by ca. 1840, "European world maps came to be acknowledged as correct by China's leading geographers."[22] This was not so much because the maps were "European," but because European cartographers had

21 Richard A. Pegg, *Cartographic Traditions in East Asian Maps* (Honolulu: University of Hawaii Press, 2014), 16–25. For a general overview and introduction to traditional Chinese mapping practices, see Richard J. Smith, *Chinese Maps: Images of "All Under Heaven"* (Oxford University Press, USA, 1996), Cordell D.K. Yee, "A Cartography of Introspection: Chinese Maps as Other than European," *Asian Art* 5, no. 4 (1992): 28–47, and Cheng Yinong, ed. *"Nonscientific" Traditional Maps of China: A Study of Traditional Chinese Mapmaking* (Singapore: Springer Nature, 2022).

22 Mosca, *From Frontier Policy to Foreign Policy*, 214.

discovered "in mathematical cartography a research method that brought all geographic data into competition on a unified field of debate."[23] A situation in which different geographic idioms—both textual and cartographic—had co-existed until the mid-nineteenth century was no longer intellectually defensible. Instead, by the late nineteenth century Europe and China had come to largely share a single geographic system dominated by coordinate geography.[24] The adoption or acceptance of this technology accelerated the global integration of space, and provided the conditions for the acceptance of territorial sovereignty as a prerequisite of the modern nation state.

In the context of the consolidation of the Chinese nation during the 1920s and 30s (the Qing having fallen in 1911), nationalist leaders of China were trying desperately to hold the new nation together in face of disintegration from within and European imperial predation from without. Mapping the geo-body of the nation was one way to assert the sovereignty and territorial integrity China's leaders sought to establish. By the early twentieth century the concept of coordinate mapping introduced to China by Ricci (via Ortelius) had become the medium in which content that is uniquely Chinese—in that it makes claims unique to China's national self-definition—was purveyed on a style of map that most of us do not recognize as particularly Chinese. Yet, it was precisely the internationally agreed upon map language of the coordinate system that made the introduction of the nine-dash line and its claims possible. Using a base map already accepted by the international community was essential to making its claims, even as the claim upset the status quo in other ways.

---

23 Mosca, *From Frontier Policy to Foreign Policy*, 45.

24 Mosca, *From Frontier Policy to Foreign Policy*, 209. Mosca's wording in describing this process is important: "Information of Western origin, known in China for centuries, was becoming a master system of geographic organization that would not replace, but rather subsume and reorganize, other geographic worldviews." In this context it is also worth noting that the variants of the Kangxi survey maps that populated the pages of the *Gujin tushu jicheng* (a well-known imperially commissioned collectanea) and the Illustrated Collected Statues of the Qing, or *Da Qing huidian tu*, these maps long went unrecognized as being anything other than the traditional Chinese maps with which they coexisted because they were (mostly) reproduced in a way that omitted indications of latitude and longitude. Cordell D.K. Yee, "Traditional Chinese Cartography and the Myth of Westernization," in *The History of Cartography* 2.2: 170–202 (Chicago: University of Chicago Press, 1994). See also Laura Hostetler, "Contending Cartographic Claims: The Qing in Manchu, Chinese, and European Maps," in *The Imperial Map: Cartography and the Mastery of Empire* (Chicago: University of Chicago Press, 2009), 124–125 and fig. 3.9, which shows the foldout map from the *Huidian tu* that does indicate latitude and longitude, unlike the maps of smaller individual jurisdictions maps shown on its regular-sized pages.

According to Li and Li, *The Map of Chinese Islands in the South China Sea* (*Zhongguo nanhai daoyu tu* 中國南海島嶼圖) that this committee published in April 1935 declared that China's southernmost boundary should extend to a latitude of 4° North. The move was entirely unprecedented; the three editions of the imperial surveys of the Qing empire made during the eighteenth century all clearly marked Hainan Island at a latitude of 18° North as the southernmost point in the empire.[25] Large stone markers, which still stand, also commemorated this fact (see Figures 10.2, 10.3, and 10.4).

In 1936, a second ROC map, *The Map of Chinese Domain in the South China Sea* (*Haijiang nan zhan hou zhi zhongguo quantu* 海疆南战后至中國全圖), showed "the Pratas Islands, the Paracel Islands, the Macclesfield Bank, and the Spratly Islands ... drawn as being within Chinese territory in the South China Sea." This map also reiterated the unprecedented claim of the southernmost national boundary for China that included all islands in the South China Sea north of 4° North.[26] In 1935, this claim must have seemed both grandiose and unrealistic to those outside of China, if they were paying attention at all. But in the context of the times, with Hitler having just come to power in Germany and Japan having recently invaded Manchuria, international attention was largely focused on other matters.[27]

Within four years the Sino-Japanese war had begun. Under these conditions there was no question of China acting as a sovereign state within the islands of the South China Sea, much less within much of the former Qing territory. By 1937 Shanghai, Beijing, and Nanjing were occupied by Japanese armies; Chiang Kai-shek's government had been forced to relocate to the western city of Chongqing in Sichuan Province, from which it carried on as best it could.

In addition to being at war with Japan, Chiang Kai-shek was also beset with the communist threat from within. After Chiang ended the GMD/CCP alliance in a surprise attack in Shanghai in 1927, the communists had gone underground. A number of small soviets had established themselves in the south, most notably the Jiangxi Soviet located on the borders of Jiangxi, Fujian, and Zhejiang provinces. Nationalist forces had encircled the group and were

25 The later editions of this map reach to 85° North, all the way to the Arctic Circle. See Laura Hostetler, "Sino-Western Cultural Relations and the Mapping of Modern China," in *Narratives from the Hinterland: Perspectives, Methodologies and Trends in the Study of Christianity in China*, ed. Xiaoxin Wu, 157–184 (Guilin: Guangxi Normal University Press, 2014) and, Laura Hostetler, "Early Modern Mapping," 15–32.

26 Li and Li, "Dotted Line," 289.

27 For a rather satirical comment on the practice of using maps to make grandiose claims, see Mark Monmonier, *How to Lie with Maps* (Chicago: University of Chicago Press, 1991), 88.

FIGURE 10.2 Detail of an engraved boulder on the southern coast of Hainan Island reading *nan tian yi zhu* 南天一柱, which translates as "Pillar of the Southern Sky." The characters to the right of the central inscription date it to the Yongzheng era. Photo: Laura Hostetler.

FIGURE 10.3 The image of the "Pillar of the Southern Sky" had long circulated on the two yuan note, which had largely dropped out of circulation by 2014. Photo: Laura Hostetler.

working to annihilate the threat it posed when communist forces broke out of the encirclement to start their famed "Long March," which would lead them over 6,000 miles on foot to the northwestern base camp of Yan'an and establish Mao Zedong's leadership of the party. During the Sino-Japanese War Chiang Kai-shek would be forced into a second alliance with the Communist Party. However, both parties knew it would not last and the communists took advantage of this opportunity to establish their own system of government in areas where Nationalist control was weak, and even behind Japanese lines. In other words, the ROC was hardly a sovereign nation at this time even within the heartland of what had long been "China."

After WWII, China experienced freedom from foreign encroachment on the mainland for the first time since the settlement of the Opium War in 1842. Aside from Hong Kong and Macao, international territorial concessions were ended. Furthermore, in 1943 the allied European powers had ended the provisions of

FIGURE 10.4 Detail of an engraved boulder on the southern coast of Hainan Island reading *tianya* 天涯, which translates as "end of the world," or "the remotest corner of the earth." This inscription is dated "first year of the Xuantong Reign," or 1909. Photo: Laura Hostetler.

extraterritoriality, which had allowed them legal jurisdiction over their residents in China. Taiwan, which had been occupied by Japan, was also returned to Chinese sovereignty. Things were finally looking up for the new nation.

In this context, the government of the ROC, beleaguered as it still was by the communist threat, took up once again the question of China's territorial extent. The government renamed the Islands in the South China Sea (a typical colonial practice[28]), "reaffirmed" Chinese control to 4° North, and asserted that it would "protect" Chinese fishermen in the area. The results were then

28 For a comparative case see Barbara E. Mundy, *The Mapping of New Spain: Indigenous Cartography and the Maps of the Relaciones Geográficas* (University of Chicago Press, 2000), 166–167. For the case of Xinjiang see James Millward, "'Coming onto the Map': 'Western Regions' Geography and Cartographic Nomenclature in the Making of Chinese Empire in Xinjiang," *Late Imperial China* 20, no. 2 (December 1999): 61–91.

published by the Geography Department of the Ministry of Internal Affairs in 1947 in an internal map entitled *The Location Map of the South China Sea Islands* (*Nanhai zhudao weizhi tu* 南海諸島位置圖). The map included eleven dashes encircling the South and East China Seas. Significantly, even before this map was published, the claims of Chinese sovereignty over the islands and shoals within the South China Sea to a latitude of 4° North had been standardized in Chinese government publications, in school curricula, and the press more generally.[29] To a domestic audience these claims were already being naturalized. In February of 1948 the claims made in this internal map were to be made available internationally as part of *The Administrative Division Map of the Republic of China* (*Zhonghua minguo xingzheng quyu tu* 中華民國行政區域圖).[30]

After the establishment of the People's Republic of China in 1949, the new government adopted the basic principles of the 1948 map. (See Fig. 10.5). In 1953 the two new dashes were, however, rescinded on either the approval or the initiative of Foreign Minister Zhou Enlai in his negotiations with Vietnam. Thereafter the Nine-Dash Map largely held sway, although it was supplemented in 2013 with a tenth dash to the east coast of Taiwan, reiterating PRC claims to the island, and also making more aggressive claims toward the Philippines.[31] Interestingly, the ROC, while it has its own maritime disputes with the PRC, also continues to stand by the principle of the nine-dash line, evincing the map's deeply nationalist roots.

In the 2010s, tensions arose in the South China Sea between China, the largest power in the area, and other countries who also have claims to its islands and territorial waters. In 2016 a tribunal of the Permanent Court of Arbitration at The Hague ruled on allegations brought by the Philippines that China's claims and activities violate the United Nations Convention on the International Law of the Sea (UNCLOS). UNCLOS stipulates that nations have exclusive maritime rights, or sovereignty, only within a range of twelve nautical miles from their shorelines.[32] Within a range of twelve to 200 miles of a shoreline, nations have sovereign rights to the continental shelf and its resources and to the water

29 Li and Li, "Dotted Line," 290.

30 Edited by Fu Jiaojin and compiled by Wang Xiguang, et al.

31 Michaela Del Callar, "China's New '10-Dash Line Map' Eats into Philippine Territory" GMA News Online. Accessed November 22, 2022. https://www.gmanetwork.com/news/topstories/nation/319303/china-s-new-10-dash-line-map-eats-into-philippine-territory/story/.

32 The United Nations Convention on the Law of the Sea (UNCLOS) was adopted in 1982, replacing several earlier treaties, none of which dated before 1958. It was not until UNCLOS that the idea of an exclusive economic zone within 200 nautical miles of shore was adopted.

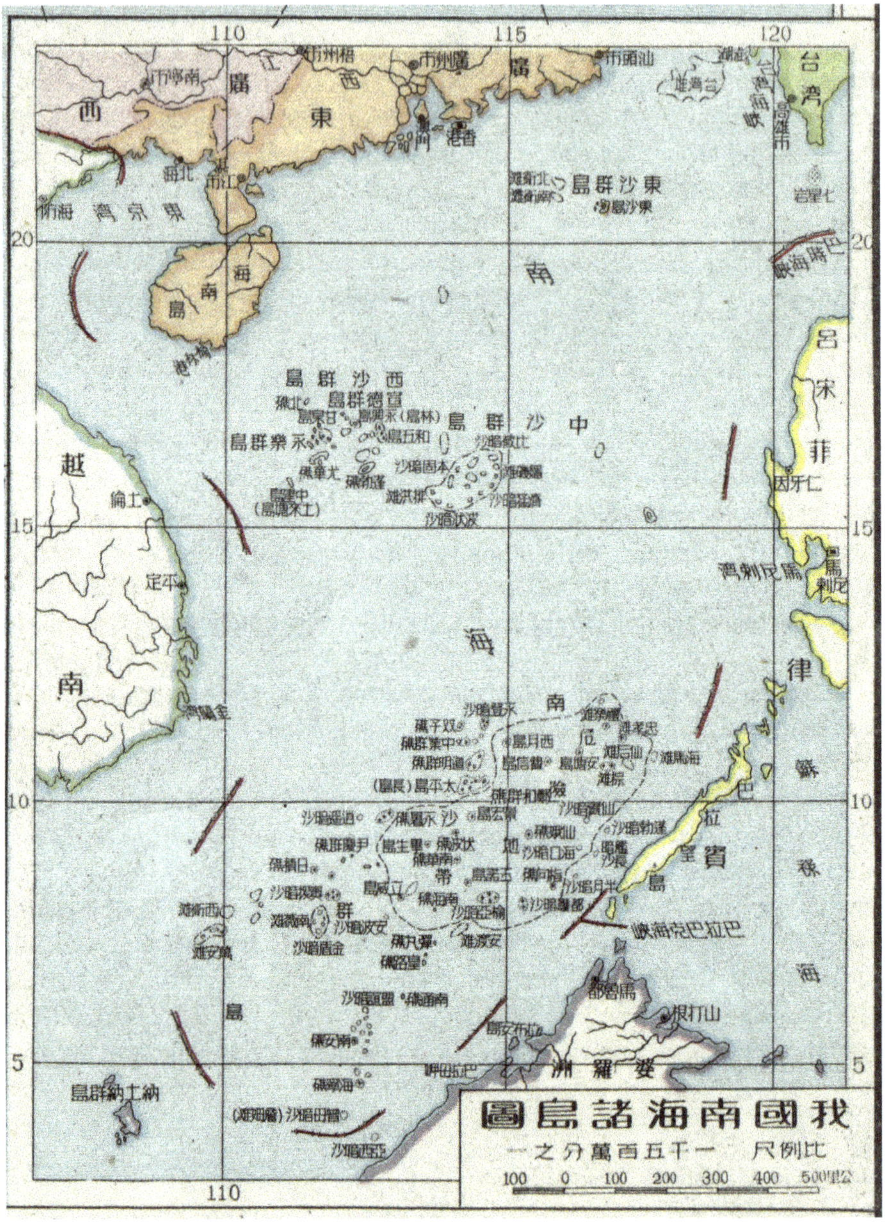

FIGURE 10.5 Detail from the *Zhonghua Renmin Gongheguo da di tu* 中華人民共和國大地圖 [Large Map of the Republic of China] showing a cut-out of the South China Sea titled *Woguo nanhai zhudao tu* 我國南海諸島圖 [Map of China's Islands in the South Sea] that clearly shows the dashed line sweeping down to 4° North. Interestingly the line does not appear on the larger map. Because the cut-out does not include more than the southern tip of Taiwan, it is unclear whether this example is modeled on a ten or eleven-dash line. Shanghai, 1952. Library of Congress.

column. This zone is referred to by international law as an Exclusive Economic Zone, or "EEZ." Within it the waters themselves are not considered territory and are open to international navigation. The distance from 200 to 350 nautical miles from shore is designated by UNCLOS as the "high seas," but neighboring countries nonetheless enjoy sovereign rights to the continental shelf under its provisions. Anything beyond 350 nautical miles from the shore is also considered the high seas, and no national rights are recognized to the seafloor.

China's claims and activities in the South China Sea reach deeply into the UNCLOS-designated Exclusive Economic Zones. The People's Republic of China does not recognize UNCLOS, the formation of which postdates the earliest of China's claims to the islands of the South and East China Seas. Growing out of treaty agreements first signed in 1958, the third meeting of UNCLOS concluded in 1982. Its provisions came into force in 1994 when the 60th nation signed on to the agreement. Currently 168 parties have ratified the agreement. An additional 14, including the United States, have signed on but not yet ratified the agreement.

In addition to the Philippines, five other nations also have territorial disputes with the People Republic of China in the area. The disputed maritime formations include the Diaoyu (in Chinese) or Senkaku (in Japanese) Islands, the Paracel Islands, the Spratly Islands, Macclesfield Bank, and James Shoal. The uninhabited Diaoyu/Senkaku Islands are located to the south of Okinawa and are contested by Japan and China. The Paracels are roughly equidistant from the coasts of Hainan Island and Vietnam and are claimed by both Vietnam and China. The Spratly Islands are located off the coasts of the Philippines and Malaysia. China, Taiwan, Vietnam, the Philippines, and Malaysia all lay claim to them and have physical toeholds in the region.[33] Macclesfield Bank consists of a series of underwater reefs and shoals east of the Paracels and north of the Spratlys. James Shoal is just off the coast of Malaysia; it is located to the south of the Spratly Islands and is sometimes grouped with them.[34]

33 Taiwan controls Itu Aba and China has a presence on Gaven Reef, Johnson South Reef, Fiery Cross Reef, and Mischief Reef, where it has built up structures at a pace radically speeded up since 2014. Vietnam has constructed a harbor and additional facilities on Southwest Cay, which it wrested from the Philippines in 1975. Malaysia has engaged in reclamation and reconstruction at Swallow Reef, which it occupied in 1983. The Philippines holds Thitu Island, where it also has plans to upgrade its defense capabilities. This information is from Derek Watkins, "Before and After: The South China Sea Transformed," *Asia Maritime Transparency Initiative*, February 18, 2015. http://amti.csis.org/before-and-after-the-south-china-sea-transformed/.

34 James Shoal was named by British surveyors in 1870. The Chinese name designated in 1933 was a transliteration of "James" to *zengmu* 曾母 (a phonetic transcription) and the character *tan* 攤 [sandbank or shoal]. In 1947 it was redesignated as a reef (*ansha* 暗沙).

Since China's President, Xi Jinping, took office and the US articulated a "Pivot toward Asia" in its foreign policy, China has been aggressively asserting its dominance in the South China Sea. More specifically, the government has been dredging sand from the ocean floor to build up reefs and then equipping the reclaimed land with military bases complete with airstrips and radar stations. Interestingly, while it may not recognize UNCLOS, China has been speedily populating the islands with permanent residents and vegetation that is able to sustain those residents, both conditions of sovereign recognition under UNCLOS.[35] The establishment of these new bases and their potential for overt militarization is a major concern for neighboring countries. In fact, China deployed HQ-9 SAMS (surface-to-air missiles) on Woody Island in the Paracel Islands, whose sovereignty is disputed with Vietnam and Taiwan, in February 2016. The built-up reefs in the more distant Spratlys are equipped with radar equipment, supply platforms, gun emplacements and, according to some reports, anti-aircraft towers.[36]

The 2016 UNCLOS ruling was ultimately in favor of the Philippines. Skirting matters of sovereignty, over which it does not have clear jurisdiction, the court instead focused on the status of various maritime features. One of the concerns that neighboring countries had was precisely how these new structures would be dealt with under UNCLOS; now that the structures built on many of the reefs are inhabited, and are clearly intended to be permanent, what will this mean for their classification in the future, especially as journalists and others begin casually to refer to them as islands? The tribunal reaffirmed that under the terms of the convention reefs and shoals that have been built up do not constitute islands, but are rather designated as low tide elevations (LTES). LTES are, by definition, shoals or reefs that do not emerge from above the water during high tide. Unlike islands, LTES do not enjoy any sort of territorial privileges with regard to the neighboring waters. From the standpoint of history, the ruling is just one more episode in a longer contestation involving maps, territory, and political power.

One could write a whole article on historical analogies for China's maritime expansion. In 2014 the Philippine President, Benigno S. Aquino III, provocatively compared China's incursions in the South China Sea to Hitler's occupation of the Sudetenland areas of Czechoslovakia in 1938.[37] From a longer-term

35 I am grateful to Craig Asberry for sharing this observation with me.

36 Derek Watkins, "Before and After," February 18, 2015. http://amti.csis.org/before-and-after-the-south-china-sea-transformed/.

37 Derek Watkins, "What China has been Building," *New York Times, Asia Pacific* section Feb. 29, 2016 and Keith Bradsher "Philippine Leader Sounds Alarm on China," *New York Times*, Feb. 4, 2014. See also accompanying interview at http://www.nytimes.com/2014

historical perspective, Portuguese assertions of power in the region in the late fifteenth and early sixteenth centuries make an appropriate comparison. To frame a historically grounded story with an open ending one could explore the phenomenon of imperial overreach followed by implosion, which may be what we are currently seeing in the case of Russia's invasion of Ukraine.[38] One might also point to the US's own aggressive moves in East Asia in the context of WWII and its aftermath. My goal here, however, is to explore the tensions in the South China Sea from the perspective of the history of cartography. By understanding how these claims came about and how China has continued to attempt to legitimize them we can historicize them, and also explore what kinds of questions this situation continues to pose for us cartographically, and geo-politically, today.

Part of the maddening nature, and strategic brilliance, of the nine-dash map is precisely the imprecision in regard to just what exactly it is claiming. A study published in the PRC in 2003 highlights the lack of clarity in the symbolism of the dashes, even while supporting China's claims—whatever they actually are. As the authors summarize in their paper, "some take the view that the dotted line relates only to the enclosed islands, others believe that it asserts Chinese sovereignty over the waters, and still others link the dotted line to a claim of historic title over the waters."[39]

This lack of clarity has served China's maritime territorial ambitions well. Those who don't recognize the dashed line as a serious cartographic symbol may simply brush it off or ignore it, allowing the map's claims to go unchallenged. Others may over-interpret what the dashed line means, as in some reproductions that transform the dashes into more of a line, suggesting—if not accepting—that the claims are over the entire maritime area as well as the islands. Also unclear from the nine-dash map is any indication of when the claims date from, leaving open the interpretation that they date not from the time of the publication of the map but rather record something that already existed. As with other examples of political rhetoric with which we are all too familiar, the outrageous can make the merely egregious seem mild, or at least less unpalatable, by comparison.

The malleability of the claim makes it difficult to know how seriously to take it, and how best, strategically, to refute it. While China's claim over the

---

/02/05/world/asia/philippine-leader-urges-international-help-in-resisting-chinas-sea-claims.html?_r=1.

38 David Atkinson, "Geopolitics, Cartography and Geographical Knowledge: Envisioning Africa from Fascist Italy," in *Geography and Imperialism, 1820–1940*, edited by Morag Bell, et al. (Manchester, UK: Manchester University Press, 1995), 265–297.

39 Li and Li, "Dotted Line," 291.

islands within the nine-dash line is unprecedented, and clearly upsetting to its neighbors, the biggest concern is that China's goal extends beyond sovereignty over the islands—and newly constructed structures on the LTEs—to maritime domination of the entire region. Analysts have identified control of the enormously lucrative trade that flows through the area as a probable goal. According to Peter Dutton, Professor and Director of the China Maritime Studies Institute at the US Naval War College:

> The logical conclusion drawn from China's adding ... islands in the southern part of the South China Sea with military-sized runways, substantial port facilities, radar platforms and space to accommodate military forces is that China's objective is to dominate the waters of the South China Sea at will ... Building the islands is therefore, in my view, a significant strategic event ... They leave the potential for the South China Sea to become a Chinese strait, rather than an open component of the global maritime commons.[40]

These aims would certainly be in alignment with Xi Jinping's Maritime Silk Road Initiative, part of his larger Belt and Road initiative.

Language, like maps, can be a slippery tool. Even the analyst's use of the word "islands" to describe the land China has raised above the sea can be taken as acknowledgment of their existence as such. As its immediate neighbors grow ever more concerned, the United States has also weighed in, sending naval vessels into the area to reinforce claims to the international navigability of these waters.[41]

Maps are persuasive documents. They seduce us with their beauty and artistry, they invite us to rely on them for finding our way in the world, and they can lead us into a sense of complacency that we know the lay of the land and that it is fixed, as on the map. Yet maps are forever changing, both in terms of what they display and in terms of the technologies—and map languages—that they use. Maps are complex persuasive documents that reflect the skill

40 "Beijing's Real Strategy in the South China Sea." *Yahoo Finance*. Accessed April 8, 2016. http://finance.yahoo.com/news/beijings-real-strategy-south-china-200053728.html. See also Peter A. Dutton, "China's Maritime Disputes in the East and South China Seas," *Naval War College Review* 67, no. 3 (2014): 4 (also at https://digital-commons.usnwc.edu/nwc-review/vol67/iss3/2/) in which he argues that China is working to expand its continental control into maritime spaces in order to extend its control and influence in the interest of political and economic benefits.

41 Another concern, which I do not explore here, is, of course, environmental.

of individual mapmakers as well as social assumptions and political exigencies. They are representative of their times, but also often active in shaping the future.

In 1933 or even in 1948 the nine-dash map was simply a claim in the way that many maps make claims on paper, however unsubstantiated. It was not a reflection of reality—either in regard to China's position in the region or in regard to the existence of competing national claims. As J.B. Harley has taught us,

> Maps are never value-free images … [b]oth in the selectivity of their content and in their signs and styles of representation maps are a way of conceiving, articulating, and structuring the human world which is biased towards, promoted by, and exerts influence upon particular sets of social relations.[42]

China is not unique in this way. It is not alone in its selectivity in regard to what it chooses to show in its maps. On any map, "[t]o avoid hiding critical information in a fog of detail, the map must offer a selective, incomplete view of reality. There's no escape from the cartographic paradox: [even] to present a useful and truthful picture, an accurate map must tell white lies." In a context within which the partial nature of the map is understood, such a "white lie" is not serious. "Because most map users willingly tolerate white lies on maps, it's not difficult for maps also to tell more serious lies," which is clearly a more serious problem.[43]

The effects of such misrepresentations are compounded in contexts where competing representations of reality through alternate maps or textual sources of information are not only unwelcome but not tolerated.[44] If

42 J.B. Harley, "Maps, Knowledge, and Power," in *The New Nature of Maps: Essays in the History of Cartography*, ed. Paul Laxton and J.H. Andrews (Baltimore: The Johns Hopkins University Press, 2001), 53.

43 Monmonier, *How to Lie with Maps*, 1.

44 How could they be, by a regime that in 2016 was simultaneously attempting full control over its own press, was putting pressure on the Hong Kong press, and publicly asserted that it can't tolerate even April Fools jokes? On the April Fools interdiction see Austin Ramzy, "No April Fooling Please, We're Chinese," *The New York Times*, April 1, 2016. http://www.nytimes.com/2016/04/02/world/asia/china-april-fools-day-xinhua.html. For repeated blockage in March 2016 of access to the *South China Morning Post* out of Hong Kong see "China's Internet Regulator Responds to Apparent Blocking of HK Newspaper Sites," Reuters, March 11, 2016. https://www.reuters.com/article/hongkong-china-newspaper-idUSL1N16J06R/. China's attempts at control of both its population and the media message have only intensified since that time.

other representations of the South China Seas and competing claims to the region were widely available in China, either on other maps or through textual sources, the nature of the nine-dash map would not be so misleading, or so powerful. However, as by far the dominant representation of the region and the one widely used in schools, a myth was created that, like the early twentieth-century concession areas, this region too was an inalienable part of the motherland worth dying for.[45]

One "lie" on the nine-dash map, to use Monmonier's term, is the implication that China's claims to the area within the dashes (however defined) are exclusive. However, Japan, Taiwan, the Philippines, Myanmar, Brunei, and Vietnam all also have longstanding claims to the islands in these waters, and feel they have both a historic and a legal right to exclusive use of the continental shelves of the 200-nautical-mile Exclusive Economic Zones articulated by UNCLOS, on which China's nine-dash line, land reclamation activities, and military buildup heavily infringe. In fact, a number of countries have been more assertive in their military development in recent years. In March of 2016 Indonesia blew up twenty-three fishing vessels (10 Malaysian and 13 Vietnamese) to show its seriousness over protecting the sovereignty of its Exclusive Economic Zone. The same week, Malaysia protested the incursion of one hundred Chinese fishing boats into its waters.[46]

If the first lie of the nine-dash map is that the sovereignty of islands within the South China Sea is uncontested, the second lie is about the historicity of China's claims to this area. While experts in their interpretation of the nine-dash line are generally careful not to allude to historical claims, in the popular press and media an idea has been fostered that China's claims to the South China Sea are historic, or even "ancient."[47]

These claims to historic control of the area are spurious, but serve several purposes.[48] First, fitting as they do into a narrative of loss and betrayal, they

45 As Mark Monmonier writes: "Map users generally are a trusting lot .... Map users seldom, if ever, question these authorities, and they often fail to appreciate that map's power as a tool of deliberate falsification or subtle propaganda." Monmonier, *How to Lie with Maps*, 1.

46 Trefor Moss, "Indonesia Blows Up 23 Foreign Fishing Boats to Send a Message," *Wall Street Journal*, April 5, 2016, sec. World. http://www.wsj.com/articles/indonesia-blows-up-23-foreign-fishing-boats-to-send-a-message-1459852007.

47 For an extensive refutation (and documentation) of such claims see Tri H. Pham, "'Since Ancient Times': China's Maritime Claims vs. History Books and Ancient Maps," *Journal of Strategy and Politics* 3 (2016): 32–177. https://strategyandpolitics.org/the-journal-of-strategy-and-politics/issue-3-autumn-2016/.

48 As an entrée into the literature on China's territorial conception of itself see Emma Teng. *Taiwan's Imagined Geography*. Teng makes the point that during the Ming anything beyond the seas (*haiwai* 海外) was not considered to be part of China. During the Qing,

drum up nationalist sentiment among Chinese young people who have been schooled in this rhetoric. Having been brought up with the nine-dash map and its claims, it is easy for those targeted by this narrative to make the leap from "I've always heard it this way" to "it always was this way." Once having internalized the fabricated truth of China's entitlement to the South China Sea they are willing to defend it—some of them with their lives, others by throwing their loyalty behind the government who will "defend" China from claims to the contrary. In aligning nationalist sentiment with its own role as defender of China's "territorial integrity" the ruling government (however unpopular in other specifics) earns the support of those who find in it their best hope for "preserving" or creating a big strong China, the dream for which so many have sacrificed so much over the last century. It is no coincidence that the biggest push for occupation of the South China Seas has come under President Xi Jinping, who has proven to be a leader ruthless in the pursuit of his political enemies and therefore very much in need of a rallying cry to unite the people in the interest of building a strong nation.

A second way in which the spurious claims to a long history of control of the region ("it's always been this way") serve the interests of the state is as a diversionary tactic. Critics of China's claims and activities in the South China Sea have spent countless hours compiling the scholarship to refute these claims.[49] But perhaps more importantly, the claim, implicit or otherwise, that these waters have always been China's obscures their nationalist origins in the 1930s. The South China Seas dispute is not about China's historic claims and their legitimacy—that is only a diversion. What is at stake in China's current claims is control over as much territory and as many resources—including shipping lanes—as possible. Projecting those claims back in time is simply an attempt to give them a type of legitimacy born only of the long durée.[50]

While the nine-dash line seems not to have generated significant response or concern in 1948 and the decades immediately following as China struggled to establish itself as a nation, the political and cartographic exigencies of the

---

the government reluctantly took control over the western portion of the island rather than have it serve as a base for rebellion. It was only in the nineteenth century, under threat of control of the eastern portion of the island by foreign powers, that it also took that territory and those populations under its direct control.

49 Ulises Granados, "The South China Sea and Its Coral Reefs during the Ming and Qing Dynasties: Levels of Geographical Knowledge and Political Control," *East Asian History* 32/33 (December 2006): 109–128, and Tri H. Pham "'Since Ancient Times'."

50 The seminal book on describing this process is Prasenjit Duara, *Rescuing History from the Nation: Questioning Narratives of Modern China* (Chicago: University of Chicago Press, 1995).

moment can, and often do, have much more serious consequences down the road. By the turn of the twenty-first century China had become an economic powerhouse with the muscle to begin to actually remake the landscape in such a way as to recreate what had been only imagined on the nine-dash map. Long-accustomed to these claims, and having only limited exposure to alternative views of the region, China's people had little reason to question Xi Jinping's aggressive moves in the region. As David Atkinson suggests,

> [T]here is … a history, especially in colonial contexts, of supposedly accurate and 'scientific' cartography mis-representing territory through … omission … of actual variation and detail. In such cases … a colonizing nation may then have few qualms over invading regions that appear on maps to be little more than vacant territory.[51]

Thus, from the perspective of the history of cartography we can see the move to bring reality into line with the map is indeed an aggressive move, and one used countless times in the colonial context.

Some will take umbrage at my implicit comparison of China's behavior in the South China Sea to that of a colonial power. The PRC prefers to understand colonialism as a Western practice of which it was a victim in the nineteenth century. (For the same reason, it casts Qing expansion into Xinjiang and Tibet as "imperial" rather than "colonial" expansion.) However, in seeking to understand the current situation we can only analyze China's actions, which speak for themselves. Whether colonial, imperial, or hyper-nationalist in its push into the South China Sea—not to mention present day Xinjiang—China's actions have been aggressive. They are not the only nationally-inspired claims in the region backed up by political bluster and military power, but they do far exceed those of other nations in the region in their scale. As a relative latecomer to the Spratlys, having constructed its first airstrip there only in 2014 well after there were other contenders to sovereignty over the region, China seems eager to make up in size for what it may lack in other ways. In this case, as occurs all too often in familial relationships, it seems the victim has grown up to become the victimizer.

In the post-WWII era, other modern nation states—whether winners or losers in that conflict—have generally accepted (with the notable recent exception of Russia) that the flip side of marking out and enjoying firm territorial borders as nation states is a tacit understanding that territorial expansion is no longer acceptable. I note specifically Britain's return of Hong Kong to Chinese

51 David Atkinson, "Geopolitics," 277–278.

control in 1997, and Portugal's return of Macao in 1999. The People's Republic of China's refusal to accept the "flip side of the coin" of enjoying territorial sovereignty marks it out as a bully on the playground. It seems at best like a nation not willing to play by the rules of international law and at worst as though it is emulating Japa's early twentieth-century wartime aspirations to create a so-called "East Asia Co-Prosperity Sphere" in the region.

Even more disturbingly China's claims, due to their potential maritime nature, are about conceptualizing territoriality and sovereignty in an entirely new way. As such they may even compete for significance with Ricci's introduction of the coordinate system into China. In the early seventeenth century, the arguments of the Dutch legal scholar Grotius (Huig de Groot) for '*The Freedom of the Seas*' won out over Portugal's attempts to claim control over the Indian Ocean and Britain's desire to control the North Sea. But, as Timothy Brook has eloquently put it, in the debate between Grotius and Selden, who represented Britain, "neither prevailed completely to the exclusion of the other. This is why the international law of the sea today consists of a mixture of the two, recognizing both freedom of movement and reasonable jurisdiction,"[52] as is presently reflected in the United Nations Convention on the Laws of the Sea. Should China's nine-dash line come to be interpreted and accepted as claiming control also over the waters as well as the lands of the South China Sea, the implications would be revolutionary and their reverberations felt for centuries to come.

Bibliography

Atkinson, David. "Geopolitics, Cartography and Geographical Knowledge: Envisioning Africa from Fascist Italy." In *Geography and Imperialism, 1820–1940*, edited by Morag Bell, et al. Manchester, UK: Manchester University Press, 1995.

Bradsher, Keith. "Philippine Leader Sounds Alarm on China." *The New York Times*, February 4, 2014, sec. World. https://www.nytimes.com/2014/02/05/world/asia/philippine-leader-urges-international-help-in-resisting-chinas-sea-claims.html.

Brook, Timothy. *Completing the map of the world: Cartographic interaction between China and Europe 全圖：中國與歐洲之間的地圖學互動*. Taibei: Institute of Modern History, Academia Sinica, Lecture Series no. 5, 2020.

Brook, Timothy. *Mr. Selden's Map of China: Decoding the Secrets of a Vanished Cartographer*. New York: Bloomsbury Press, 2013.

52 Timothy Brook, *Mr. Selden's Map of China: Decoding the Secrets of a Vanished Cartographer* (New York: Bloomsbury Press, 2013), 38.

Del Callar, Michaela. "China's New '10-Dash Line Map' Eats into Philippine Territory." *GMA News Online*. Accessed November 22, 2022. https://www.gmanetwork.com/news/topstories/nation/319303/china-s-new-10-dash-line-map-eats-into-philippine-territory/story/.

Cams, Mario. *Companions in Geography: East-West Collaboration in the Mapping of Qing China (c.1685–1735)*. Leiden: Brill, 2017.

Cams, Mario. "Reimagining Qing Space: Yongzheng's Eurasian Atlas (1727–8)." *Late Imperial China* 42, no. 1 (2021): 93–129.

Cheney, Amanda. "Tibet's Twenty Years Crisis: Indeterminacy, International Systems Change and Sino-Tibetan Relations, 1919–1937," Paper Presented at the Annual Meeting of the Association of Asian Studies, March 2016.

Cheng Fangyi. "Pleasing the Emperor: Revisiting the Figured Chinese Manuscript of Matteo Ricci's Maps." *Journal of Jesuit Studies* 6, no. 1 (March 11, 2019): 31–43.

Day, John D. "The Search for the Origins of the Chinese Manuscript of Matteo Ricci's Maps." *Imago Mundi* 47, no. 1 (January 1995): 94–117.

Du Halde, J.-B. *Description Géographique Historique, Chronologique, Politique, Et Physique De l'Empire De La Chine Et De La Tartarie Chinoise: Enrichie Des Cartes Générales Et Particulières De Ces Pays, De La Carte Générale & Des Cartes Particulières Du Thibet, & De La Corée, & Ornée D'un Grand Nombre De Figures & Vignettes Gravées en Taille-Douce*. Paris: Chez P.G. Le Mercier, 1735.

Duara, Prasenjit. *Rescuing History from the Nation: Questioning Narratives of Modern China*. Chicago: University of Chicago Press, 1995.

Fontana, Michela. *Matteo Ricci: A Jesuit in the Ming Court*. Lanham, MD: Rowman & Littlefield Publishers, Inc, 2011.

Foss, Theodore N. "A Western Interpretation of China: Jesuit Cartography." In *East Meets West: The Jesuits in China, 1582–1773*, 209–251. Chicago: Loyola University Press, 1988.

Gaubil, Antoine. *Correspondance De Pékin, 1722–1759*. 1st éd. Genève: Libraire Droz, 1970.

Granados, Ulises. "The South China Sea and Its Coral Reefs during the Ming and Qing Dynasties: Levels of Geographical Knowledge and Political Control." *East Asian History* 32/33 (July 2006): 109–128.

Harley, J.B. "Maps, Knowledge, and Power." In *The New Nature of Maps: Essays in the History of Cartography*, edited by Paul Laxton and J.H. Andrews, 51–81. Baltimore: The Johns Hopkins University Press, 2001.

Hostetler, Laura. "Contending Cartographic Claims: The Qing in Manchu, Chinese, and European Maps." In *The Imperial Map: Cartography and the Mastery of Empire*, edited by James Akerman, 93–133. Chicago: University of Chicago Press, 2009.

Hostetler, Laura. "Imperial Competition in Eurasia: Russia and China." In *The Cambridge World History. Volume VI: The Construction of a Global World, 1400–1800 CE*,

Part 1, Foundations, edited by Jerry H. Bentley, Sanjay Subrahmanyam, and Merry E. Wiesner-Hanks, 297–322. Cambridge: Cambridge University Press, 2015.

Hostetler, Laura. "Mapping, Registering, and Ordering: Time, Space, Knowledge." In *Oxford World History of Empire*, 288–317. Oxford University Press, 2020.

Hostetler, Laura. *Qing Colonial Enterprise: Ethnography and Cartography in Early Modern China*. Chicago: University of Chicago Press, 2001.

Hostetler, Laura. "Sino-Western Cultural Relations and the Mapping of Modern China." In *Narratives from the Hinterland: Perspectives, Methodologies and Trends in the Study of Christianity in China*, edited by Xiaoxin Wu, 157–84. Guilin: Guangxi Normal University Press, 2014.

Jami, Catherine. *The Emperor's New Mathematics: Western Learning and Imperial Authority During the Kangxi Reign*. Oxford: Oxford University Press, 2012.

Li Jinming, and Li Dexia. "The Dotted Line on the Chinese Map of the South China Sea: A Note." *Ocean Development and International Law* 34, no. 3/4 (2003): 287–295.

Konvitz, Josef. *Cartography in France, 1660–1848: Science, Engineering, and Statecraft*. Chicago: University of Chicago Press, 1987.

Millward, James. "Coming onto the Map." *Late Imperial China* 20, no. 2 (December 1999): 61–91.

Monmonier, Mark, and H.J. de Blij. *How to Lie with Maps*. 2nd edition. Chicago: University of Chicago Press, 1996.

Mosca, Matthew. *From Frontier Policy to Foreign Policy: The Question of India and the Transformation of Geopolitics in Qing China*. Stanford University Press, 2013.

Moss, Trefor. "Indonesia Blows Up 23 Foreign Fishing Boats to Send a Message." *Wall Street Journal*, April 5, 2016, sec. World. https://www.wsj.com/articles/indonesia-blows-up-23-foreign-fishing-boats-to-send-a-message-1459852007.

Mundy, Barbara E. *The Mapping of New Spain: Indigenous Cartography and the Maps of the Relaciones Geográficas*. University of Chicago Press, 2000.

Parker, Charles H. *Global Interactions in the Early Modern Age, 1400–1800*. Cambridge; New York: Cambridge University Press, 2010.

Pegg, Richard A. *Cartographic Traditions in East Asian Maps*. Honolulu: University of Hawaii Press, 2014.

Pratte, Anne-Sophie. "Mapping the Steppe: The Politics of Cartography in Qing Mongolia, 1780–1911." PhD, Harvard University, 2021.

Ramzy, Austin. "No April Fooling Please, We're Chinese." *The New York Times*, April 1, 2016, sec. World. https://www.nytimes.com/2016/04/02/world/asia/china-april-fools-day-xinhua.html.

Rosenfeld, Everett. "Beijing's Real Strategy in the South China Sea." Accessed July 19, 2021. http://finance.yahoo.com/news/beijings-real-strategy-south-china-200053728.html.

Smith, Richard J. *Chinese Maps: Images of "All Under Heaven."* Oxford University Press, 1996.

Song Gang. "Re-Locating the 'Middle Kingdom': A Seventeenth-Century Chinese Adaptation of Matteo Ricci's World Map." In *Mapping Asia: Cartographic Encounters between East and West*, eds. Martijn Storms, Mario Cams, Imre Josef Demhardt, and Ferjan Ormeling. Cham: Springer International Publishing, 2019.

Teng, Emma. *Taiwan's Imagined Geography: Chinese Colonial Travel Writing and Pictures, 1683–1895*. Harvard East Asian Monographs 230. Cambridge, MA: Harvard University Asia Center, 2004.

Thongchai Winichakul. *Siam Mapped: A History of the Geo-Body of a Nation*. Honolulu: University of Hawaii Press, 1994.

Watkins, Derek. "Before and After: The South China Sea Transformed." *Asia Maritime Transparency Initiative*, February 18, 2015. https://amti.csis.org/before-and-after-the-south-china-sea-transformed/.

Watkins, Derek. "What China Has Been Building in the South China Sea." *The New York Times*, July 30, 2015.

Yee, Cordell D.K. "A Cartography of Introspection: Chinese Maps as Other than European." *Asian Art* 5, no. 4 (1992): 28–47.

Yee, Cordell D.K. "Traditional Chinese Cartography and the Myth of Westernization." In *The History of Cartography*, 2.2: 170–202. Chicago: University of Chicago Press, 1994.

POSTLUDE

Reflections on the Curation of Cartographic Knowledge

CHAPTER 11

Writing Technologies and Special Collections: Agents and Arbiters of Change through the Transmission of Knowledge

Marguerite Ragnow

Maps. Mention maps to anyone, young or old, and watch their eyes light up. Maps fascinate us; they excite our imaginations and inspire wonder. On a panel of his 1602 world map, *Kunyu wanguo quantu* 坤宇萬國全圖, Jesuit missionary Matteo Ricci (1552–1610), lamenting the size of the world and the impossibility of seeing and hearing everything in it first hand, had printed:

> That's why maps and history are precious; history records and maps disseminate [knowledge]: scholars from all over the world can see them. The Ancients record and posterity observes, decreasing effortlessly its ignorance and increasing its wisdom. Oh! How great are the merits of maps and history![1]

As described by Ricci, maps both preserve and transmit knowledge, in this case, historical knowledge. The 1602 Ricci map is a xylograph, a print made from an engraved wood block. This form of writing technology was ancient in China by that time, dating back to the third century CE.[2] It joins other writing technologies such as the creation of alphabets and hieroglyphs, the stylus and wax tablet, as well as the invention of moveable type as representatives of significant and singular change in the transmission of knowledge around the pre-modern globe, sometimes concurrent with oral culture, but often leading to its demise. This happened not because the written word is intrinsically

1 "Ecco perché le carte et la storia sono preziose; la storia annota e le carte diffondono; gli studiosi di tutto il mondo possono vederle. Gli Antichi registrano e i posteri osservano, diminuendo senza sforzo la loro ignoranza e aumentando la loro sapienza. Oh! quanto grandi sono i meriti delle carte e della storia!" Filippo Mignini, ed., *La Cartografica di Matteo Ricci* (Rome: Liberia dello Stato, Istituto poligrafico e Zecca dello Stato, 2013), 181. Translation from Chinese to Italian by Huang Ping and Filippo Mignini. Author's translation of Italian into English.

2 See for example, Roderick Whitfield, Anne Farrer, Shelagh J. Vainker, and Jessica Rawson eds., *Caves of the Thousand Buddhas: Chinese Art of the Silk Road* (London: George Braziller, 1990).

© MARGUERITE RAGNOW, 2024 | DOI:10.1163/9789004684782_013
This is an open access chapter distributed under the terms of the CC BY-NC-ND 4.0 license.

better than the spoken word, but because of how the written word travels, across both time and space.

Prior to writing, oral cultures transmitted information from person to person, and while individuals could travel and thus share their information, the scope was extremely limited. Writing changed that. The written word could travel thousands of miles in many directions at the same time, and equally important, it could be preserved across centuries. And, of course, as writing changed the scope of knowledge transmission, so, too, did printing; the survival of texts often depended on both manuscript and print. In his 2011 book, *The Swerve*, Stephen Greenblatt tells the story of Renaissance humanist and book hunter Poggio Bracciolini (1380–1459). During his quest to track down manuscripts of lost texts of the classical world, Bracciolini stumbled across a manuscript copy of *De rerum natura* [*On the Nature of Things*], the only surviving work by the Roman philosopher Lucretius (ca. 99 BCE–ca. 55 BCE), thought to have been lost for a thousand years.[3] The poem, designed to explain Epicurean philosophy to a Roman audience, survives in a few early medieval copies.[4] The copy discovered by Bracciolini, which reintroduced the poem to the scholarly world, has not survived, but a copy of it made by a fellow humanist, Niccolò di' Niccoli (ca. 1365–1437), has been preserved in the Laurentian Library in Florence.[5] It was first printed in 1473, and in his book, Greenblatt traces its impact from Galileo to Freud. Thomas Jefferson (1743–1826), who owned five Latin editions, as well as translations in English and other languages, was influenced profoundly by the poem and Epicurean ideas found their way into the American Declaration of Independence.[6]

Another example of the printed word serving to preserve and transmit knowledge across time and space is the work *De re Metallica* by Georg Bauer, another Renaissance humanist, better known as Georgius Agricola (1494–1555). A German scholar of Greek, chemistry, medicine, and physics, he was appointed, upon the receipt of his doctorate, as town physician of Joachimsthal, an important center of mining and smelting in Bohemia that now is part of the Czech Republic. There he applied the habits of mind developed from his earlier training in Greek and philosophy to better understand minerals and their role or potential role in the treatment of disease and other

3 Stephen Greenblatt, *The Swerve. How the World Became Modern* (New York and London: W.W. Norton and Company, 2011).

4 Re the purpose of the poem, see John Barker Stearns, "Lucretius and Memmius," *The Classical Weekly* 25 (1931): 67–68.

5 Codex Laurentianus 35.30.

6 Greenblatt, *The Swerve*, 243–263.

human ailments.[7] During his lifetime, he published several works on a variety of subjects, but *De re Metallica*, completed in 1550 but published a year after his death, in 1556, is his most famous work and the one that had the most lasting impact. It is a systematic treatise on mining and extractive metallurgy, which owes a debt to classical works such as Pliny the Elder's *Historia Naturalis*, among others, but goes much further.[8] According to the editors John Carter and Percy H. Muir of *Printing and the Mind of Man*, "*De re Metallica* embraces everything connected with the mining industry and metallurgical processes, including administration, prospecting, the duties of officials and companies, and the manufacture of glass, sulphur, and alum."[9] The work's first English translators, Herbert Hoover (1874–1964), a mining engineer and later thirty-first president of the United States, and his wife, Lou Henry Hoover, a geologist and Latinist, published their translation by private subscription in 1912.[10] Hoover wrote of the text that it was

> the first important attempt to assemble systematically in print the world-knowledge of mining, metallurgy, and industrial chemistry. It was the great textbook of those industries for two centuries and had dominated thought and practice all that time. In many mining regions and camps, including the Spanish South American, it was chained to the church altar and translated by the priest to the miners between religious services.[11]

The practical knowledge to be found in this sixteenth-century text not only preserved knowledge from the ancient world but added to it; the preservation

---

7 For more on Agricola, see Herbert Clark Hoover and Lou Henry Hoover, "Introduction," in *Georgius Agricola De re metallica: tr. From the 1st Latin ed. of 1556, with biographical introduction, annotations and appendices upon the development of mining methods, metallurgical processes, geology, mineralogy & mining law, from the earliest times to the 16th century* (London, 1912; reprinted New York: Dover Publications, 1950), v–vii, as well as Abraham Wolf, *A History of Science, Technology, and Philosophy in the 16th and 17th Centuries*, 2nd ed., vol. 2 (New York: Harper Torchbooks, 1959).

8 The creation of these types of compendia of knowledge were popular in early modern Europe, but also flourished in China. See, for example, Benjamin Elman, "Collecting and Classifying: Ming Dynasty Compendia and Encyclopedias (Leishu)," *Extrême-Orient, Extrême-Occident*, hors série, (2007): 131–157.

9 John Carter and Percy H. Muir, eds., *Printing and the Mind of Man, a descriptive catalogue illustrating the impact of print on the evolution of Western civilization during five centuries* (New York: Holt, Rinehart & Winston, 1967), 79.

10 See note 6, above.

11 Herbert Hoover, *The Memoirs of Herbert Hoover*, vol. 1: *Years of Adventure, 1874–1920* (New York: The Macmillan Company, 1952), 117.

of Agricola's text transmitted that knowledge across both centuries and continents. The Hoovers' translation, which includes copious footnotes dealing with the difficulties of Agricola's handling of German technical terminology in Latin, opened the door to a variety of new translations in other languages, greatly broadening its 20th-century audience.

Numerous other examples of the impact of the transmission of knowledge through manuscripts and printed works could be cited. However, there are many more manuscripts, documents, and books whose value and impact as knowledge transmitters is more complicated to trace. Ricci's *Kunyu wanguo quantu* [Map of the Ten Thousand Countries of the Earth] is one such example. The essays in this volume help us approach these complicated questions from a number of angles. Part Two: "The Production of Early Modern Geographical Knowledge in Europe" explores the complex and burgeoning state of geographical knowledge production in the European context where Ricci received his early scholarly training. The essays in Part Three: "The Reception and Legacy of Ricci's Maps in East Asia" explore the intriguing questions of what can happen when knowledge traditions from one part of the world intersect with those of another. Finally, there are questions that emerge pertaining to the genre of the map itself, that perspectives from the History of Science—and in particular the study of mapmaking—can help us to unpack, which I would like to touch on here. In the case of Ricci's map, as with all maps, certainly there was knowledge there to transmit, but as with all early modern maps, it was subjectively and purposefully constructed. And while this is the case with all knowledge creation, modern consumers tend to see maps, whether created 500 years ago or yesterday, as objective documents based on unchallengeable science and technology that are designed to get one from point A to point B.[12]

Kunyu wanguo quantu, the oldest surviving world map in Chinese that includes the Americas, was created in 1602 in Beijing, China, during the reign of the Wanli emperor (1563–1620, r. 1572–1620), under Ricci's direction. It measures roughly 12 feet by 5 feet in six panels and was printed using six blocks per panel on native paper. The map depicts China and the Pacific Ocean in the center, with the Americas to the east and Europe and Africa to the far west. It was acquired for the benefit of the Bell Library by the James Ford Bell Trust in 2009, one of only six known exemplars that have survived into the twenty-first century, and became a permanent part of the Bell Library's collection in 2020.[13]

12 Whenever I make the point about the subjectivity of maps, both in the past and today, my students find it extremely unsettling.

13 The James Ford Bell Trust gifted the Ricci map to the University of Minnesota in June 2020 for the Bell Library collection.

According to Ricci's diary, it was designed to fit on a folding screen, so one could contemplate the world from the comfort of one's armchair.[14]

In 1582, Matteo Ricci joined fellow Jesuit Michele Ruggieri (1543–1607) at Macao, a bustling commercial center and enclave for Europeans, as well as a Roman Catholic archdiocese from 1583. At this time, Macao was an island off the coast of Guangdong province. Later, in the seventeenth century, land reclamation turned it into a peninsula, although it shares only nineteen miles of coastline with the mainland. Travel from Macao to the mainland was strictly regulated—this was the period when China forbade most interactions with foreigners. The Jesuits were allowed to join European merchants on the mainland just twice each year for commercial fairs. The goal of the missionaries was to establish themselves on the mainland, however, and this they accomplished in 1583. Ruggieri and Ricci received permission to construct a mission at Zhaoqing, an important administrative center and seat of the viceroy of Guangdong and Guangxi—it was the first Catholic mission on mainland China.[15]

During a visit to the mission, the district administrator, Wang Pan, noticed a "cosmographical chart" hanging on the wall of the mission house. It was the first map he had seen that showed the New World. He asked Ricci for a copy and to make the map "speak Chinese." Ricci complied, with the aid of local Chinese, and in 1584 his first Chinese-language world map was printed. In his diary, Ricci wrote that he had brought with him to China the 1570 atlas of Abraham Ortelius and many scholars who work on Ricci believe that it was Ortlelius' world map, *Typus Orbis Terrarum*, that hung on the mission house wall and which formed the basis of this first Ricci world map.[16] By the time Ricci finally arrived in Beijing, in 1601, at the invitation of the Wanli emperor who was interested in his scientific knowledge, he had acquired more recent publications, including Dutch cartographer Petrus Plancius' 1590 atlas and maps by Gerardus Mercator.[17] Ricci's wall map was a composite product

14 Ricci's diary was published after his death, expanded and translated from the original Italian by fellow Jesuit Nicolas Trigault, under the title *De Christiana expeditione apud Sinas suscepta ab Societate Jesu* … (Augsburg: Apud Christoph. Mangium, 1615). An English translation was undertaken by Louis J. Gallagher, S.J., and published under the title *China in the Sixteenth Century: The Journals of Matthew Ricci, 1583–1610* (New York: Random House, 1953). This concept was most likely derived from Abraham Ortelius' 1570 atlas, in which he suggests the same activity.

15 Ricci, *De Christiana expeditione*.

16 See the critical edition by Mignini, *La cartografia di Matteo Ricci*.

17 John Day, "The Search for the Origins of the Manuscript Copies of the Maps of Matteo Ricci," *Imago Mundi* 47 (1995): 94–117.

of Western and Chinese cartographic and geographic knowledge drawn by Chinese artists.

While we can document at least some of the geographical sources that Ricci had at his disposal in constructing his 1602 map, its subjective nature is demonstrated most clearly by the text printed on the it. In addition to place names, Ricci populated Europe, in particular, with an ethno-mythology that provided Chinese readers of the map with few actual facts about its people, choosing instead to relate legends and stories and, in some cases, deliberate distortions of the truth, attempting to alleviate Chinese fears about the intentions of the Europeans toward China.[18] Religious disputes, particularly the wars of religion that wracked Europe for most of the sixteenth century, are not mentioned; Europe is deliberately portrayed as a peaceful bucolic place, where grapes are grown and excellent wine is made.

What effect did this "knowledge" about the West have on the *literati* at the Ming court for whom the map was created, some of whom assisted with its creation? Did it affect how they treated Ricci and his fellow missionaries? Ethno-mythology also was provided by the Chinese, especially about the people living north of the Great Wall. Did the map and its stories create a camaraderie between men from different sides of the globe who lived and worked in small but brilliant pockets of civilization surrounded by barbarians? The testimonial passages printed on the map certainly indicate a high level of mutual respect. Did the ethno-mythological descriptions of Scandinavia and elsewhere enter the Chinese geographic lexicon?

That this map and his previous maps were desired acquisitions among Ricci's Chinese peers is reflected in what we know about their distribution. The government official, Wang Pan—who had asked Ricci to create his first map, a Chinese-language version of a European map, published in 1584—saw the map as an object of desire. It meant little to him in its original state: it had text in a language he couldn't read and images of land masses and places with which he was unfamiliar; some his own culture deemed imaginary. The same map with Chinese text would affect his understanding, but also enhance his cultural capital. According to Ricci, Wang Pan printed copies to give to all of his friends and colleagues, and then sent copies to other provinces. Ricci believed that the Chinese had never seen anything like this, particularly the cartographic depiction of the Americas from which the Chinese were obtaining large quantities of silver via the Manila galleon trade with the Spanish. To have this knowledge in the form of an exotic map in the Chinese language

18 Ricci, *De Christiania expeditione*.

greatly enhanced Wang Pan's prestige.[19] When Ricci reached Nanjing in 1599, he was confronted with a "Chinese" map printed on native paper very similar to his own map of 1584. J.F. Baddeley contended that such maps, of which this was only one of many, were meant to show that the Chinese were not without geographical knowledge and could produce maps for themselves. However, he argued, once the map was proven to be merely a version of Ricci's own map, it greatly increased Ricci's reputation.[20]

During Ricci's sojourn in Nanjing, an important mandarin, Wu Zhuohai, requested that Ricci make a new version of his map with even more annotations than the one he had made in Guangdong. This revised, enlarged map, with errors corrected and new "annotations and declarations" was cut into stone tablets from which it was then printed, with a preface dedicated to Wu Zuohai. According to Filippo Mignini, Wu Zuohai's interest in the revised edition of Ricci's first map indicates "how much the country's leading intellectuals understood the importance of the document and how much interest they had in making it an interest of publication education."[21] Trigault relates the reaction of one Chinese scholar, Wang Zhongming, president of the Board of Rites in Nanjing:

> The President took great pleasure in studying this tablet, wondering that he could see the great expanse of the world depicted on such a small surface, and that it contained the names of so many new kingdoms and a list of their customs. He would examine it over and over again and very attentively, in an effort to memorize this new idea of the world.[22]

The cultural capital of the map continued to increase and broaden over time. In part, this was due to Ricci's ongoing collaborations with Chinese scholars; several works were published as "written" by Chinese scholars "as told by Matteo Ricci."[23] In addition to the copies of the 1602 map printed for the Ming court by Zhang Wentao, to which Ricci affixed Jesuit seals, Ricci's diary informs

19 Matteo Ricci and Nicolas Trigault, *Histoire de l'expedition chrestienne en la Chine, entreprise par les peres de la Compagnie de Iesvs* (Paris: P. Le-Mvr, 1618).

20 J.F. Baddeley, "Father Matteo Ricci's Chinese World Maps, 1584–1608," *The Geographical Journal* 50, no. 4 (Oct. 1917): 262–263.

21 Filippo Mignini, "Introduzione," in *La Cartagrafia di Matteo Ricci*, ed. Filippo Mignini, xxiii. My translation.

22 Gallagher, *Journals of Matteo Ricci*, 301.

23 Theodore N. Foss, "Ricci's World Map. The 1602 *Kunyu Wanguo Quantu*," in *China at the Center: Ricci and Verbiest World Maps*, ed. Natasha Reichle (San Francisco: Asian Art Museum, 2016), 19.

us that the engravers created two sets of woodblocks and operated a sort of pay-as-you-go business from late 1602 to 1607, printing illicit copies of the map to anyone willing to pay the price. Apparently, hundreds of maps were sold in this way.[24] We probably will never know the effect that "knowledge" had on the Chinese people who purchased and read copies of the map.

Although the Ricci map is considerably derived from maps created by western Europeans, it does contain information not available in the West in 1602, the time of the map's creation, derived from Chinese or other geographical sources.[25] Li Zhizao (1565–1630), a prominent mathematician and geographer, assisted Ricci with the East Asian regions of the map and introduced him to the work of Ma Tuan-lin (1245–1322), a Yuan dynasty encyclopedist and historical writer. His *Comprehensive Examination of Literature* (*Wenxian tongkao*) provided Ricci with material for additional explanatory passages related to the East.[26] Ricci also was influenced by Luo Hongxian (1504–1564), whose 1555 *Atlas of Universal Land* (*Guang yu tu*), an expanded and revised atlas based on the work of Zhu Siben (1273–1333), emphasized geographic accuracy over the more casual depictions in many Chinese gazetteers.[27]

There is a bit of text on an island at the top of the third panel of the Ricci map that reads as follows:

> In this region the cold is so excessive that the sea turns into ice, and the natives pass over it with horses and carts. They scoop out holes in the ice and catch large fish in great numbers. As the five grains will not grow in their country, they eat fish to appease their hunger, burn fish oil in their lamps, and use fish bones to make their houses, boats, and carts.[28]

---

24 Gallagher, *Journals of Matteo Ricci.*

25 Ricci's preface to the map states that this version is richer than the previous versions published at Guangdong and Nanjing.

26 Foss, "Ricci's World Map," 22. See Matteo Ricci, *Storia dell'introduzione del Cristianesimo in Cina*, ed. Pasquale M. d'Elia (Rome: La Libreria dello Stato, 1942–1949).

27 Hui-hung Chen, "The Human Body as a Universe: Understanding Heaven by Visualization and Sensibility in Jesuit Cartography in China," *The Catholic Historical Review* 93, no. 7 (2007): 517–552.

28 According to the legends on the map, this island is home to the Yakshas (nocturnal devils) and vagrant spirits. Further description seems to reference early Inuit, who lived in partially underground houses: "the inhabitants live in holes, clothe themselves in skins, and do not know how to ride." Translations from the Italian in Pasquale M. D'Elia S.J., *Il Mappamondo Cinese del P. Matteo Ricci SJ* (Cittá del Vaticano: Biblioteca apostolica Vaticana, 1938) by Marguerite Ragnow with checks against the Chinese of the map by Professor Ann Waltner, University of Minnesota.

Europeans were not yet aware of Alaska, much less any islands in the northern Pacific or in that part of the Arctic when the Ricci map was created in 1602. However, since the seventeenth century, there had been an awareness, probably beginning with the Russians, that the Chukchi peoples of Siberia knew of Arctic islands and "Great Lands" or continents "beyond."[29] Therefore, it is probable that the Ricci map reflects some knowledge from Siberia of the Chukchi and their legends. Linguist Michael Krauss believes that Wrangel Island, located between the Chukchi Sea and the East Siberian Sea was a way station on a trade route that linked the Inuit peoples of North America with the northern Siberian coast and that the island, itself, was populated with North American Inuits.[30] Krauss, however, points to early eighteenth-century European maps as the first evidence of this connection; he was not aware of the Ricci map—and in all probability, neither were the European mapmakers to whom he referred. We must also ask whether the Jesuits at the Ming court who undertook the creation and dissemination of the Ricci map paid any attention to this early notice of the Inuit peoples. And what about the many Chinese who purchased copies of the Ricci map? Was the curious bit of information on this map about the Inuit already known to them or was it news? If so, what did they make of it? How many of the Chinese who purchased the map could read the text?

The case of this Arctic island on the Ricci map points to an issue not often discussed when we consider the transmission of knowledge within an historical context via writing technology: was knowledge actually transmitted?[31] Because for knowledge to be transmitted, the text must be read and understood. The images must be understood as well as viewed. Often when we consider a text to be widely disseminated—where we have records of extant copies, multiple printings, and translations—we believe we have a sense of how widely the content of the text was known. But in many instances, we really only know how widely the text was sold, how widely it was distributed, not how many people actually read it, or even could read it.[32]

Scholars of the Middle Ages, beginning in the 1960s, began revising upward the number of literate people in parts of Europe and the number continues

29 Michael E. Krauss, "Eskimo Languages in Asia, 1791 on, and the Wrangel Island-Point Hope Connection," *Études/Inuit/Studies* 29, no. 1–2 (2005): 171.

30 Krauss, "Eskimo Languages in Asia, 1791 on," 163–185, esp. 171–178.

31 To say nothing of "false" or inaccurate information.

32 See, for example, Andrew Pettegree, *The Book in the Renaissance* (New Haven: Yale University Press, 2010).

to climb.[33] Because texts in the Middle Ages were in manuscript form, more difficult to produce and therefore not widely available, reference to one text in another text suggests readership by a higher percentage of the literate population than we can infer from printed materials in the early modern period. Additionally, books were read aloud, often to entire households, so that a single copy of a book could reach many more people than its owner. In the Middle Ages and in the early modern period, we also must account for the fact that book ownership was prestigious; for some it was more important to own the book than to read it or even to be able to read it. The question of how many people in the pre-modern world were literate—including visual literacy, the ability to "read" images in books, figures carved on the portals of churches, maps—has not yet been satisfactorily answered by scholars. Nor are we able to determine how many people purchased books and maps as potential sources of knowledge or just as precious objects meant to impress, or even as décor.[34] It may be possible that we can chart the knowledge that was "available" to be transmitted and trace its spread through some small segment of the population, but that requires data about information accrual and how it might manifest itself. Does knowledge accrue prior to transmission or through the process of transmission. Or both? Knowledge acquisition often is unique to an individual. The ability to understand knowledge requires the accrual of earlier acquired knowledge in a seemingly endless cycle that may not be traceable except in the instance of a very well-documented individual.

And yet, however small the percentage of the population at any given time obtained information from written or printed sources directly, we know that knowledge was transmitted because in some cases it changed the world, as noted in the Bracciolini example at the beginning of this chapter. It was transmitted to the right people at the right time and in such a form that attention

33 James Westfall Thompson, *The Literacy of the Laity in the Middle Ages* (New York: Franklin Classics, 1960); Michael T. Clanchy, *From Memory to Written Record: England 1066–1307* (Cambridge: Harvard University Press, 1979); Armando Petrucci, *Writers and Readers in Medieval Italy*, ed. and trans. Charles M. Radding (New Haven: Yale University Press, 1995); Ian Frederick Moulton, ed., *Reading and Literacy in the Middle Ages and the Renaissance* (Turnhout, Belgium: Brepols, 2004).

34 On printed books, see Stephen Füssel, *Gutenberg and the Impact of Printing*, trans. Douglas Martin (Aldershot, Hampshire, UK: Ashgate: 2005), orig. published in German as *Gutenberg und seine Wirkung* (Frankfurt am Main: Insel, 1999); Miriam Usher Chrisman, *Lay Culture, Learned Culture: Books and Social Change in Strasbourg, 1450–1599* (New Haven: Yale University Press, 1982); Pettegree, *The Book in the Renaissance*, and Flavia Bruni and Andrew Pettegree, eds., *Lost Books: Reconstructing the Print World of Pre-Industrial Europe* (Leiden: Brill, 2016); as well as the several essays in Michael F. Suarez, S.J., and H.R. Woudhuysen, eds. *The Book: A Global History* (Oxford: Oxford University Press, 2013).

was paid. In other cases, there are many more questions. Was this knowledge transmitted orally or in writing? In manuscript or in print? Scholars often use the Columbus and Vespucci letters as examples of the broad dissemination and transmission of knowledge that wrought lasting changes.[35] Seventeen editions of Columbus' letter were published from 1493 to 1497, with additional manuscript versions surviving in whole or in part.[36] Vespucci's letters, although their authenticity is highly disputed, saw at least twenty-seven print editions, possibly more.[37] If Columbus' and Vespucci's letters had not been printed and sold throughout Europe, how long would it have taken for entrepreneurs and governments to willingly invest in trans-Atlantic enterprises? How important to the exploration and exploitation of the New World was the printing press? Where would New World cartography be without the early printed work of Martin Waldseemüller, whose 1507 globe and planisphere—both woodcut maps like the Ricci world map—were the first to depict the name "America"?

The answers to these questions are unknown at this time; perhaps they will never be known. It is fortunate that the Waldseemüller maps are better documented than most maps from the early sixteenth century. Waldseemüller and his collaborator, Martin Ringmann, wrote a little book documenting their creation, explaining why they gave the name America to the New World

35 Columbus' letter was first printed in Spanish at Barcelona in 1493, followed by a Latin edition in Rome about a month later. Eleven different editions were published that year, with six more editions published from 1494 to 1497. See Cecil H. Clough, "The New World and the Italian Renaissance," in *The European Outthrust and Encounter: The First Phase c. 1400–c. 1700: Essays in Tribute to David Beers Quinn on his 85th Birthday*, ed. Cecil H. Clough and P.E.H. Hair (Liverpool: Liverpool University Press, 1994), 291–328; and Matthew H. Edney, "The Columbus Letter: The Diffusion of Columbus' Letter Through Europe, 1493–1497," Osher Map Library, University of Southern Maine. http://www.oshermaps.org/special-map-exhibits/columbus-letter/iv-diffusion-columbuss-letter-through-europe-1493-1497 (accessed July 12, 2017).

36 Cesare De Lollis, ed., *Scritti di Cristoforo Colombo*, vol. 1, part 1, of *Raccolta di Documenti e Studi pubblicati dalla Reale Commissione Colombiana pel quarto Centenario della scoperta dell'America*. (Rome: Ministero della pubblica istruzione, 1894). On the new manuscript letter discovered in 1985, see David Henige, "Finding Columbus: Implications of a Newly Discovered Text," in *European Outthrust and Encounter*, ed. Clough and Hair, 141–166. For context and a translation, see Margarita Zamora, *Reading Columbus* (Berkeley: University of California Press, 1993).

37 On Vespucci, see Germán Arciniegas, *Amerigo and the New World: The Life and Times of Amerigo Vespucci*, trans. Harriet de Onis (New York: Knopf, 1955), orig. published in Spanish, 1952; Luciano Formisano, ed., *Letters from A New World: Amerigo Vespucci and the Discovery of America*, trans. David Jacobsen (New York: Marsilio, 1992); and Felipe Fernándo-Armesto, *Amerigo: The Man Who Gave His Name to America* (New York: Random House, 2007).

as they knew it, and explaining how they intended the maps to be used.[38] Documentary evidence even survives that tells us that the maps and the book were at one time sold as a set. The circumstances of the creation of the Ricci map, too, are adumbrated by Ricci's journal, as well as text on the map itself. For most maps this is not the case. Even when the creator is believed to be known, the location where the map was created, the reason behind its creation, and why the cartographer chose to include or exclude particular information, may not be evident. And so, we speculate, we infer.

Some historians of cartography invest a lot of time in trying to determine issues of chronology.[39] Was this map created before that map? Did that map influence this map? Many people who study maps believe that a straight line can be drawn from one map to another through an examination of the map, its style, its content. While it is known that many premodern cartographic workshops used models or templates, and that maps were copied, the copies disseminated and then copied again, some believe that lineage can be determined. Rarely is it posited that several maps might stem from a common, as yet unknown, "Ur" map, nor is the possibility considered that even maps believed to be first in a line of later maps might, themselves, be compilations of cartographic knowledge from a variety of sources. According to J.B. Harley, the crowns of both Spain and Portugal engaged in campaigns of disinformation with respect to the Americas, creating maps with false information and sending them out into the world and, whenever possible, into the hands of agents of their rivals.[40] How many maps in the canon of cartographic historiography are based on the false maps created by the Spanish and Portuguese in the sixteenth and early seventeenth centuries? How many of these false maps now hold pride of place in museum and library collections?

Pre-modern maps pose additional challenges to today's readers. For one thing, familiar places often are not in familiar locations. The pre-modern world had no global positioning system to accurately locate specific places; personal experience and observation, manual surveying techniques, often with inadequate instrumentation, were the tools of the early cartographer. While scholars generally agree that maps are subjectively-created cultural objects—reflections of a particular point of view at a particular point in time and space—it is not

38 Martin Waldseemüller, *Cosmographiae introductio* (St. Dié, France: Walter and Nikolaus Lud, 1507). See John W. Hessler, *The Naming of America: Martin Waldseemüller's 1507 World Map and Cosmographiae Introductio* (London: Giles, 2008).

39 See the excellent book by Matthew H. Edney, *Cartography: The Ideal and Its History* (Chicago: University of Chicago Press, 2019).

40 J.B. Harley, "Silences and Secrecy: The Hidden Agenda of Cartography in Early Modern Europe," *Imago Mundi* 40 (Jan 1988): 57–76.

true that pre-modern maps generally represent the state of geographic knowledge at the time they were created. Their very subjectivity precludes that, as does the fact that maps were often used as political tools in territorial disputes and bids for expansion.[41] Pre-modern maps also frequently included places that people "knew" existed, the source of that knowledge either lost in the ever popular mists of time or drawn from legends and stories that most scholars now believe to be fictional.[42]

In the collection of the James Ford Bell Library at the University of Minnesota is a portolan chart dated 1424 and attributed to Venetian cartographer Zuane Pizzigano.[43] It was part of the collection of Sir Thomas Phillips (1792–1872), perhaps the greatest private collector of manuscripts in modern times. During his lifetime he amassed a collection of approximately 60,000 items, primarily vellum manuscripts of historical interest.[44] A substantial part of the Phillips collection, including this map, was acquired in 1946 by William H. Robinson, Ltd. of London, very distinguished dealers in rare books and manuscripts in the mid-twentieth century. It was as the firm began to sell the items in the Phillips collection that the Pizzigano chart first came to the attention of scholars;[45] it was acquired by the James Ford Bell Library in 1954, less than a year after the library's founding.

This nautical chart is hand drawn on parchment (measuring 88 × 56 cm) and is undecorated. Rather than focusing on the Mediterranean as do most

41 In addition to Harley, "Silences and Secrecy," see: Philip E. Steinberg, "Insularity, Sovereignty, and Statehood: The Representations of Islands on Portolan Charts and the Construction of the Territorial State," *Geograpfiska Annaes* Series B: *Human Geography* 27.4 (2005): 253–265; and Valerie A. Kivelson, "'Between All Parts of the Universe': Russian Cosmographies and Imperial Strategies in Early Modern Siberia and Ukraine," *Imago Mundi* 60 (2008): 166–181.

42 Edward Brooke-Hitching's recent book, *The Phantom Atlas: The Greatest Myths, Lies, and Blunders on Maps* (London: Simon & Schuster Ltd, 2016), offers an interesting take on these, but should be read with caution. It is too easy to dismiss things as myths, lies, and blunders when we do not know their origin or the motivation behind their first inclusion on maps.

43 John Parker, *Antilia and America; A Description of the 1424 Nautical Chart and the Waldseemüller Globe Map of 1507 in the James Ford Bell Collection at the University of Minnesota* (Minneapolis: James Ford Bell Collection, 1955).

44 It was listed as item No. 25924; the printed catalog of the Phillips collection reached only to No. 23837.

45 Armando Cortesão, "The North Atlantic Nautical Chart of 1424," *Imago Mundi* 10 (1953): 1–13; followed by ibid, *The Nautical Chart of 1424 and the early discovery and cartographical representation of America: a study on the history of early navigation and cartography*, foreword by Maximino Correia (Coimbra and Minneapolis: University of Coimbra, 1954). These are the first and only major studies of this map.

pre-Columbian portolans,[46] its focus is the western ocean; it depicts the western coast of Europe and North Africa, along with England and Ireland and the recently discovered Canaries and Madeira Islands, as well as the Azores, which would not be discovered "officially" by the Portuguese for another few years, but which had already made a few appearances on fourteenth-century maps, including the famous Catalan atlas.[47] A few of the more famous so-called legendary islands also appear, including that of Brazil or Hy Brazil, off the western coast of Ireland. This map's claim to fame, however, rests in the earliest known depiction of what is often called the Antilia group of islands, improbably large islands in the western Atlantic, named Antiylia (Antilia), Saya, Satanazes, and Ymana. It was seeing these islands on the Pizzagano chart that inspired former British submariner Gavin Menzies to speculate on their discovery and "true" location in the popular book, *1421: The Year China Discovered America*. Menzies posited that these islands were discovered in the Caribbean by one of the Chinese treasure ships sent out into the world to collect tribute in the first quarter of the fifteenth century.[48] Menzie's theory sparked world-wide controversy and an increased interest in exploration history, with many scholars finding his claims unsupported by viable evidence.[49]

---

46 On early portolan charts, see Philipp Billion, *Graphische Zeichen auf mittelalterlichen Portolankarten—Ursprünge, Produktion und Rezeption bis 1440* (Marburg: Tectum, 2011); Ramon J. Pujades i Bataller, *Les cartes portolanes: la representació medieval d'una mar solcada* (Barcelona: Institut Cartogràfic de Catalunya ..., 2007); and Corradino Astengo, "The Renaissance Chart Tradition in the Mediterranean," in *Cartography in the European Renaissance*, vol. 3 of *The History of Cartography*, ed. David Woodward (Chicago: University of Chicago Press, 2007), part 1, 174–262. The foundational works remain Tony Campbell, "Census of Pre-Sixteenth-Century Portolan Charts," *Imago Mundi* 38 (1986): 67–94 and ibid., "Portolan Charts from the Late Thirteenth Century to 1500," in *Cartography in Prehistoric, Ancient and Medieval Europe and the Mediterranean*, vol. 1 of *The History of Cartography*, ed. J.B. Harley and David Woodward (Chicago: University of Chicago Press, 1987), 371–463.

47 Armando Cortesão, *The Nautical Chart of 1424*, suggests that they may have been found "accidentally" during the mapping expedition that discovered the Canaries.

48 Gavin Menzies, *1421: The Year China Discovered America* (New York: William Morrow, 2003); originally published as *1421: The Year China Discovered the World* (London: Bantam, 2002).

49 See, for example, historian Robert Finlay, "How Not to (Re)Write World History: Gavin Menzies and the Chinese Discovery of America," *Journal of World History* 15.2 (2004): 229–242; Leo K. Shin's review in *The Dalhousie Review* 84.2 (2004): 319–321; and Du Gangjian 杜钢建, 中国人远洋美洲的若干历史问题——读孟席斯《1421：中国发现世界》*Zhongguoren yuanyang meizhou de ruogan lishi wenti—du Meng Xisi "1421: Zhongguofaxian shijie"* [Some Historical Issues of the Chinese Colonization of the Americas—Reading Menzies "1421: China's Discovery of the World"], *Taipingyang xuebao* 太平洋学报 21.3 (2013): 86–95, among many others on both sides of the issue.

The northernmost of the two largest islands is Satanaze: Satan's or Devil's Island. Many researchers have focused their attention on the southernmost island, Antilia, which eventually gave its name to the Greater and Lesser Antilles in the Caribbean; few have written about Satanaze. Part of the reason for the lack of attention paid to this island is that it has frequently been linked on the 1424 chart with Antilia as part of the Antilia island group. However, a closer look at these islands reveals that they are really two separate groups. Satanaze and the smaller islands near it are actually quite a bit north of Antilia and the small island of Ymana to its east: between approximately 45° N and 50° N.[50]

In 2015, University of Minnesota graduate student Joshua Marcotte discovered that this island, Devil's Island or Isle of Demons, was also noted on a world map by Joken Nishikawa, published in 1708 and titled (roughly translated) "Map of 10,000 Countries on the Spherical Earth," a title almost identical to Ricci's "Map of the Ten Thousand Countries of the Earth."[51] An examination of the Ricci map quickly led to the discovery that an Isle of Demons was also depicted on it, in approximately the same location: off the coast of Nova Scotia between 45° N and 50° N.

As noted above, Ortelius' 1570 *Typus Orbis Terrarum* accompanied Matteo Ricci to China and was a likely source for his maps. This map, too, depicts an Isle of Demons in roughly the same spot, as does Mercator's 1569 world map, which Ortelius reproduced, as well as Petrus Plancius' 1590 *Terris Orbis*, also sources for the Ricci map. Ortelius listed among his sources the 1562 map of Diego Gutiérrez, published by Hieronymus Cock in Antwerp, but that map does not depict the Isles of Demons.[52] However, the 1565 map of the New World by Paolo Forlani, reprinted in 1566 by Bologino Zaltieri, does include it.[53] Moving

50 For a survey of maps depicting Satanaze, see Cortesão, *The Nautical Chart of 1424*, 134. For a recent survey and discussion of Antilia, see the self-published Jerald Fritzinger, *Pre-Columbian Trans-Oceanic Contact* (Morrisville, NC: LuLu, 2016), ch. 5. James E. Kelley, Jr. was one of the earliest to consider a North American location for Santanze: "The Oldest Portolan Chart in the New World," *Terrae Incognitae* 9 (1979): 22–48.

51 Joken Nishikawa, *Zōho Kai tsūshō kō* 西川如見, 5 vols. (Kyōto: n.p., 1708); revised and expanded edition of *Kai tsūshō kō*, 2 vols. (Kyōto, 1695).

52 Diego Guitterez, *Americae sive quartae orbis partis nova et exactissima descriptio* (Antwerp, 1562). See https://www.loc.gov/resource/g3290.ct000342 (accessed June 7, 2015). For Ortelius' sources, see Abraham Ortelius, *Theatrum orbis terrarum* (Antwerp: Apud Aegid Coppeneim Diesth, 1570).

53 Paolo Forlani, *Il disegno del discoperto della nova Franza, il quale s'e hauuto ultimamente dalla novissima nauigatione de' Franzesi in quel luogo* (Venice, 1565). See David Woodward, "Paolo Forlani: Compiler, Engraver, Printer, or Publisher?" *Imago Mundi* 44 (1992): 45–64. Bolognino Zaltieri apparently acquired Forlani's plate and reprinted the map in 1566 with

back to portolan charts, the 1463 Benincasa chart, for example, includes the Antilia group of islands, depicted similarly to the Pizzigano portolan, but in a more stylized form. This form, which represents most of the portolan depictions of the islands also can be seen on the 1466 chart of Majorcan cartographer Petrus Roselli, as well as on the 1489 chart of Albinus de Canepa.[54] Martin Behaim's 1492 globe includes Antilia, but not Satanaze. Moving into the sixteenth-century and the post-Columbian world, the Isle of Demons is clearly labeled in a location close to that of the others on the 1507/1508 world map of Johannes Ruysch.[55]

The 1424 Pizzigano chart is the earliest extant map to show the Satanaze and Antilia groups of islands. However, it is not possible to say that knowledge of these island groups was transmitted directly from the Pizzagano chart to the other portolan charts noted above. What we can say is that knowledge of the islands was depicted on some portolan charts and not others. We also can state that at least one cartographer believed that Satanaze was located, not in the Caribbean, but off the coast of Nova Scotia. That information was transmitted via the maps of Ortelius, Mercator, and Plancius, and possibly others, to China, leading Matteo Ricci to include the Isle of Demons on his 1602 map.

The case of the Nishikawa map is more complicated because little has been written about Nishikawa in English and it is difficult to find Japanese sources that discuss his work. It is likely, however, that the Ricci map could be a possible source, but so, too, could seventeenth-century Dutch maps. The Ricci map was widely disseminated and, in fact, of the six surviving copies of the 1602 printing, three reside in Japan, along with numerous later renditions. Nishikawa was born and educated in Nagasaki, the entry point for Europeans into Japan during a period of strictly regulated contact, which functioned not unlike Macao did in relation to China. In the 1580s, due to the instability of the government, Nagasaki was virtually a Jesuit colony and it is believed that European books and maps were in wide circulation. By 1602, other Catholic missionaries also were in evidence. Portuguese, Dutch, and English merchants traded there. However, fear that the Christians were becoming too powerful

---

his own name added to the cartouche. Both the Newberry Library and the John Carter Brown Library hold the 1565 state of the map; the James Ford Bell Library holds the 1566. The 1582 map by Michael Lok, prepared for Richard Hakluyt, *Illustri Viro, Domino Philippo Sidnaeo* (London: Thomas Woodcocke, 1582), is another example that includes the Isle of Demons. It also shows St. Brendan's Isle at the same latitude.

54 Both the Roselli and the Canepa portolan charts are part of the James Ford Bell Library collection.

55 Ruysch's map was published in the 1507 and 1508 editions of Ptolemy's *Geographia*, published at Rome.

led to the expulsion of the missionaries in 1614 and the banning of Catholicism. The Shimabara Rebellion of 1637, although not incited by European Christians, was found to include Japanese Christian rebels with close ties to the Portuguese among its ringleaders. The Portuguese were expelled and forbidden access. The Dutch were allowed to move into a specially-built island in Nagasaki harbor, Dejima, but contact with the Japanese people was extremely limited and Dutch books were banned until 1720. As a scholar, it is possible that Nishikawa had access to Dutch materials that others did not.[56] His illustration of Europeans, which appeared in the same book as the world map, was so clearly that of a Dutchman and his wife that it became iconic and was reused throughout Japan.

Scholars have mixed views of the influence of the West on Asian cartography. However, it seems certain that the West, in the persons of Matteo Ricci and other Jesuit missionaries who promoted the creation of world maps as part of their missionary efforts, introduced China to the geographic depiction of the Americas as well as to map projections based on the use of latitude and longitude. The Ricci map enabled the Chinese to see for the first time where the Americas were in relationship to China. And it was through western maps that the Isle of Demons was introduced as part of that new American geography.

As this brief discussion suggests, understanding the transmission of knowledge in the pre-modern world, particularly via maps, is a complex process, fraught with minefields and pitfalls. How many maps should one examine to trace the transmission of a particular place name? Without knowing more about each individual cartographer's resources and training what level of scholarly value can such examinations provide?

Another key element to the transmission of knowledge is the preservation of the artifacts—books, maps, manuscripts—that contain said knowledge. Collectors like Bracciolini and Phillips, noted above, are integral to the survival of these artifacts. So, too, are institutional repositories: libraries, museums, archives. The James Ford Bell Library at the University of Minnesota, home of the 1602 Ricci map, as well as the Pizzigano and the 1507 Waldseemüller globe gores, is not a map library. The collection holds about 300 flat, unbound maps, and about half of them were once bound into a volume of some kind. Thanks to a grant from the National Endowment for the Humanities, the James Ford Bell Library at the University of Minnesota, where I serve as Curator, has been able to document more than 22,400 maps in its books, which it is making

56 For this period in Japanese science, see Masayoshi Sugimoto and David L. Swain, *Science and Culture in Traditional Japan* (Rutland, VT and Tokyo: Charles E. Tuttle Co., 1989), esp. part 4, 225–290.

available freely online through digital surrogates. With the flat maps, there are about 23,000 maps in the collection altogether, yet we are not a map library. It is neither our specialty nor our focus, although it is generally the maps that get the most press.

All of our maps, as well as everything else in the Bell Library collection, are a testament to the preservation of knowledge through writing technologies. We collect materials based on a rather broad criteria: every item we acquire must be a product of the pre-modern world and have something to do with the history of trade and cultural exchange and their impact. Not counting the maps found in our books, the collection houses close to 40,000 items ranging from government documents, treatises on scurvy, and polemical tracts arguing for or against slavery to geographies, works of natural history and science, ethnographies, and instructions and plans for building ships. It contains the records of a sugar plantation operated by the Jesuits in Mexico, the archives of the Swedish East India Company, account books for a Portuguese merchant firm doing business in eighteenth-century Peru, the fifteenth-century correspondence of a prominent Venetian merchant family, and a set of atlases once owned by Pope Clement XI (1649–1721), famous for their provenance as well as for their creator, the renowned Dutch publisher and cartographer Johannes Blaeu (1596–1673).

This collecting scope is broadly conceived and at one time the Bell Library had the funds to purchase as many editions of a particular text as were available, to create the depth necessary for textual comparison. For example, the collection includes twenty-eight different editions of Ptolemy's famous *Cosmographia*, including the first one from 1475,[57] published without maps, as well as other books that incorporate parts of the *Cosmographia*. This focus on depth is no longer possible. While some additional editions are added occasionally in areas in which the library already has an investment, the collecting focus since 2005 has been on manuscript materials—materials that will set the Bell Library apart from other such repositories. What is added to the collection, whether manuscript or print, first depends on what is available to add, what has come on the market. Then, of these materials available that fit the library's collection scope, we identify those that fit the current budget. That short list is then ranked by scarcity (even manuscript materials can be of a type that is widely available even if individually unique), their importance to scholarship generally, importance to the research and teaching interests of the faculty and graduate students at the University of Minnesota, and then the importance to

57 Claudius Ptolemy, *Cosmographia*, trans. Jacopo D'Angelo, ed. Angelus Vadius and Barnabus Picardus (Vicenza: Hermannus Liechtenstein, 1475).

the collection: whether the materials fill a gap, or augment existing materials, and so forth.

These types of decisions are made by curators and acquisitions librarians at institutions all over the world. And in making these acquisition decisions, they influence what is available for scholarly research. If an item on the market is purchased by an institution that makes the item available in some way, that item is available for study. If, on the other hand, the item is purchased by a private collector, it may never be seen again by anyone but the purchaser and their immediate circle. The scholarly world will know it exists through the evidence of the dealer's catalog, but it will not have access to it. If the dealer's catalog is available only online, that record will eventually fade from view. If the dealer also published a printed catalog, then there is a much better chance that at least knowledge of the item will not be lost forever to scholarship, provided at least one person saves the dealer's catalog. In either case, in 100 years, should the family wish to sell the item, it may cause quite a stir in scholarly circles.

The survival of material objects created by writing technologies that transmit knowledge (sculpture, glass, and other art objects have their own stories of transmission to tell) is random at best. We interpret the survival of comparatively numerous quantities of a single item to reflect the importance or popularity of the work, but we cannot know that our interpretation is correct without corroborating evidence. A fire that destroys one print shop but not another influences survival, as do war, natural disasters, and economic upheavals. The famous Anglo-Saxon epic Beowulf survives but in a single manuscript that dates, based on paleographic analysis, to somewhere from the late tenth to the early eleventh century. Many scholars believe that the epic, itself, was first composed in the late seventh or early eighth century. The survival of a single epic poem, more than 1,200 years old, in a single manuscript, itself more than 1,000 years old, offers a striking example of a different type of interpretation of survival—to have survived at all the item *must* be important. Or was it? What we do know is that because it has survived, it is extremely important to early medieval scholarship in the modern age. It is the basis, along with a handful of other texts, for our knowledge of the Anglo-Saxon language. It is taught widely around the globe; indeed, the standard edition of the text was created by the University of Minnesota's own Frederick Klaeber in the early twentieth century.[58] The first recorded owner of the Beowulf manuscript was

---

58 Frederick Klaeber, *Beowulf and the Fight at Finnsburg* (Boston, New York: D.C. Heath & Co., 1922; 3rd ed Boston: Heath, 1950). A new fourth edition with substantial introduction was published by Robert Dennis Fulk, Robert E. Bjork, and John D. Niles, eds., *Klaeber's Beowulf* (Toronto: University of Toronto Press, 2008).

Laurence Nowell, a sixteenth-century pioneer of the study of Old English or Anglo-Saxon. It was then acquired by Sir Robert Cotton, who built up a significant personal library and whose heirs, in the early eighteenth-century, bequeathed it to the British nation. The Cotton Library was one of the foundational collections of the British Museum when it was founded in 1753, now part of the British Library. However, prior to the founding of the British Museum, the Cotton manuscripts were housed in Ashburnham House at Westminster. According to the British Library, fire broke out on the night of 21 October 1731 and a great many manuscripts were destroyed. Although the Beowulf manuscript survived, it was damaged and suffered even greater loss along the edges through handling over the years until, in the mid-nineteenth century, it was placed in paper frames and access to it became more restricted.[59]

Fate determined that the Beowulf manuscript survived, but what has been lost to fire and other means of destruction over the centuries? If Cotton had not acquired the Beowulf manuscript for his collection, it may have been lost to scholarship much earlier. The role of collectors, both individuals and institutions, cannot be underestimated when it comes to the survival of texts and the transmission of knowledge. When the Roman Empire collapsed and the preservation of knowledge through the copying of texts fell to the Catholic Church in the early Middle Ages, much was lost to the limited selection criteria of what would be copied and what would not. In addition to some philosophically-based decisions, it really was a matter of where to place one's investment dollar, so to speak. With limited time and resources, the texts that supported the mission of the Church—the conversion of everyone to orthodox Christianity—took precedence. Ricci's map, created, in part, as a tool in his campaign to convert the Chinese to Christianity, exists for a similar reason. The copies that are extant in the west survived primarily due to their connection to the Catholic Church and the Jesuit order. The 1602 map now owned by the James Ford Bell Library was acquired not only and not primarily for its cartography, but for its contribution to the Bell Library's significant collection of Jesuit-related materials.

Today, we are faced with a similar dilemma that may have a significant effect on the transmission of knowledge in the future. While creators make decisions about what information to include in their work, decisions that affect knowledge transmission at the source so to speak, today we are faced with the decision—again based on where to invest limited resources—on whether or not to digitize a book, map, or other document. Here again, special collection

59 Fulk, Bjork, and Niles, *Klaeber's Beowulf*, xxv–xxxv. Also see the British Library, https://www.bl.uk/collection-items/beowulf (accessed March 4, 2016).

and rare book library curators play an integral role. The Google Book project describes its goal as:

> simple: make it easier for people to find relevant books—specifically books they wouldn't find any other way such as those that are out of print—while carefully respecting authors' and publishers' copyrights. Our ultimate goal is to work with publishers and libraries to create a comprehensive, searchable, virtual card catalog of all books in all languages that helps users discover new books and publishers discover new readers.[60]

What is most disturbing about this statement is their use of the word "relevant." Who decides? What happens to books or other historical materials that are deemed not relevant? The University of Minnesota was an early partner in the Google Books project and it, along with other academic partners, was instrumental in the formation of the Hathi Trust, a consortium of academic repositories that have agreed to fund the maintenance, in perpetuity, of digital surrogates created as part of the Google Books project as well as those created by other means. Should Google cease to exist, the material scanned by Google from these institutions would be preserved through the Trust.[61]

Such consortiums are becoming increasingly necessary. In libraries and archives across the country, and indeed around the world, there is a storage crisis. Some universities have negotiated cooperative storage arrangements. For example, certain members of the coalition agree to hold in perpetuity a certain number of copies of a particular book so that other members can free up space. If the item is scanned, then one or more members of the coalition commit to maintaining the digital surrogate. It is a complex problem and there are no easy solutions.[62] For curators of special collections, it is a more complicated issue. The costs of scanning rare materials is considerably greater than for modern or less valuable materials; rare books cannot be disbound to be scanned. They and other rare materials must be photographed expertly with an overhead digital camera. Large-scale items, like the Ricci map, must be scanned in sections, which are then stitched together using an image editor like Adobe Photoshop. In fact, creating a digital surrogate was one of the

---

60 https://www.google.com/googlebooks/library/ (accessed February 6, 2016).

61 https://www.hathitrust.org/ (accessed February 6, 2016).

62 For a discussion of these changes and more, see Lorcan Dempsey, Constance Malpas, and Brian Lavoie, "Collection Directions: The Evolution of Library Collections and Collecting," *portal: Libraries and the Academy* 14.3 (2014): 393–423.

priorities for the University of Minnesota. The map was a previously unknown copy that had been put on the market by a private collector, and we are committed to ensuring its survival, both physically and digitally.

Curators and librarians make decisions everyday about what to scan and what not to scan and then how to make those scans available and under what circumstances. They decide whether an item can be studied digitally on the other side of the world, or if it can be studied only by people who can find their way into a particular library. Here, as much as when acquisition decisions are made, curators and librarians are arbiters of the transmission and preservation of knowledge. Academic libraries, in particular, are changing their focus from collectors to disseminators, pushing out affordable content whenever possible. Special collections curators are struggling to find a balance between preserving and growing their collections and sharing them digitally. In each case, what scholars have available to study is in their hands.[63]

According to Walter Ong, "Fully literate persons can only with great difficulty imagine what a primarily oral culture is like, that is, a culture with no knowledge whatsoever of writing or even of the possibility of writing. Try to imagine a culture where no one has ever 'looked up' anything. In a primarily oral culture, the expression 'to look up something' is an empty phrase: it would have no conceivable meaning."[64] Will future generations find it impossible to imagine what it was like to hold a codex? To feel the paper or parchment? To physically turn a page? When people decry the demise of the book that is generally what they mean: the demise of the codex. In the face of Google Earth and other satellite-based mapping systems, the demise of the paper map, seems close behind. However, the "book" or "map" is the contents, not the container. The book and the map will survive, whether written on a papyrus scroll, on illuminated or decorated parchment, printed and bound in leather or mass-produced on cheap paper, or encoded and viewed on a computer, a tablet, or even a phone. The question before us should be, "which books and which maps?"

63 In addition to the Dempsey et al. article noted above, see James G. Neal, "The Entrepreneurial Imperative: Advancing from Incremental to Radical Changes in the Academic Library," *portal: Libraries and the Academy* 1.1 (2001): 1–13.

64 Walter J. Ong, *Orality and Literacy: The Technologizing of the Word* (London and New York: Methuen, 1982), 31.

Bibliography

Arciniegas, Germán. *Amerigo and the New World: The Life and Times of Amerigo Vespucci*. Trans. Harriet de Onis. New York: Knopf, 1955.

Astengo, Corradino. "The Renaissance Chart Tradition in the Mediterranean." In *Cartography in the European Renaissance*. Ed. David Woodward. Part 1, 174–262. In Vol. 3 of *The History of Cartography*. Chicago: Chicago University Press, 2007.

Baddeley, J.F. "Father Matteo Ricci's Chinese World Maps, 1584–1608." *The Geographical Journal* 50, no. 4 (Oct. 1917): 262–263.

Billion, Philipp. *Graphische Zeichen auf mittelalterlichen Portolankarten—Ursprünge, Produktion und Rezeption bis 1400*. Marburg: Tectum, 2011.

Brooke-Hitching, Edward. *The Phantom Atlas: The Greatest Myths, Lies, and Blunders on Maps*. London: Simon & Schuster, Ltd., 2016.

Bruni, Flavia, and Andrew Pettegree. *Lost Books: Reconstructing the Print World of Pre-Industrial Europe*. Leiden: Brill, 2016.

Campbell, Tony. "Census of Pre-Sixteenth-Century Portolan Charts." *Imago Mundi* 38 (1986): 67–94.

Campbell, Tony. "Portolan Charts from the Late Thirteenth Century to 1500." In *Cartography in Prehistoric, Ancient and Medieval Europe and the Mediterranean*. Ed. J.B. Harley and David Woodward, 371–463. In Vol. 1 of *The History of Cartography*. Chicago: Chicago University Press, 1987.

Carter, John, and Percy H. Muir, eds. *Printing and the Mind of Man, a descriptive catalogue illustrating the impact of print on the evolution of Western civilization during five centuries*. London and New York: Holt, Rinehart & Winston, 1967.

Chen, Hui-hung. "The Human Body as a Universe: Understanding Heaven by Visualization and Sensibility in Jesuit Cartography in China." *The Catholic Historical Review* 93, no. 7 (2007): 517–552.

Chrisman, Miriam Usher. *Lay Culture, Learned Culture: Books and Social Change in Strasbourg, 1450–1599*. New Haven: Yale University Press, 1982.

Clanchy, Michael T. *From Memory to Written Record: England 1066–1307*. Cambridge: Harvard University Press, 1979.

Clough, Cecil H. "The New World in the Renaissance." In *The European Outthrust and Encounter. The First Phase c. 1400–c. 1700: Essays in Tribute to David Beers Quinn on his 85th Birthday*. Ed. Cecil H. Clough and P.E.H. Hair, 291–328. Liverpool: Liverpool University Press, 1994.

Codex Laurentianus 35.30. Biblioteca Medicia Laurenzian, Florence.

Cortesão, Armando. *The Nautical Chart of 1424 and the early discovery and cartographical representation of America: a study on the history of early navigation and cartography*. Foreword by Maximino Correia. Coimbra and Minneapolis: University of Coimbra, 1954.

Cortesão, Armando. "The North Atlantic Nautical Chart of 1424." *Imago Mundi* 10 (1953): 1–13.

Day, John. "The Search for the Origins of the Manuscript Copies of the Maps of Matteo Ricci." *Imago Mundi* 47 (1995): 94–117.

D'Elia, Pasquale M. *Il Mappamondo Cinese del P. Matteo Ricci* SJ. Cittá del Vaticano: Biblioteca apostolica Vaticana, 1938.

Dempsey, Lorcan, Constance Malpas, and Brian Lavoie. "Collection Directions: The Evolution of Library Collections and Collecting." *Portal: Libraries and the Academy* 14, no. 3 (2014): 393–423.

Edney, Matthew H. *Cartography: The Ideal and Its History*. Chicago: University of Chicago Press, 2019.

Edney, Matthew H. "The Columbus Letter: The Diffusion of Columbus' Letter Through Europe, 1493–1497." Osher Map Library, University of Southern Maine. http://www.oshermaps.org/special-map-exhibits/columbus-letter/iv-diffusion-columbuss-letter-through-europe-1493-1497. Accessed July 12, 2017.

Elman, Benjamin. "Collecting and Classifying: Ming Dynasty Compendia and Encyclopedias (Lieshu)." *Extrême-Orient, Extrême-Occident*, hors série (2007): 131–157.

Fernández-Armesto, Filipe. *Amerigo: The Man Who Gave His Name to America*. New York: Random House, 2007.

Forlani, Paolo. *Il disegno del discoperto della nova Franza, il quale s'e hauuto ultimamente dalla novissima navigatione de' Franzesi in quel luogo*. Venice, 1565.

Formisano, Luciano. *Letters from a New World: Amerigo Vespucci and the Discovery of America*. New York: Marsilio, 1992.

Foss, Theodore N. "Ricci's World Map. The 1602 *Kunyu Wanguo Quantu*." In *China at the Center: Ricci and Verbiest World Maps*. Ed. Natasha Reichle. San Francisco: Asian Art Museum, 2016.

Fritzinger, Jerald. *Pre-Columbian Trans-Oceanic Contact*. Morrisville, NC: Lulu, 2016.

Fulk, Robert Dennis, Robert E. Bjork, and John D. Niles, eds. *Klaeber's Beowulf*. Toronto: University of Toronto Press, 2008.

Füssel, Stephen. *Gutenberg and the Impact of Printing*. Trans. Douglas Martin. Aldershot, Hampshire, UK: Ashgate, 2005.

Gallagher, S.J., Louis J. *China in the Sixteenth Century: The Journals of Matthew Ricci, 1583–1610*. New York: Random House, 1953.

Greenblatt, Stephen. *The Swerve. How the World Became Modern*. New York and London: W.W. Norton and Company, 2011.

Guitterez, Diego, *Americae sive quartae orbis partis nova et exactissima descriptio*. Antwerp, 1562. https:/www.loc.gov/resource/g3290.ct000342.

Harley, J.B. "Silences and Secrecy: The Hidden Agenda of Cartography in Early Modern Europe." *Imago Mundi* 40 (1988): 57–76.

Henige, David. "Finding Columbus: Implications of a Newly Discovered Text." In *The European Outthrust and Encounter. The First Phase c. 1400–c. 1700: Essays in Tribute to*

David Beers Quinn on his 85th Birthday. Ed. Cecil H. Clough and P.E.H. Hair, 141–166. Liverpool: Liverpool University Press, 1994.

Hessler, John W. *The Naming of America: Martin Waldseemüller's 1507 World Map and Cosmographiae Introductio*. London: Giles, 2008.

Hoover, Herbert. *The Memoirs of Herbert Hoover*. Vol. 1: *Years of Adventure, 1874–1920*. New York The Macmillan Company, 1952.

Hoover, Herbert Clark and Lou Henry Hoover. *Georgius Agricola De re metallica: tr. From the 1st Latin ed. of 1556, with biographical introduction, annotations, and appendices upon the development of mining methods, metallurgical processes, geology, mineralogy & mining law, from the earliest times to the 16th century*. London, 1912; reprinted New York: Dover Publications, 1950.

Kelley, Jr., James E. "The Oldest Portolan Chart in the New World." *Terrae Incognitae* 9 (1979): 22–48.

Kivelson, Valerie A. "'Between All parts of the Universe': Russian Cosmographies and Imperial Strategies in Early Modern Siberia and Ukraine." *Imago Mundi* 60 (2008): 166–181.

Klaeber, Frederick. *Beowulf and the Fight at Finnsburg*. 3rd ed. Boston: Heath, 1950.

Krauss, Michael E. "Eskimo Languages in Asia, 1791 on, and the Wrangel Island-Point Hope Connection." *Études/Inuit/Studies* 29, no. 1–2 (2005): 163–185.

Lollis, Cesar De, ed. *Raccolta di Documenti e Studi pubblicati dalla Reale Commissione Columbiana del quarto Centenario della scoperta dell'America*. Vol. 1, part 1: *Scritti di Cristoforo Colombo*. Rome: Ministero della pubblica istruzione, 1894.

Menzies, Gavin. *1421: The Year China Discovered America*. New York: William Morrow, 2003.

Mignini, Filippo, ed. *La Cartografica di Matteo Ricci*. Rome: Liberia dello Stato, Istituto poligrafico e Zecca dello Stato, 2013.

Moulton, Ian Frederick, ed. *Reading and Literacy in the Middle Ages and the Renaissance*. Turnhout, Belgium: Brepols, 2004.

Neal, James G. "The Entrepreneurial Imperative: Advancing from Incremental to Radical Changes in the Academic Library." *portal: Libraries and the Academy* 1, no. 1 (2001): 1–13.

Nishikawa, Joken. *Zōho Kai tsūshō kō* 西川如見 5 vols. Kyōto: n.p., 1708.

Ong, Walter. *Orality and Literacy: The Technologizing of the Word*. London and New York: Methuen, 1982.

Ortelius, Abraham. *Theatrum orbis terrarum*. Antwerp: Apud Aegid. Coppeneim Diesth, 1570.

Parker, John. *Antilia and America: A Description of the 1424 Nautical Chart and the Waldseemüller Globe Map of 1507 in the James Ford Bell Collection at the University of Minnesota*. Minneapolis: James Ford Bell Collection, 1955.

Petrucci, Armando. *Writers and Readers in Medieval Italy*. Ed. and trans. Charles M. Radding. New Haven: Yale University Press, 1995.

Pettegree, Andrew. *The Book in the Renaissance*. New Haven: Yale University Press, 2010.

Ptolemy, Claudius. *Cosmographia*. Trans. Jacopo D'Angelo. Ed. Angelus Vadius and Barnabus Picardus. Vincenza: Hermannus Liechtenstein, 1475.

Pujades I Bataller, Ramon J. *Les cartes portolanes: la representació medieval d'una mar solcada*. Barcelona: Institut Cartogràfic de Catalunya …, 2007.

Ricci, Matteo. *De Christiana expeditione apud Sinas suscepta ab Societate Jesu* …. Trans. and ed. Nicolas Trigault. Augsburg: Apud Christoph. Mangium, 1615.

Ricci, Matteo. *Storia dell'introduzione del Christianesimo in Cina*. Ed. Pasquale M. d'Elia. Rome: La Libreria dello Stato, 1942–1949.

Ricci, Matteo, and Nicolas Trigault. *Histoire de l'expedition chrestienne en la Chine, entreprise par les peres de la Compagnie de Iesus*. Paris: P. Le-Mvr, 1618.

Steinberg, Philip E. "Insularity, Sovereignty, and Statehood: The Representations of Islands on Portolan Charts and the Construction of the Territorial State." *Geograpfiska Anaes* Series B: *Human Geography* 27.4 (2005): 253–265.

Stearns, John Barker. "Lucretius and Memmius." *The Classical Weekly* 25 (1931): 67–68. https://doi.org/10.2307/4389660.

Suarez, Michael F., and H.R. Woudhuysen, eds. *The Book: A Global History*. Oxford: Oxford University Press 2013.

Sugimoto, Masayoshi, and David L. Swain. *Science and Culture in Traditional Japan*. Rutland, VT and Tokyo: Charles E. Tuttle Co., 1989.

Thompson, James Westfall. *The Literacy of the Laity in the Middle Ages*. New York: Franklin Classics, 1960.

Waldseemüller, Martin. *Cosmographie introductio*. St. Dié, France: Walter and Nikolaus Lud, 1507.

Whitfield, Roderick, Anne Farrer, Shelagh J. Vainker, eds. *Caves of the Thousand Buddhas: Chinese Art of the Silk Road*. London: George Braziller, 1990.

Wolf, Abraham. *A History of Science, Technology, and Philosophy in the 16th and 17th Centuries*, 2nd ed. vol. 2. New York: Harper Torchbooks, 1959.

Woodward, David. "Paolo Forlani: Compiler, Engraver, Printer, or Publisher?" *Imago Mundi* 44 (1992): 45–64.

Zamora, Margarita. *Reading Columbus*. Berkeley: University of California Press, 1993. http://ark.cdlib.org/ark:/12020/ft009nbocv/. Accessed August 2, 2017.

CHAPTER 12

East Asian Map Collections in the Library of Congress: A Unique Source for the Study of Cartography and East–West Cultural Exchange

Ralph E. Ehrenberg

Maps and atlases have been a vital part of the collections of the Library of Congress since its establishment in 1800 when a congressional committee purchased three maps and an atlas from a London dealer. From this modest beginning, the Library's cartographic holdings have grown during the past two centuries to almost six million maps and atlases in all formats, which are held primarily in the Geography and Map Division located in the James Madison Building on the Library's Capitol Hill Campus. Included among these vast holdings is one of the most extensive collections of Asian cartographic materials outside of China, Japan, and Korea, totaling more than 370,000 cartographic materials dating from the Ming Dynasty (1368 to 1644 CE) in China to the present day.

Established in 1897 as the Hall of Maps to serve Congress and United States government agencies, the Geography and Map Division now functions as the National Map Library.[1] Its primary purpose is developing the Library's cartographic collections, which includes responsibility for all aspects of map curatorship, ranging from acquisitions, processing, preservation, and storage to cataloging, reference services, and exhibitions. An average of thirty to forty thousand cartographic items are acquired yearly through government agency deposits, transfers of superseded maps from United States government libraries, copyright deposits, domestic and international exchanges, purchases, and donations.

More than 400,000 cartographic materials have been cataloged by Division staff, including some 19,000 Chinese, Japanese, and Korean map and atlas titles, with their bibliographic records available online.[2] As a major service to the map library community, the Geography and Map Division also establishes,

1 An overview of the Geography and Map Division is found on the Geography and Map Division's online website http://www.loc.gov/rr/geogmap/.

2 Access to the Library of Congress Online Catalog keyword search functions by title, author/creator, or subject is https://catalog.loc.gov.

© RALPH E. EHRENBERG, 2024 | DOI:10.1163/9789004684782_014
This is an open access chapter distributed under the terms of the CC BY-NC-ND 4.0 license.

maintains, and disseminates national standards for classifying and cataloging cartographic materials under the direction of Min Zhang and her team of eight catalogers. In addition, two illustrated bilingual cartobibliographies provide detailed descriptions of the Division's pre-1900 Chinese holdings.[3] The Division's East Asia collections are available for research to scholars and the general public through personal visits, distance research, and online websites.

The Division's public reading room is open daily, Monday thru Friday, from 8:30 AM to 5:00 PM. The value of the Division's Asian collection has long been recognized for its research value. One of the earliest requests recorded was from the Japanese Embassy in 1934 concerning "a large-scale map of Manchuria and adjacent portions of China."[4] More recently, Chinese, Japanese, and Korean scholars have spent extended periods of time personally examining maps of their regions. Two professors from Osaka University, for example, identified several early manuscript maps in the Title Collection of topographic surveys conducted in Korea by the Japanese Imperial Land Survey. These appear to be the only copies that have survived!

An array of digital scanners, including book-to-net and roller feed scanners, are available for scanning and reproducing maps.

For those scholars unable to visit the Library in person, distance research is available through "Ask a Librarian," an online reference service that allows researchers to submit reference questions directly to the Division's reference librarians, and the Library's Online Digital Map Collection that contains more than 1,000 East Asian, Chinese, Japanese, and Korean titles.[5] In addition, a Chinese language website, *Meiguo guohui tushuguan diancang zhi Zhongguo xiangguan ditu wenku chaxun xitong* 美國國會圖書館典藏之中國相關地圖文獻查詢系統 [Inventory of China Related Historical Maps Archived at the Library of Congress], provides access to some 79,000 digital map files of Chinese maps in the Geography and Map Division. This website was created by Taipei's Academia Sinica Digital Center between 2004 and 2008 as part of a joint program between Academia Sinica and the Library of Congress to make

3 Li Xiaocong 李孝聪, ed. *Meiguo guo hui tu shu guan cang Zhong wen gu di tu xu lu* 美国国会图书馆藏中文古地图叙录 [Descriptive catalogue of ancient Chinese maps in the collection of the Library of Congress] (Beijing: Wenwu Publishing House, 2004); and Lin Tianren 林天人 and Zhang Min 張敏, eds., *Reading Imperial Cartography: Ming-Qing Historical Maps in the Library of Congress* [*Huangyu suolan: Meiguo guohui tushuguan suo cang Ming Qing yutu* 皇輿搜覽：美國國會圖書館所藏明清輿圖] (Washington, DC: Library of Congress in association with the Academia Sinica Digital Center, 2013).

4 Geography and Map Division Annual Report 1934 (Typescript in Geography and Map Division, Library of Congress).

5 Online access to Ask a Librarian is: http://www.loc.gov/rr/askalib/ask-geogmap2.html; access to the Library's Online Map Collections is: http://www.loc.gov/maps/.

"more broadly known the rich holdings of the Chinese maps in the Geography and Map Division's collections."[6]

1 Development of the Library's Early East Asian Map Collection

The Library's East Asian Map Collection developed gradually, initially through international exchanges between the Chinese and American governments. These were made possible by Qing Emperor Daoguang, who sent the tenth president of the United States John Tyler a letter in 1843 that read: "Now that the English barbarians have been allowed to trade, whatever other countries there are, the United States and others, should naturally be permitted to trade without discrimination, in order to show our tranquilizing purpose."[7] Shortly thereafter, President Tyler appointed former American Congressman and United States Army General Caleb Cushing, "Commissioner, Envoy Extraordinary and Minister Plenipotentiary of the United States to China." Cushing's stay in China led to the Wangxia Treaty of 1845, the first formal trading agreement between the United States and China.

The first maps acquired by the Library that were produced in East Asia accompanied a donation by the Qing Dynasty's Tongzhi emperor in 1869. This gift was part of an international exchange authorized by the US Congress two years earlier.[8] Although the bulk of the donation was devoted to the Confucian classics and science, a rare set of administrative maps for 26 provinces were included, compiled by the Hubei Provincial Publishing Office (Hubei guanshuju 湖北官書局) in the 3rd year of the Qing Tongzhi reign (1864).[9] An interesting element of the maps comprising this set is that a Chinese cartographic grid is used to indicate distance, with each grid side representing 100 *li* (approximately 33 miles).[10] This grid may also provide a guide for copying maps. In comparison, the "Complete Map of Hubei Province," a woodcut prepared by provincial governor Yan Shusen, and printed in Wuchang in 1862

6 Online access to the Chinese website *Inventory of China Related Historical Maps Archived at the Library of Congress* (美國國會圖書館典藏之中國相關地圖文獻查詢系統) is: http://webgis.sinica.edu.tw/map_loc/.

7 Quoted in Warren I. Cohen, *America's Response to China: A History of Sino-American Relations* (New York: Columbia University Press, 2010), p. 11.

8 *Library of Congress Asian Collections: An Illustrated Guide* (Washington, DC: Library of Congress, 2000) https://catalog.hathitrust.org/Record/004120975, p. 14.

9 For descriptions and illustrations of some of these maps, see: *Reading Imperial Cartography*, pp. 134–153.

10 See for example, Hubei Guan Shu Ju, *Jiangsu quan tu* 江蘇全圖 [Complete Map of Jiangsu Province], Wuchang, 1864. http://hdl.loc.gov/loc.gmd/g7823j.ct002578.

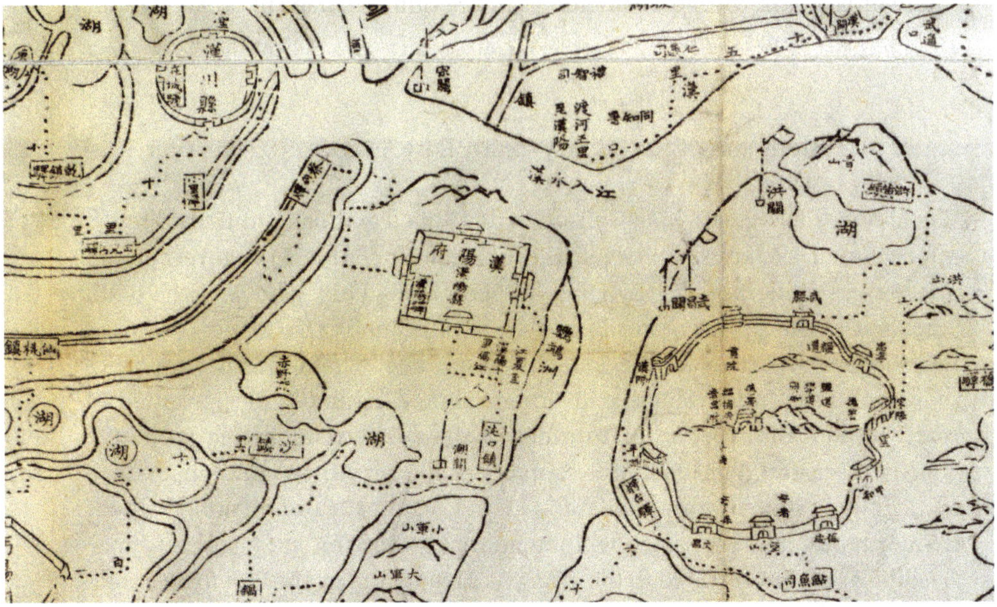

FIGURE 12.1 Detail from *E sheng quan tu* 鄂省全图 [Complete Map of Hubei Province], Tongzhi Reign. Wuchang, 1862. Wood block print, 123 × 197 cm. Library of Congress, Geography and Map Division, Call Number G7823.H64 1862. Y3. http://hdl.loc.gov/loc.gmd/g7823h.ct002593.

is drawn in the traditional Chinese manner without a grid or measurement guide. Urban centers, roads, and streams are displayed planimetrically, like those by the Hubei Provincial Publishing Office, using a variety of symbols, but the provincial capital, mountains, and coastlines are presented pictorially with an exaggerated perspective typical of Chinese maps of that era.[11] (Figure 12.1).

During his stay in China, Ambassador Cushing, who had become proficient in the Chinese language, selectively collected some 2,500-works devoted to the Chinese classics, history, poetry, medicine, and other standard works. Included among these was the *Nong zheng quan shu* 農政全書 [Encyclopedia of Agriculture Among the Chinese], which was compiled by Xu Guangqi (1562–1633), a notable Ming scholar who worked closely with the Italian Jesuit missionary Matteo Ricci. A Roman Catholic convert, Xu was the first Chinese to translate European works into Chinese.[12] The Library of Congress purchased

---

11 Yan Shusen, *E sheng quan tu* 鄂省全图 [Complete Map of Hubei Province], 1862. http://hdl.loc.gov/loc.gmd/g7823h.ct002593. Lin and Zhang, *Reading Imperial Cartography*, 107; Li, *Library of Congress Ancient Chinese Map Descriptive Catalog*, 61–62.

12 *Library of Congress Asian Collections: An Illustrated Guide*, 14–15.

this rare collection from the Cushing family in 1869 for $4,000, which represented almost half of the Library's annual acquisition budget.

Other American counsels and missionaries soon followed, some of whom donated maps. Examples include a Chinese map of Fuzhou, the provincial capital city of Fujian province, annotated in English by Thomas Dunn, the American counsel there from 1858–1873.[13] Drawn in the traditional Chinese pictorial style, three-mast American or European ships anchored in the provincial port are surrounded by Chinese junks. Chinese walled cities and mountains are drawn in relief. In contrast are maps drawn by missionaries in the western style. These include an administrative map of Jieyang county in Guangdong province, drawn by a missionary in 1875, which was donated to the Library by the American Baptist National Society. Mountains are depicted by hachures, a cartographic device consisting of a series of short lines that follow the direction of the slope. Hachuring was used extensively in Europe and American during the 19th century to illustrate terrain planimetrically rather than panoramically.

During the first four decades of the twentieth century, the Library's Chinese map collection continued to grow with exchanges and donations. In 1904, the Chinese government donated to the Library 198 works in 1,965 volumes on a variety of subjects, including the *Gujin tushu jicheng* 古今图书集成 [Imperial Encyclopedia of China]. A Shanghai official donated 291 volumes of Chinese books and four maps in 1909 in exchange for official government documents.

United States Department of Agriculture botanist Dr. William Tennyson Swingle donated a large number of scientific books, maps, and official local gazetteers that he collected during extensive visits to China between 1913 and 1939 to study Chinese herbs and plants. During Swingle's long career in China, he introduced the Chinese soybean, tung tree oil, and some twenty-two citrus strains to the United States, making possible entirely new agricultural industries. Swingle became close friends with Librarian of Congress Herbert Putnam and assisted him in achieving his dream to expand the Library from a "national" to a "universal" library.[14]

Walter B. Nance, a professor of philosophy and later president of Suzhou University in Jiangsu province, visited the Library of Congress in 1912 and

---

13 Anonymous, *Fu sheng quan tu* 福省全圖 [Map of Fujian Province], between 1842–1884. http://hdl.loc.gov/loc.gmd/g7824f.ct003181; Lin and Zhang, *Reading Imperial Cartography*, 118–119.

14 *Library of Congress Asian Collections: An Illustrated Guide*, 17.

donated two military maps made during the Qing Dynasty that were captured a year earlier and presented to the American mission by a Christian soldier.[15]

The largest single collection of rare early maps in the Chinese language was acquired through the efforts of Arthur W. Hummel, a distinguished Sinologist who led the Library's Orientalia Division (now Asian Division) from 1928 to 1964, and the generosity of philanthropist and art collector Andrew W. Mellon, who funded Hummel's later acquisition trips. Hummel began collecting Chinese maps shortly after he started teaching English for the American Board of Commissioners in the Foreign Mission School in Fenyang, Shansi, in 1915. At the end of each day while teaching English at the mission school, Hummel recalled, "he visited shops in downtown Fenchow, a trading center of importance on a camel trail from Northwest China, where … old maps were for sale at extremely low prices."[16] He once told a friend that the ten years he spent teaching in the interior of China were "the most interesting, and most important, years of my life."[17]

Hummel's collection of some 200 rare manuscript and wood engraved maps and atlases was donated in two parts. Andrew Mellon funded the first donation in 1928–1930, and Hummel personally deposited the remaining maps and atlases in 1962.[18] The collection consists of some eighty-five scrolls, wall maps, atlases, and maps dating from the late Ming Dynasty (1368–1644) to the end of the Qing Dynasty (1644–1911). It includes a wide variety of themes typical to classic Chinese government activities, including administrative and boundary issues, military campaigns, transportation, entombment, and flood control.

A few examples of Hummel's cartographic treasures illustrate the depth and breadth of this incredible collection. The earliest work, dating from the late Ming period, is an atlas titled *Guang yu tu* [Atlas of Heaven and Earth], which consists of a celestial chart, a map of the Ming Dynasty, and 18 provincial maps.[19]

---

15 Geography and Map Division Annual Report 1912 (Typescript in Geography and Map Division, Library of Congress).

16 Edwin Beall and Janet Beall, "Arthur W. Hummel," *Journal of Asian Studies* 35 (Feb. 1976), 265–276.

17 Edwin Beall and Janet Beall, "Arthur W. Hummel," *Journal of Asian Studies* 35 (Feb. 1976), 265–276.

18 US Library of Congress, *Report of the Librarian of Congress for the Fiscal Year Ending June 30, 1907* (hereafter cited *Librarian of Congress AR*), pp. 360–361; *Librarian of Congress AR*, 1962, p. 28.

19 Zuo Junheng, *Guang yu tu* 廣輿圖 [Atlas of Heaven and Earth], 1601. http://hdl.loc.gov/loc.gmd/g7820m.gct00225; Lin and Zhang, *Reading Imperial Cartography*, 50–53; Li, *Library of Congress Ancient Chinese Map Descriptive Catalog*, 7–8.

One of Hummel's most significant donations was the rare double hemisphere world map prepared by the Jesuit missionary Ferdinand Verbiest that was published in Beijing on eight scrolls in 1674[20] (Figure 7.1). Of the 475 pre-1900 individual Chinese maps in the Geography and Map Division, more than 140 or about thirty percent were donated by Hummel.

Scroll maps are well represented. A late seventeenth-century silk scroll measuring almost 20 feet in length depicts, in the form of a landscape painting, four important frontier regions of the Manchu dynasty, including one illustrating a clash between Russian and Manchurian troops on the Amur River (Figure 12.2).[21] Other scroll maps, measuring 60 and 80 feet long, depict in detail market towns, villages, temples, inns, courier stations, monuments, rivers, rapids, cliffs, and caves.

Coastal defense was a prominent theme of early Chinese mapping, as illustrated by an early eighteenth-century pen-and-ink watercolor copy of an original map drawn during the reign of the Jiajing emperor in about 1524 to aid officials in curbing piracy and smuggling. Titled "Illustrated Map of Qing Empire Coastal Fortifications," this twelve-foot manuscript on folded boards displays military garrisons along the Grand Canal and the Chinese coast from Hainan Island to the Liaodong Peninsula.[22] Mountains and islands are rendered as miniature landscape paintings in the Chinese tradition with flagpoles designating military garrisons.

Chinese road maps form another important segment of Hummel's donation. Particularly fascinating is an ink and color manuscript titled *Shan jing Shu dao tu* [Annotated Road Map from Shaanxi to Sichuan], compiled about 1750 (Figure 12.3). Formatted as a scroll map measuring one foot high by nearly 55 feet long, it depicts a strategic trans-mountain route that linked Central and Southwest China during the Qing dynasty.[23] A variety of urban and rural centers, residences of public officials, temples, inns, courier stations, and

---

20 Ferdinand Verbiest, *Kun yu quan tu* 坤輿全图 [World Map], Beijing, 1674. http://hdl.loc.gov/loc.gmd/g3200.ct004397gh.

21 Anonymous, *Aihun, Luosha, Taiwan, [Nei] Menggu tu* 艾渾，羅刹，台灣，蒙古圖 [Pictorical Maps of Aihun, Russia, Taiwan, and (Inner) Mongolia], between 1697–1722. http://hdl.loc.gov/loc.gmd/g7801r.ct002202a. Lin and Zhang, *Reading Imperial Cartography*, 154–157.

22 Tan Jiuchou, *Wan li hai fang tu shuo* 万里海防圖说 [Illustrated Map of Qing Empire Coastal Fortifications], 1725. https://www.loc.gov/item/gm71005020/. Lin and Zhang, *Reading Imperial Cartography*, 260–263.

23 Anonymous, *Shan jing Shu dao tu* 陝境蜀道圖 [Annotated Road Map from Shaanxi to Sichuan], 1750. http://hdl.loc.gov/loc.gmd/g7823sm.gct00272. Lin and Zhang, *Reading Imperial Cartography*, 300–307; Li, *Library of Congress Ancient Chinese Map Descriptive Catalog*, 131.

FIGURE 12.2 Detail from scroll map of Luosha, depicting siege of Russian forces by Qing army along the Heilongjiang (Amur) River in 1686. From *Aihun, Luosha, Taiwan, Menggu tu* 艾渾，羅刹，台灣，蒙古圖 [Pictorial Maps of Aihun, Russia, Taiwan, and (Inner) Mongolia]. Library of Congress, Geography and Map Division, Hummel Collection 1930-6. http://hdl.loc.gov/loc.gmd/g7801r.ct002202a.

monuments are displayed along the route with rivers and terrain features. In form and content, this eighteenth-century scroll map is similar to one of the Division's Japanese road maps displaying the legendary Tōkaidō Road that connected Edo to Kyoto and Nagasaki (Figure 12.4).[24] Similarly, both these landscape itinerary maps depict settlements, courier stations, inns for rest and food, temples for meditation, and bridges. These road maps with their vivid details were voyages in themselves. Japanese map historian Unno Kazutaka explains: "Without moving from his chair, one can gaze at misty mountains in the distance and then pass on to inspect the remains of famous sites along the road; he can visit soaring castles and come right down to the travelers on their way and the boatmen rowing across the water."[25]

Silk fans were another popular device for displaying maps in both China and Japan. One example from Hummel's donation is an ink and watercolor silk fan from the late Qing period that displays that dynasty's territory on one side, and a star chart on the other side.[26] Contemporary European and American pocket globes served a similar purpose.

William Woodville Rockhill was another major map donor to the Geography and Map Division. An early explorer in China and Tibet, and author of the US Open Door Policy for China, Rockhill was one of America's leading experts on that country at the turn of the twentieth century. He first visited Beijing as personal secretary to John Russell Young, who served as the American minister to China from 1882 to 1885. Rockhill donated his entire collection of some 6,000 Chinese books and maps as a memorial to John Russell Young, who was the Librarian of Congress from 1897 to 1899.[27]

In addition to donating several early 17th century atlases of Imperial China, Rockhill furnished the Library with an embankment map showing dikes along the Yellow River, drawn shortly after the river was flooded in 1855.[28] Hummel

24 Anonymous, *Tōkaidō bunken emaki* 東海道分間絵図 [Detailed pictures of the Tōkaidō road], 17–?, http://hdl.loc.gov/loc.gmd/g7964tm.gct00271a. See also: Ralph E. Ehrenberg, *Mapping the World* (Washington, DC: National Geographic Society: 2006), pp. 132–133.

25 Unno Kzutaka, "Notes to the Plates," *Old Maps in Japan*, edited by Nanba Matsutaro, Muroga Nobuo, and Unno Kazutaka; translated by Patricia Murray (Osaka: Sogensha Inc., 1973), 187.

26 Anonymous, *Da Qing yi tong tian di quan tu* 大清一统天地全图 [The Great Qing Dynasty's Complete Map of All under Heaven], 1890? http://hdl.loc.gov/loc.gmd/g7820.ct003395r. Lin and Zhang, *Reading Imperial Cartography*, 94–95; Li, *Library of Congress Ancient Chinese Map Descriptive Catalog*, 30.

27 *Library of Congress, Asian Collections: An Illustrated Guide*, p. 16.

28 Anonymous, *Caozhou Fu Shan Xian nan di dong hao tu* 曹州府單縣南堤東濠圖 [Embankment map of Shan County in Caozhou Prefecture], 1856–1866. http://hdl.loc.gov

FIGURE 12.3 Detail from an eighteenth-century Chinese road map. *Shan jing Shu dao tu* 陝境蜀道圖 [Route Map from Shaanxi to Sichuan], after 1750. Detail from Scroll 1 (31 cm × 1,672 cm) Library of Congress, Geography and Map Division, Hummel Collection 1930-16. http://hdl.loc.gov/loc.gmd/g7823sm.gct00272.

FIGURE 12.4 Detail from seventeenth-century Japanese road map. *Tōkaidō bunken emaki* 間絵東海分間絵図 [A Pictorial Map of the Stages of the Route along the Eastern Sea] by Moron Hishikawa and Ochikochi Dōin, ca. 1618–1694. One map on two scrolls, 26 cm × 3,575 c Detail is from scroll 2 (Image 12). Library of Congress, Geography and Map Division, Ca Number G7964.T7P2 17-- .T6. http://hdl.loc.gov/loc.gmd/g7964tm.gct00271b.

also donated several levee maps. The Geography and Map Division's annual report noted that in 1928 two manuscript scroll maps and an atlas that illustrated artificial levees along portions of the Huanghe or Yellow River were of such interest to members of a congressional committee in the United States Congress, that copies were provided to them by the Geography and Map Division for their discussion of a 1928 Mississippi River flood control bill.[29]

The bulk of the Library's collection of early rare Korean maps and atlases was acquired from two eminent educators, archaeologist and East Asian historian Langdon Warner and geographer and diplomat Shannon McCune. Warner established the American School of Archaeology in Peking for the Smithsonian Institution and later was associated with the Fogg Museum of Harvard University.

Warner's donation in 1929 ranged from a hand-colored woodblock print of Korea dated 1822 to several atlases (and one hand-held fan) depicting the eight administrative circuits of Korea during the eighteenth and nineteenth centuries from different political points-of-view.[30]

Another item donated by Warner is the *Great Ming Dynasty's Atlas of Mountains and Seas*, based on data derived from the *Da Ming yi tong zhi* 大明一統志 [Comprehensive Gazetteer of China in the Ming Period] originally compiled in 1461 CE.[31] Compiled in 1721 by a Korean student named Won Hak-saeng in an effort to recreate the geography of China, Korea, and Manchuria during the Ming period, it is believed to be the earliest cartographic work in the Geography and Map Division drawn by a Korean mapmaker.[32]

Won Hak-saeng's preface to his great atlas still resonates today with all armchair geographers:

/loc.gmd/g7823s.ct003230. Lin and Zhang, *Reading Imperial Cartography*, 165; Li, *Library of Congress Ancient Chinese Map Descriptive Catalog*, 130.

29 *Librarian of Congress AR*, 1928, p. 92; Lawrence Martin, "Annual Report of the Chief of the Division of Maps Library of Congress for the Fiscal Year Ending June 30, 1928," Typescript, p. 5.

30 For example, see *Tongguk chido*, (176–?), which displays prefectural and district towns and roads as well as rivers and mountain ranges. https://www.loc.gov/item/83675015/.

31 Won Hak-saeng?, *Da Ming yi tong shan he tu* 大明一统山河图 [The Great Ming Dynasty's Atlas of Mountains and Seas], 1721–1724. http://hdl.loc.gov/loc.gmd/g7820m.gct00230; Lin and Zhang, *Reading Imperial Cartography*, 59; Li, *Library of Congress Ancient Chinese Map Descriptive Catalog*, 10–11. The Library of Congress holds four copies of *Da Ming yi tong zhi*, dated from 1461 to 1559.

32 Arthur W. Hummel, "Chinese and Other East Asiatic Books Added to the Library of Congress, 1929–30," *Librarian of Congress AR*, 1930, p. 361; Li, *Library of Congress Ancient Chinese Map Descriptive Catalog*, 11.

> A scholar who never travels but stays at home is not worthy to be accounted a scholar. From my youth on I had the ambition to travel, but could not afford to wander over the 300 counties of Korea, much less the whole world. So, carrying out an ancient practice, I drew a geographical atlas. And while gazing at it for long stretches at a time I feel as though I were carrying out my ambition … Morning and evening, while bending over my small study table, I meditate on it and play with it—and there in one vast panorama are the districts, the prefectures and the four seas, and endless stretches of thousands of miles.[33]

Shannon McCune was born in Korea to American missionary parents. A trained geographer, he later served as the civil administrator of the Ryukyu Islands (1962–64), president of University of Vermont, and director of the American Geographical Society. His extensive acquisition, acquired in 1982, includes both manuscript maps and wood block impressions, which generally are rarer and more valuable than Korean manuscript copies.

A distinctive achievement of Korean cartography was the hand atlas, which came into general use in the sixteenth century as a reference source for reading Chinese classics and elementary geographical instructions. The Division's Korean hand atlases range in date from the mid-eighteenth century to 1896 and are representative of those produced during the Chosŏn (Yi) dynasty (1392–1910).

Korean hand atlases generally included a circular world map or *ch'ŏnhado*, popularly known as a "wheel map." Uniquely Korean in concept and design, the *Ch'ŏnhado* [Map of all Under Heaven] reflected a vision of the world that embraced Taoist and Confucian traditions as well as Chinese and Korean geographical concepts.[34] It placed the "Hermit Kingdom" near the center of the world along with China.

The Library's Japanese collections began slowly in the 1870s with books and maps obtained through a United States government documents exchange program with Japan initiated by Congress's Joint Committee on the Library and administered by the Smithsonian Institution.[35] By 1871, the Smithsonian

33 Quoted in *Librarian of Congress AR*, 1929, 149–150; Li, *Library of Congress Ancient Chinese Map Descriptive Catalog*, 11.

34 Yŏ, On, *Ch'ŏnhado*, 18–? http://hdl.loc.gov/loc.gmd/g7900m.gct00240. See also Ralph E. Ehrenberg, *Mapping the* World, 124–125.

35 Andew Kuroda, "A History of the Japanese Collection in the Library of Congress, 1874–1941," *Senda Masao Kyōju Koki Kinene Toshokan Shiryō Ronshu* (Senda Masao Kyōju Koki Kinenkai, 1970), pp. 283–284.

Institution acknowledged the receipt of "maps and prehistoric relics" along with unidentified books, specimens, and gold coins.[36]

Early donors included Crosby Stuart Noyes, editor of the *Washington Evening Star*, in 1905, and, particularly, Prof. Kan'ichi Asakawa of Yale University, who was commissioned by the Library in 1906 to obtain significant titles relating to Japanese history, literature, and the arts.[37] A graduate of Waseda University in Tokyo and Dartmouth College, Dr. Asakawa obtained his PhD from Yale University in 1902. During Asakawa's year-long acquisition tour of Japan, his acquisition of an estimated 45,000 volumes (3,160 titles) laid the foundation for the Library's Japanese collection.[38] Included among this total were two categories that he made a special effort to obtain: "old maps of some of the larger cities" as well as "old and new books on the geography and historical geography of different localities, many of which, printed from wooden plates, are becoming rarer every year."[39]

Shiho Sakanishi, the first Japanese native-born area specialist in the Library of Congress, nearly tripled the size of the Japanese language collection from 1930 to 1941. A graduate of Tokyo Women's Christian College and Wheaton College, Norton, Massachusetts, Dr. Sakanishi earned her PhD degree from the University of Michigan in 1929. Before she was deported to Japan at the outset of WWII, she tripled the size of the Library's Japanese Collection.[40] The Library of Congress now holds probably the largest Japanese language collection in the world outside of Japan. Regrettably no cartobibliography of its early Japanese map collection is available, but a large number are cataloged.

One of the oldest examples is a world map by the Japanese Buddhist scholar-priest Hotan that provides an image of the Buddhist vision of world geography in terms of its cosmology. Titled *Nansenbushū bankoku shōka no zu* [Map of the Universe], it displays India, where Buddha was born, and China as the center of the world.[41] Printed from woodcuts in 1710, this prototype map was popular in Japan until the mid-nineteenth century. The Geography and Map Division holds two versions.

Most of the Division's early Japanese maps date from the nineteenth century. The most significant is Inō Tadataka's original survey of the coast of

36 Kuroda, pp. 293–294.

37 *Library of Congress Asian Collections: An Illustrated Guide*, pp. 31–32.

38 Kuroda, p. 305.

39 "Kan'ichi Asakawa Report, Sept. 27, 1907," in *Librarian of Congress AR*, 1907, p. 25; Kuroda, p. 302.

40 Kuroda, pp. 318, 323–324.

41 Sōshun, *Nansenbushū bankoku shōka no zu* 南瞻部洲萬國掌菓之圖 [Map of the Universe], Kyoto, 1710. http://hdl.loc.gov/loc.gmd/g3200.ct002184.

Japan, *Dai Nihon enkai yochi zenzu* [Full View of the Coast of Japan], which was completed in 1819.[42] Compiled at a scale of 1:216,000, Ino's original maps were destroyed in a fire fifty years later. Only a few sheets were known to have survived until a Japanese researcher chanced upon the Division's set of 207 of the original 214 sheets some seventeen years ago. Japanese geographers and historians only knew about this collection through Inō's diaries. It was a great discovery noted Osamu Nishikawa, professor emeritus at the University of Tokyo: "It is epochal in terms of enabling historical researchers to compare the situation in various locations in Japan 200 years ago and today, with this extremely detailed large-scale map."[43] As an indication of its value to Japan, the Geospatial Information Authority of Japan, in cooperation with the Library of Congress, digitized the collection.[44] These images were then printed along with the other known copies and displayed in Japan through large floor exhibitions measuring nearly 200 × 100 feet, such as one at the Nagoya Baseball Dome Stadium, which was viewed by some 24,000 Japanese over two days. Since 2009, more than ten floor exhibitions have been held throughout Japan.[45]

Each of Ino's survey sheets measures about 5¼ feet, or the size of a Japanese tatami mat, as one Japanese commentator observed. These copies survived because they were hand copied by the Imperial Japanese Army General Staff to be used as a framework for its 1:200,000 scale map of Japan, according to one authority associated with the INOH Tadataka Society. Following World War II and the allied occupation of Japan, this collection along with other major map series that were produced or acquired by Japanese military agencies were transferred to the US Army Map Service and later to the Library of Congress when Japanese military units were disbanded following demobilization.

42 Inō Tadataka, *Dai Nihon enkai yochi zenzu* 大日本沿海輿地全図 [Full View of the Coast of Japan], 1816–1819; recreated by the Japanese Army in 1884. (214 manuscript sheets). http://hdl.loc.gov/loc.gmd/g7960m.gct00032.

43 "Copy of Ino Tadataka's Map of Japan Found in U.S.," *Kyodo News International* (July 4, 2001). https://www.thefreelibrary.com/Copy+of+Ino+Tadataka's+map+of+Japan+found+in+U.S.-a076664695.

44 "Rare Japanese Maps to be Preserved," *Library of Congress Gazette* (January 10, 2002), p. 10.

45 Y. Hoshino, M. Kitani, J. Suzuki, M. Horino, and I. Watanabe, *Floor Exhibition of Inoh's Map, The Oldest Scientifically Surveyed Map in Japan*. https://icaci.org/files/documents/ICC_proceedings/ICC2011/Oral%20Presentations%20PDF/D4-Digital%20technologies%20and%20cartographic%20heritage/CO-396.pdf.

2 Atlas Collection

The Library's atlas collection is comprised of some 80,000 titles primarily related to the United States, World, and Europe, but East Asian coverage is not insignificant. The following examples provide a brief introduction to the range and scope of the Division's East Asian holdings in addition to those already described:

1. Illustrative of early European atlases of Asia are Martino Martini's *Novus atlas Sinensis* (Amsterdam: J. Blaeu, 1665), probably the best known early atlas of Asia published in Europe; and Jean Baptiste Bourguignon d'Anville's *Atlas général de la Chine, de la Tartarie chinoise, et du Tibet: pour servir aux différentes descriptions et histoires de cet empire* (Paris: Dezauche, 1790).[46]
2. Representative of an English language atlas published in China is *The New Atlas and Commercial Gazetteer of China, a Work Devoted to its Geography & Resources and economical & commercial development* (Shanghai: The North-China News & Herald, 1917). Edited by John Dingle, it includes 25 bilingual maps that he claimed to be compiled and translated from the latest and most authoritative surveys and records. In 1918, the chief of the Geography and Map Division described the work as the largest and most comprehensive English language survey of China published to date.[47]
3. Examples of thematic atlases include *Chōsen chishitsuzu* [Geological Atlas of Korea] (Seoul: Geological Survey of Korea, 1929–1932), and *Japanese Pre-War Colonization* (Washington, DC by the Federal Bureau of Investigation, 1935?), which displays the distribution of Japanese nationals worldwide and their agricultural and mineral production on the eve of World War II.[48]
4. Ting Wen-chiang's *New Atlas of the Republic of China* (*Zhonghua minguo xin ditu*) (Shanghai, 1934) exemplifies a category of atlas that was translated from Chinese, Japanese, or Korean to English.[49] While this work documented many geographical changes in China that were not

46 Six images of the Martini atlas are available online at https://www.loc.gov/resource/g7820m.gct00076/?sp=2; Four plates from d'Anville's *Atlas de la Chine* are online: http://hdl.loc.gov/loc.gmd/g7820m.gct00075.

47 Geography and Map Division Annual Report, July 1918, p. 6.

48 The Geological Atlas of Korea, *Geological Atlas of Chosen NO-8* is available online at http://hdl.loc.gov/loc.gmd/g7901cm.gct00307; The FBI atlas of Japanese Pre-War Colonization is available online at http://hdl.loc.gov/loc.gmd/g3201em.gct00493.

49 Ding Wenjiang 丁文江, *Zhonghua Minguo xin di tu* 中華民國新地圖 [New Atlas of the Republic of China] (Shanghai, 1934).

previously known outside that country following the establishment of the Republic in 1911, its usefulness was limited until the Geography and Map Division published a translation in 1949, *A Supplementary Key to Accompany the V.K. Ting Atlas of China* (Edition of 1934).[50]

5. Finally, a typical facsimile atlas is represented by Walter Fuch's *Der Jesuiten-Atlas der Kanghsi-Zeit: China und die Aussenlaender* [The Jesuit Atlas of the Kanghsi Period: China and the Foreign Countries] is representative of facsimile atlases. Published in Beijing by the Catholic University (Katholischen Universität) in 1941, it reproduces 35 provincial woodcut maps from the original edition, *Kangxi huangyu quanlan tu* 康熙皇輿全覽圖 [Kangxi Atlas of all Surveyed] (Beijing, 1721). These maps were based on a nationwide field survey conducted between 1708 and 1717 by Qing officials and Jesuits Jean Baptiste Regis and Pierre Jartoux that employed European survey techniques and Nicolas Sanson's projection.[51]

3 Large-Scale Multi-sheet Map Series

The largest collection of East Asia Maps is found among the Multi-Sheet Map series, which represents sixty-four percent of the Division's entire East Asian Collection. This series consists of medium or large-scale maps that were produced from field surveys at uniform size and scale with standardized symbols. Numbering some 236,000 map sheets, it encompasses such diverse subjects such as topography, geology, land-use, and census data; large-scale plans of cities; transportation maps; aeronautical charts; and hydrographic or nautical charts. The maps date from Inō Tadataka's Survey to the present.

Our major source for maps of China during the post-World War II period was the US Army Map Service (now the National Geospatial-Information Agency), which had acquired thousands of large-scale topographical maps that had been prepared during the first half of the twentieth century by Chinese, Japanese, and American military units. These included more than 100,000 large-scale topographic maps produced by the Chinese General Staff Land Survey from about 1917 to the end of World War II that depict topography, roads, and drainage for much of east China and Sichuan province.

50 L.W. Wadsworth, comp., "A supplementary key to accompany *Zhonghua Minguo xin di tu*" [the V.K. atlas of China] (edition of 1934). (Washington, DC: Library of Congress, Reference Department, Division of Maps, 1949).

51 Walter Fuchs, ed., *Der Jesuiten-Atlas der Kanghsi-Zei: China und die Aussenlaender* (Peking: Katholischen Universität, 1941). http://hdl.loc.gov/loc.gmd/g7820m.gct00265.

Among these transfers were a special category of maps prepared by Japanese Army topographers for areas outside the Japanese homeland that were known collectively as *gaihōzu* maps (外邦図), or "maps of outer lands." They were compiled by the Japanese Land Survey Department of the General Staff Headquarters between the Meiji era and the end of WWII at scales ranging from 1:25,000 to 1:500,000. Geographical coverage stretches from Alaska southward to Australia, and from America westward to parts of Pakistan and Afghanistan. The maps were prepared from field surveys conducted by Army surveyors, often secretly, or copied from existing maps and then updated with the aid of Japanese aerial photographs.[52]

Generally compiled for military purposes, most *gaihōzu* were classified as secret and large numbers were destroyed or confiscated during demobilization. Those that were acquired by the United States military were used during the Cold War for geographical information not available elsewhere, and today remain important for their scientific and historical research data delineating changing landscapes from the late nineteenth century to the first half of the twentieth century.

The Division holds some 20,000 to 25,000 *gaihōzu*, namely topographical maps of former Japanese colonies, covering mostly China and Korea. Cartographic Materials Cataloger Setsuko Means is working closely with colleagues from Stanford, Osaka, and Tōhoku universities to create a "Union Database" for this important series.

A core group of Japanese scholars have been studying the Division's collection of *gaihōzu*, particularly with respect to those related to China and Korea during the 1880s.[53] These maps were based on traverse surveys of the main

52 Min Zhang and Setsuko Means, "Maps of Outer Lands," *Philip Lee Phillips Map Society of the Library of Congress* Newsletter 14.3 (Summer 2016): 3.

53 S. Imazato and T. Hisatake, "Japanese overseas maps in the Library of Congress and the Library of American Geographical Society," in *Japanese Modern Cartography concerning Asia-Pacific Areas: Approaches to Gaihōzu*, ed. S. Kobayashi, 55–69, Osaka University Press (in Japanese); R. Watanabe, K. Yamachika, and S. Kobayashi, "Map making of Korean Peninsula by Japanese army officers during 1880s: The study of the maps in the Library of Congress, Washington, DC," *Map, Japan Cartographers Association* 47.4 (2009), 1–16 (in Japanese); K. Yamachika, R. Watanabe, and S. Kobayashi, "The route maps of the Korean Peninsula drawn by Japanese army officers during 1880s," *Proceedings of the 14th International Conference of Historical Geographers, Kyoto 2009* (Kyoto University Press, 2010), 307–308; S. Kobayashi, R. Watanabe, and K. Yamachika, "The travel and surveying for the preparation of maps in East Asia by Japanese army officers during the 1880s," *The Shirin* (*The society of Historical Research*) 93 (2010), 473–505. (in Japanese); K. Yamachika, "Japanese Imperial Maps of the Meiji Era in the Library of Congress Collection: An Analysis of the Explanatory Notes and Legends," unpublished paper presented at

routes in these two countries by young Japanese Army officers who measured distances by pacing and direction by compass bearings. Nearly 600 maps have been scanned for their database.[54]

Korea is also well represented by large-scale topographic maps and city plans. The US Army Corps of Engineers began compiling topographic maps of South Korea at 1:50,000 and 1:250,000 scales beginning in 1945 using Japanese Imperial Land Survey maps, Japanese Hydrographic charts, and Korean General Staff maps which were updated and revised using aerial photographs.[55]

A large set of city maps of Pyongyang, North Korea, produced by the Japanese government for colonization of Korea and Manchuria in 1936 is representative. These were probably designed for potential Japanese investors in Korea. The author showed this item by chance to a South Korean businessman some years ago. He paused after looking at it, and then, pointing to a spot on one of plats, noted that that was where he was born. Later he wrote thanking me for the tour. He then went on to apologize, saying that he could not speak for a few moments upon seeing the map of his birthplace because he was overcome by emotion. When we speak about the power of maps, I will always remember that encounter.

4 Small-Scale Single Sheet Collection

Equally important for research is the Geography and Map Division's small-scale single-sheet collection for East Asia, which totals some 71,000 maps. About a third of this collection has been cataloged while the remainder constitutes a collection that is commonly known as the "Titled Collection." Although there is no comprehensive listing of the individual maps in the Titled Collection, a basic geographic arrangement was developed in the 1890s when the Hall of Maps was established. This geographical scheme was expanded as the holdings grew during the early twentieth century and is still in use today. Each map is foldered and labeled (or "titled") with basic bibliographic information including geographic area, date, scale, and author.

"Japanese Imperial Maps as Sources for East Asian History: A Symposium on the History and Future of the Gaihozu," Stanford University, October 8, 2011.

54 Shigeru Kobayashi, Hitoshi Miyazawa, and Kenta Yamamoto, "Three Databases of Japanese Imperial Map." http://www.let.osaka-u.ac.jp/geography/gaihouzu/newsletter11/pdf/n11_s9_3.pdf.

55 Examples of US Army Map Service topographic maps of Korea are available online: 1:50,000 http://hdl.loc.gov/loc.gmd/g7900m.gct00023; 1:250,000 http://hdl.loc.gov/loc.gmd/g7900m.gct00024.

These two collections are the best source for late nineteenth and twentieth century maps and are essential resources for understanding the more recent political, economic, and cultural history of East Asia. Diligent researchers will find a wide range of geographic coverage and subject content, such as a stunning pictorial wall map of China. Titled *Pictogram Map of the Republic of China* (*Xiangxing Zhonghua Minguo renwu yu di quantu* 象形中華民國人物與地全圖), it is a large wall map (7 × 5 feet) that was prepared for Chinese and English-speaking middle and high school students in Harbin, China, in 1931 by Russian mapmakers and published by the Northern Trading Co. Ltd. Drawings of local animals, men and women, and buildings reflecting local customs and economics just prior to mass movements of Chinese during World War II blanket the map.[56] Other ethnographic maps include a map of Manchuria and Mongolia, published by the Korean Daily News in Seoul, 1932, that shows the distribution of Koreans at a critical period in their history and a Japanese Army map of Central and East Asia, dated April 22, 1938, that displays seven different ethnic groups: Tungusic peoples (Tungus), Chinese (Han), Turks, Indo-Aryans, Mongols, Hmong, and Dravidians (in South India).[57] International relations are strongly represented, including one map that shows Chinese harbors open to foreign powers in 1929.[58]

Military maps from the Sino-Japanese War to the Korean War are extensive and found in all formats and languages. These include, for example, a map titled *Theater of Operations, Japan-China War 1894–5*, prepared by the Office of Naval Intelligence (Washington, DC: Navy Department); a map of the Russian-Japanese War, printed in Yiddish by the Jewish Press Publishing Company in 1904; and a series of newspaper maps by the *New York Times* and the *Washington Evening Star* relating to Japanese and allied battles in China during WWII.

Communications maps are available. Chinese postal maps were acquired by the US Post Office and then transferred to the Geography and Map Division beginning in 1905. For example, *Carte du District Postal de Kwangsi / Postal Map of Kwangsi District*, printed in 1924 at a scale of 1:600,000, displays 11 levels of mail service from rural courier to steamer and railway. An undated nautical harbor chart depicts the submarine telegraph cable network at Shanghai and in the approaches to the Yangtze River. Seven cables are named, including

56 The "Map of China" can be viewed online: http://hdl.loc.gov/loc.gmd/g7821a.ct007520.

57 From Library of Congress, Geography and Map Division Title Collection, "East Asia/ Ethnography," 1932 and 1938.

58 Zhong hua guo chi di tu (Beijing, 1929) is available at http://hdl.loc.gov/loc.gmd/g7821f .ct002301.

"American to Manila," "German to Tsingtao," and "Chinese (Really Danish) to Japan & North."

Maps showing foreign concessions in Chinese cities provide another valuable resource for analysis. These include a blueprint "Map of Peking" in 1900 compiled by the British Intelligence and Survey Office from sketches supplied by the various foreign nations that administered concessions; a detailed consolidated map of Shanghai, *Shi ce Shanghai cheng xiang zu jie tu* 十冊上海城鄉組截圖 (Shanghai, 1913), showing the Chinese walled city, French and English settlements, the city wharf, and an inset including street names in Chinese and English[59] and a 1942 "Map of Tientsin" depicting foreign concessions (Japanese, British, and French).[60] An example of the international intrigue sometimes associated with cartography is represented by a Japanese city map of Beijing (*Saishin Pekin shigai chizu: tsuketari Pekin Tenshin fukin chizu*) that was originally published in Tokyo before WWII and then reproduced ten years later in 1948 by a US Army intelligence unit from the original Japanese plates.[61]

5 Hydrographic Charts

The final major category of East Asian cartographic materials held by the Geography and Map Division is hydrographic charts. Maritime activity in East Asia was dominated by Japanese and its military during the first half of the nineteenth century and this is reflected in our holdings. The Geography and Map Division began acquiring modern Japanese nautical charts in the 1920s as Japanese naval power began to grow. The Japanese Navy was the third largest in the world in 1927, the year in which the US Coast and Geodetic Survey transferred 523 Japanese charts to the Division, which was part of a transfer of 1,858 hydrographic charts made by government institutions from 15 foreign countries.[62] The Japanese transfer was by far the largest, and over the next several years an additional 200 charts were received. Today, the Division holds some 16,000 hydrographic charts produced by the Japanese Navy, 600 by Korean naval authorities, and 120 Chinese charts.

59 A Consolidated Map of City of Shanghai is available online: http://hdl.loc.gov/loc.gmd/g7824s.ct003616.

60 Tientsin concession map is available online: http://hdl.loc.gov/loc.gmd/g7824t.ct003615.

61 The Japanese/American map of Beijing is available online: http://hdl.loc.gov/loc.gmd/g7824b.ct001951.

62 *Librarian of Congress AR*, 1927, p. 86.

Early Japanese charts followed the standard form and format established by the British Admiralty and included English titles as well as Japanese. A beautiful example is a large-scale approach chart of Kamaishi Bay on the East Coast of Japan based on surveys conducted by officers of the Imperial Japanese Majesty's Ship *Kasuga*, in 1872 (the 4th year of the Meiji era). It was published at the Hydrographic Office of the Japanese Naval Department on August 15 (5th year of Meiji), 1875. Soundings are denoted in fathoms while topography is displayed by hachures.[63]

A Chinese Maritime Customs chart of "Hankow Harbour," drawn at a little larger scale, provides detailed information about the businesses located along the Yangtze River wharf and those associated with the Japanese, French, Russian, and British concessions. It was based on a survey conducted by the Marine Department of the Chinese Maritime Customs, Chinese government surveys, and a British Admiralty chart. Soundings are indicated in feet. A cautionary note reads: "Owing to the constant shifting of the channels, no dependence can be placed on this chart as a navigational guide."[64]

Charts from this era reflect the fact that even after a century of organized chart making, the world's waters were still imperfectly known during the first half of twentieth century. For example, both American and Japanese charts of the island of Tarawa, site of one of the most lethal World War II Pacific Island land battles, were based on old surveys initially made by the US Exploring Expedition commanded by Charles Wilkes in 1841 and updated with just minor corrections by the British Admiralty in 1925![65] While the Library of Congress has traditionally focused on collecting analogue maps, it has been acquiring digital maps for the past decade. One of the most significant acquisitions for East Asian scholars is the "The Digital Atlas of China by Industries," which was created by the All China Marketing Research Co., Ltd. in Beijing and distributed by the University of Michigan China Data Center in 2011. To our knowledge the Library of Congress is the only library that is allowed to make this 180GB file available to researchers. Based on China's 2004 economic census,

63 Yanagi Narayoshi, *Rikichū no Kuni Kamaishikō no zu* [Kamaishi Bay of Rikuchu] (Tokyo: Dai Nihon Kaigun Suiroryō, Meiji 4 [1871]). https://lccn.loc.gov/2016589624. In recent years, major tsunamis in 1960 and 2011 have devastated the city of Kamaishi.

64 "Chinese Maritime Customs Chart, Published by Order of the Inspector General of Customs. China-Yangtze River. Hankow Harbour ... 1919. No. 16." Library of Congress, Geography and Map Division, Hydrographic Collection.

65 Ralph E. Ehrenberg, John A. Wolter, and Charles A. Burroughs, "Surveying and Charting the Pacific Basin," in Herman J. Viola and Carolyn Margolis, eds., *Magnificent Voyagers, the U.S. Exploring Expedition, 1838–1842* (Washington, DC: Smithsonian Press, 1985), p. 187. The Japanese chart is part of a set of ten islands on one sheet, titled "Gilbert Islands: Military Secret." https://www.loc.gov/resource/g9482m.gct00150/?so=l.

it provides geo-referenced information on more than three million Chinese companies. This includes such data as company name, address, contact information, year of establishment, ownership, primary products or services produced, industries, employee, and revenue scales. The data are in GIS shapefile format. Thus far, this atlas has been used in the Geography and Map Division Research Center by researchers from the World Bank, State Department, Johns Hopkins University, and the London School of Economics.

Finally, the Geography and Map Division is playing a lead role in developing a Capitol Hill Geographic Hosting Environment that will use geographical information systems or GIS to store, manipulate, and manage all types of geospatial data. As part of this program, a pilot project was begun to create geo-referenced digital indexes. One of the tests involved scanning a recently acquired set of some 6,000 Russian geological maps of China.

Acknowledgement

The author acknowledges with gratitude the assistance of Geography and Map Division staff members Min Zhang, Setsuko Means, and Tammy Wong in the preparation of this essay, and Diane Schug-O'Neill for digitizing the relevant map files.

Bibliography

Aihun, Luosha, Taiwan, [Nei] Menggu tu 艾渾，羅剎，台灣，蒙古圖 [Pictorical Maps of Aihun, Russia, Taiwan and (Inner) Mongolia]. 1697–1722. http://hdl.loc.gov/loc.gmd/g7801r.ct002202a.

Anville, Jean Baptiste Bourguignon d'. *Atlas général de la Chine, de la Tartarie chinoise, et du Tibet : pour servir aux différentes descriptions et histoires de cet empire*. Paris: Dezauche, ca. 1790. http://hdl.loc.gov/loc.gmd/g7820m.gct00075.

Beall, Edwin and Janet Beall, "Arthur W. Hummel, 1884–1975." *Journal of Asian Studies* 35 (February 1976): 265–276.

Caozhou fu Shan xian nan di dong hao tu 曹州府單縣南堤東濠圖 [*Embankment Map of Shan County in Caozhou Prefecture*]. 1856–66. http://hdl.loc.gov/loc.gmd/g7823s.ct003230.

"Chinese Maritime Customs Chart, Published by Order of the Inspector General of Customs. China-Yangtze River. Hankow Harbour ... 1919. *No. 16*." Shanghai: Maritime Dept., Chinese Maritime Customs, 1919.

Cohen, Warren I. *America's Response to China: A History of Sino-American Relations*. New York: Columbia University Press, 2010.

"Copy of Ino Tadataka's Map of Japan Found in U.S." *Kyodo News International* (July 4, 2001). https://www.thefreelibrary.com/Copy+of+Ino+Tadataka's+map+of+Japan+found+in+U.S.-a076664695.

Da Ming yi tong shan he tu 大明一统山河图 [The Great Ming Dynasty's Atlas of Mountains and Seas]. 1721–1724. http://hdl.loc.gov/loc.gmd/g7820m.gct00230.

Da Qing yi tong tian di quan tu 大清一统天地全图 [The Great Qing dynasty's Complete Map of All under Heaven]. ca. 1890. http://hdl.loc.gov/loc.gmd/g7820.ct003395r.

Ding Wenjiang 丁文江, *Zhonghua Minguo xin di tu* 中華民國新地圖 [New Atlas of the Republic of China]. Shanghai, 1934.

Ehrenberg, Ralph E., and National Geographic Society (US). *Mapping the World: An Illustrated History of Cartography*. Washington, DC: National Geographic, 2006.

Ehrenberg, Ralph E., John A. Wolter, and Charles A. Burroughs. "Surveying and Charting the Pacific Basin." In *Magnificent Voyagers, the U.S. Exploring Expedition, 1838–1842*, edited by Herman J. Viola and Carolyn Margolis, 165–187. Washington, DC: Smithsonian Press, 1985.

Fu sheng quan tu 福省全图 [Complete Map of Fujian Province]. 1842–1884. http://hdl.loc.gov/loc.gmd/g7824f.ct003181.

Fuchs, Walter, ed. *Der Jesuiten-Atlas der Kanghsi-Zeit: China und die Aussenlaender*. Peking: Verlegt bei der Katholischen Universität, 1941. http://hdl.loc.gov/loc.gmd/g7820m.gct00265.

Geological Atlas of Chosen. Seoul: Government General of Chosen, Geological Survey, 1929–1932. http://hdl.loc.gov/loc.gmd/g7901cm.gct00307.

Hebei Sheng. *Zhong hua guo chi di tu* 中華國恥地圖 [Map of China's National Humiliation]. Beijing: Gong shang ting, 1929. http://hdl.loc.gov/loc.gmd/g7821f.ct002301.

Hoshino, Y., M. Kitani, J. Suzuki, M. Horino, and I. Watanabe. *Floor Exhibition of Inoh's Map, The Oldest Scientifically Surveyed Map in Japan*. Internal Cartographic Association, CO-396. https://icaci.org/files/documents/ICC_proceedings/ICC2011/Oral%20Presentations%20PDF/D4-Digital%20technologies%20and%20cartographic%20heritage/CO-396.pdf.

Imazato S., and Hisatake T. "Japanese overseas maps in the Library of Congress and the Library of American Geographical Society." In *Kindai Nihon no Chizu Sakusē to Ajia Taiheiyō Chiiki: 'Gaihōzu' eno Apurōchi* [Japanese Modern Cartography Concerning Asia-Pacific Areas: Approaches to Gaihōzu], edited by Kobayashi Shigeru, 55–69. Osaka: Osaka University Press, 2009.

Inō Tadataka. *Dai Nihon enkai yochi zenzu* [Full View of the Coast of Japan]. 1816–1819; recreated by Japanese Army in 1884. http://hdl.loc.gov/loc.gmd/g7960m.gct00032.

Jiangsu quan tu 江蘇全图 [Complete Map of Jiangsu Province]. Wuchang: Hubei guan shu ju, 1864. http://hdl.loc.gov/loc.gmd/g7823j.ct002578.

Kobayashi S., Watanabe R., and Yamachika K. "The travel and surveying for the preparation of maps in East Asia by Japanese army officers during the 1880s." *Shirin* 93 (2010): 473–505.

Kuroda, Andrew Y. *A History of the Japanese Collection in the Library of Congress, 1874–1941*. Washington, 1970.

Li Xiaocong 李孝聪, ed. *Meiguo guo hui tu shu guan cang Zhong wen gu di tu xu lu* 美国国会图书馆藏中文古地图叙录 [Descriptive Catalogue of Ancient Chinese maps in the Collection of the Library of Congress]. Beijing: Wenwu chubanshe, 2004.

Library of Congress. *Annual report of the Librarian of Congress for the Fiscal Year Ended ...* Washington, DC: Library of Congress, 1907.

Library of Congress. *Annual report of the Librarian of Congress for the Fiscal Year Ended ...* Washington, DC: Library of Congress, 1927.

Library of Congress. *Annual report of the Librarian of Congress for the Fiscal Year Ended ...* Washington, DC: Library of Congress, 1928.

Library of Congress. *Annual report of the Librarian of Congress for the Fiscal Year Ended ...* Washington, DC: Library of Congress, 1929.

Library of Congress. *Annual report of the Librarian of Congress for the Fiscal Year Ended ...* Washington, DC: Library of Congress, 1930.

Library of Congress. *Annual report of the Librarian of Congress for the Fiscal Year Ended ...* Washington, DC: Library of Congress, 1962.

Library of Congress. *Asian Collections at the Library of Congress: An Overview*. Washington, DC: Library of Congress, 2020. http://www.loc.gov/rr/asian/guide2007/guide-chinese.html.

Library of Congress. *Ask a Librarian: Maps, Geography*. Washington, DC: Library of Congress. https://ask.loc.gov/map-geography/.

Library of Congress. *Geography and Map Division Annual Report July 1918*. Typescript in Geography and Map Division, Library of Congress. Washington, DC: Library of Congress, 1918.

Library of Congress. *Geography and Map Division Annual Report 1934*. Typescript in Geography and Map Division, Library of Congress. Washington, DC: Library of Congress, 1934.

Library of Congress, Geography and Map Division Title Collection. "East Asia/ Ethnography," 1932 and 1938.

Library of Congress. *Geography and Map Reading Room*. Washington, DC: Library of Congress. Geography and Map Division. http://www.loc.gov/rr/geogmap/.

Library of Congress. *Library of Congress Asian Collections: An Illustrated Guide*. Washington, DC: Library of Congress, 2000. https://catalog.hathitrust.org/Record/004120975.

Library of Congress. *Library of Congress Catalog*. Washington, DC: Library of Congress. https://catalog.loc.gov.

Library of Congress. "Rare Japanese Maps to be Preserved." *The Gazette / Library of Congress*. Washington, DC: Library of Congress (January 10, 2002): 10.

Lin Tianren 林天人 and Zhang Min 張敏, eds. *Reading Imperial Cartography: Ming-Qing Historical Maps in the Library of Congress* [*Huangyu suolan: Meiguo guohui tushuguan suo cang Ming Qing yutu* 皇輿搜覽:美國國會圖書館所藏明清輿圖]. Washington DC: Library of Congress in association with the Academia Sinica Digital Center, 2013.

Martini, Martino. *Pecheli, sive Peking, mperii sinarvm, provincia prima* [Six Images of the Martini Atlas]. Amsterdam: J. Blaeu, 1655. https://www.loc.gov/resource/g7820m.gct00076/?sp=2.

Meiguo guohui tushuguan dian cang zhi Zhongguo xiang guan ditu wenxian chaxun xitong 美國國會圖書館典藏之中國相關地圖文獻查詢系統 [Inventory of China-Related Historical Maps Archived at the Library of Congress]. http://webgis.sinica.edu.tw/map_loc/.

Ochikochi Dōin and Moronobu Hishikawa. *Tōkaidō bunken emaki* 東海道分間絵図 [Detailed Pictures of the Tōkaidō Road]. Japan: ca. 1700. http://hdl.loc.gov/loc.gmd/g7964tm.gct00271a.

Primakoff, G., cartographer, John A. Diakoff, compiler, and P. Sergeeff, engraver. *Xiang xing Zhonghua Minguo ren wu yu di quan tu* 象形中華民國人物輿地全圖 [The "Map of China"]. Harbin: Northern Trading Co. Ltd. & Mr. V.F. Yao-hsiun, 1931. http://hdl.loc.gov/loc.gmd/g7821a.ct007520.

Shan jing Shu dao tu 陝境蜀道圖 [Annotated Road Map from Shaanxi to Sichuan], 1750. http://hdl.loc.gov/loc.gmd/g7823sm.gct00272.

Shang wu yin shu guan. *Shi ce Shanghai cheng xiang zu jie tu* 实测上海城厢租借圖 [A Consolidated Map of the City of Shanghai]. Shanghai, 1913. http://hdl.loc.gov/loc.gmd/g7824s.ct003616.

Shigeru Kobayashi, Hitoshi Miyazawa, and Kenta Yamamoto. "Three Databases of Japanese Imperial Maps." http://www.let.osaka-u.ac.jp/geography/gaihouzu/newsletter11/pdf/n11_s9_3.pdf.

Sōshun. *Nansenbushū bankoku shōka no zu* 南瞻部洲萬國掌菓之圖 [Map of the Countries in Jambudvipa]. Kyoto: Bundaiken Uhei, Hōei, 1710. http://hdl.loc.gov/loc.gmd/g3200.ct002184.

Tan Jiuchou, *Wan li hai fang tu shuo* 万里海防圖说 [Illustrated Map of Qing Empire Coastal Fortifications]. Qing Yongzheng: 1725. https://www.loc.gov/item/gm71005020/.

Tianjin di tu 天津地圖 [Map of Tientsin]. Tianjin, China: Chung-Tung Litho Works, 1912. http://hdl.loc.gov/loc.gmd/g7824t.ct003615.

Tōkyō Atorasusha, and United States Far East Command. Intelligence Division. *Saishin Pekin shigai chizu: tsuketari Pekin Tenshin fukin chizu* 最新北京市街地圖 [Updated Street Map of Beijing]. Washington, DC: The Intelligence Div. of FEC, 1948. http://hdl.loc.gov/loc.gmd/g7824b.ct001951.

Tongguk chido 通國地圖 [Map of Korea (by Province)], ca. 1760. https://www.loc.gov/item/83675015/.

United States Army Map Service. *Munsan-Ni, 1950* [Korea 1:250,000]. Washington, DC: Army Map Service, U.S. Army, 1950. http://hdl.loc.gov/loc.gmd/g7900m.gct00023.

United States Army Map Service. *Kwangju, 1950* [Korea 1:50,000]. Washington, DC: Army Map Service, U.S. Army, 1950. http://hdl.loc.gov/loc.gmd/g7900m.gct00024.

United States Federal Bureau of Investigation. *Japanese Pre-War Colonization.* Washington, DC: Federal Bureau of Investigation, ca. 1935. http://hdl.loc.gov/loc.gmd/g3201em.gct00493.

Unno Kzutaka. "Notes to the Plates." In *Old Maps in Japan*, edited by Nanba Matsutaro, Muroga Nobuo, and Unno Kazutaka; translated by Patricia Murray, 187. Osaka: Sogensha Inc., 1973.

Verbiest, Ferdinand. *Kun yu quan tu* 坤輿全圖 [World Map]. Beijing: 1674. http://hdl.loc.gov/loc.gmd/g3200.ct004397gh.

Wadsworth, L.W., comp. "A supplementary key to accompany Zhonghua Minguo xin di tu" [The V.K. Atlas of China] (edition of 1934). Washington, DC: Library of Congress, Reference Department, Division of Maps, 1949.

Watanabe R., Yamachika K., and Kobayashi S. "Map Making of Korean Peninsula by Japanese Army Officers during 1880s: The Study of the Maps in the Library of Congress, Washington, DC," *"Map," Japan Cartographers Association* [*Chizu* 全図] 47, no. 4 (2009): 1–16. (In Japanese.)

Yamachika, K., R. Watanabe, and S. Kobayashi. "The route maps of the Korean Peninsula drawn by Japanese army officers during 1880s." In *Proceedings of the 14th International Conference of Historical Geographers, Kyoto 2009*, edited by Akihiro Kinda, 307–308. Kyoto: Kyoto University Press, 2010.

Yamachika, K. "Japanese Imperial Maps of the Meiji Era in the Library of Congress Collection: An Analysis of the Explanatory Notes and Legends." Unpublished paper presented at "Japanese Imperial Maps as Sources for East Asian History: A Symposium on the History and Future of the Gaihozu." Stanford University, October 8, 2011.

Yan Shusen. *E sheng quan tu* 鄂省全图 [Complete Map of Hubei Province]. Wuchang: [s.n.], Qing Dongzhi yuan nian, 1862. http://hdl.loc.gov/loc.gmd/g7823h.ct002593.

Yanagi Narayoshi. *Rikichū no Kuni Kamaishikō no zu* [Kamaishi Bay of Rikuchu]. Tokyo: Dai Nihon Kaigun Suiroryō, Meiji 4 (1871). https://lccn.loc.gov/2016589624.

Yŏ On. *Ch'ŏnhado* 輿地圖 [Terrestrial Map]. ca. 1800. http://hdl.loc.gov/loc.gmd/g7900m.gct00240.

Zhang Min and Setsuko Means. "Maps of Outer Lands." In *Philip Lee Phillips Map Society of the Library of Congress* 14, no. 3 (Summer 2016): 3. https://www.loc.gov/rr/geogmap/pdf/plp/plpnewsletters/2016/news2016sum.pdf.

Zuo Junheng. *Guang yu tu* 廣輿圖 [Enlarged Territorial Atlas]. Wanli: 1601. http://hdl.loc.gov/loc.gmd/g7820m.gct00225.

Index